Lecture Notes in Physics

For information about Vols. 1–131, please contact your bookseller or Springer-Verlag.

Lecture Notes in Physics

Edited by H. Araki, Kyoto, J. Ehlers, München, K. Hepp, Zürich
R. Kippenhahn, München, H. A. Weidenmüller, Heidelberg
and J. Zittartz, Köln

210

Cellular Structures in Instabilities

Proceedings of the Meeting
"Structures cellulaires dans les instabilités –
périodicité, défauts, turbulence de phase"
Held at Gif-sur-Yvette, France, June 20–22, 1983

Edited by J. E. Wesfreid and S. Zaleski

Springer-Verlag
Berlin Heidelberg GmbH 1984

Editors

José Eduardo Wesfreid
Laboratoire d'Hydrodynamique et Mécanique Physique
E.R.A. No. 1000 C.N.R.S.
Ecole Supérieure de Physique et Chimie de Paris
10, rue Vauquelin, 75231 Paris, Cedex 05, France

Stéphane Zaleski
Groupe de Physique des Solides
L.A. No. 17 C.N.R.S., Ecole Normale Supérieure
24, rue Lhomond, 75231 Paris, Cedex 05, France

ISBN 978-3-540-13879-2   ISBN 978-3-540-39073-2 (eBook)
DOI 10.1007/978-3-540-39073-2

2153/3140-543210

PREFACE

The need for a colloquium on the cellular structures in large aspect
ratio systems in instabilities has been considered to be necessary for
some years. Both the dynamic and fast development of investigations in
this subject and its largely pluridiscipline character add to the interest of such a meeting. The objective of this colloquium was to spread
the new findings from the last few years.

In particular the study of thermoconvective flows in large aspect
ratios has permitted new themes to be seen, both experimentally and
theoretically: These include pattern selection, driving of means flows,
defects, and phase turbulence. But if many of the results presented in
this colloquium were initially motivated especially by Rayleigh-Bénard
convection, it is possible to see in the contributions a wide range of
instabilities: instabilities in shear flows, liquid crystals, crystal-
line growth, reactive systems with spatial patterns, flames, and elastic
buckling. One important characteristic of this colloquium was the balance
in contributions on theoretical and experimental papers.

The meeting was organized in coordination with the steering committee
of the "chaos" divisional meeting of the French Society of Physics
(Grenoble 1983). We thank the members of this committee for their help
and advice. P. Bergé and the CEA should be particularly thanked for
allowing us the use of the Maison d'Hotes in Gif-sur-Yvette. E. Guyon
and all the members of the laboratory L.H.M.P. of the Ecole Supérieure
de Physique et Chimie de Paris have made an important contribution to
the organization of this meeting. We thank the staff of the Groupe de
Physique des Solides of the Ecole Normale Supérieure for assistance in
the editing of these proceedings. We wish finally to acknowledge the
DRET for their financial support (grant no. 83/1346).

J.E. Wesfreid

S. Zaleski

TABLE OF CONTENTS

CELLULAR STRUCTURES IN INSTABILITIES : AN INTRODUCTION

J.E. Wesfreid[+] and S. Zaleski[*]

[+]LHMP - ERA N°1000 - C.N.R.S.
Ecole Supérieure de Physique et Chimie de Paris
10 rue Vauquelin, 75231 Paris Cedex 05, France

[*]Groupe de Physique des Solides de l'Ecole Normale Supérieure
24 rue Lhomond, 75231 Paris Cedex 05, France

Appearance of cellular structures is a very wide-spread phenomenon among non linear systems becoming unstable. Among the most investigated ones are the Rayleigh-Bénard rolls or hexagons, the Taylor-Couette vortex system, as well as many other cellular patterns in hydrodynamical, mechanical, diffusive or physico-chemical instabilities. Many such instabilities have shown remarkable spatial patterns, such as those observed in boundary layer instabilities, buckling of long plates and shells, plane fronts or crystal growth. The experimental results as well as the theoretical approaches have often been compared to results known in other contexts. Many important results have been realized since the pioneering works of Bénard (1) and Rayleigh (2) for thermoconvective instabilities. Other advances were made primarily through the study of hydrodynamical instabilities. From those various origins a number of common features appear to be valid for a very large variety of instabilities.

As such cellular structures appear in a very wide range of systems out of equilibrium [3-4], they are of fundamental importance in many natural phenomena and technological developments [5]. Solar granulation [6-7], continental drift motion [8-9], geothermal reservoirs [10] are related to convection phenomena. It is not necessary to emphasize the technological importance of instabilities in liquid crystals [11], solidification fronts [5] [12], boundary layer instability [13], or buckling instability of plates [14-15].

This colloquium was particularly devoted to the so called "large box" [16] problems. In a Rayleigh-Bénard experiment, for instance, when the aspect ratio (i.e. ratio of horizontal dimensions to the vertical one) is large, the number of cells observed is correspondingly large. Approximate periodic patterns then appear and some classical questions of crystallography can be asked, using the concepts of optimal wavenumber, elasticity, dislocation, and phase motions. The theoretical investigations of these problems requires a deeper insight into the non linearity of the governing equations. For large aspect ratio systems, several useful methods have been developed only recently. The corresponding experiments must involve the simultaneous observation of a large number of rolls. These experiments benefited from the development of new experimental methods [17]-[26].

There has been no previous review specifically devoted to the subject of large cellular structures. General introductory volumes dealing with several instabilities can be found [27]-[31]. A general overview of convection was written for non specialists in [32] . Reviews, mostly issued in the last few years, on the hydrodynamical and convective instabilities, can be found in [33]-[42].For crystal growth instabilities, see [43]-[44]. Other instabilities were also considered in the colloquium, and we refer the reader to each particular contribution for an introduction and for references to review papers.

In this introduction we shall give first a general presentation of the various instabilities showing large cellular structures. Then we review the general properties of these patterns. A special emphasis is given to convective instabilities, as most theoretical and experimental work was done in this case.

1. INSTABILITIES WITH LARGE PERIODIC PATTERNS.

a. Thermal convection

The Rayleigh-Bénard convection occurs in horizontal layers of fluid heated from below. When the control parameter R_a (the Rayleigh number proportional to the difference of temperature across the cell) is smaller than the critical value R_{ac} the fluid is at rest and we may label this state as "laminar" or conducting. For $R_a > R_{ac}$ the system transits to the "cellular" state with thermal convection of fluid, i.e. organized motion in cellular patterns.

The thermal expansion of the fluid is at the origin of this instability. Fluctuations create small displacements of fluid, giving unbalanced pressure, and Archimedes'buoyancy further displaces the fluid. This destabilizing action is opposed by stabilizing dissipative actions as viscosity damping and temperature homogenization by thermal diffusion.

The structure of the non-dimensional parameter - the Rayleigh number: $R_a = \alpha \, g \, \Delta T \, d^3 \, / (\nu \, \kappa)$- reflects these competing effect. α, ν, κ are physical constants of the fluid : respectively expansion coefficient, kinematical viscosity, and thermal diffusivity. ΔT is the difference of temperature across the layer of depth d. This number may

be written as a combination of characteristic stabilizing and destabilizing times. The destabilizing time $\tau_a = (d/(\alpha\ g\Delta T))^{1/2}$ is typical for
buoyancy advection. The two typical stabilizing times are : the thermal
diffusion time $\tau_{th} = d^2/\kappa$, and the vorticity diffusion time $\tau_\nu = d^2/\nu$.
Indeed, the Rayleigh number may be written as $R_a = \tau_{th}\ \tau_\nu/\tau_a^2$, and it is
easy to understand that if the geometrical average of the relaxation
times is smaller than τ_a ($R_a <$ cte), the fluid can not develop convective
motion.

The main results of stability for the primary flow - at rest -
come from the analysis of the full set of equations describing the system. These are : the state equation $\rho = \rho_0$ $(1 - \alpha\ \Delta T)$, the Navier-
Stokes equation with the term of force due to buoyancy, and the heat or
energy equation. Quiescient fluid ($\vec{v} = 0$) is a stationary solution of
this system, when temperature varies with a linear law in the vertical
direction and with hydrostatic pressure variation. As is classical,
stability is studied by imposing small disturbances in the velocity $\vec{v}$,
and in the deviation from the linear profile in temperature θ, and
pressure p.
With suitable non-dimensional variables, the equations for disturbances
are :

Navier-Stokes equation : $P_r^{-1}\left|\dfrac{\partial\vec{v}}{\partial t} + (\vec{v}.\vec{\nabla})\vec{v}\right| = -\vec{\nabla}p + \Delta\vec{v} + R\theta\hat{z}$ (1)

heat equation   : $\dfrac{\partial\theta}{\partial t} + \vec{v}.\vec{\nabla}\theta = \Delta\theta + v_z$ (2)

continuity equation   : $\vec{\nabla}.\vec{v} = 0$ (3)

where the Prandtl number $P_r = \nu/\kappa$, gives the ratio between the two characteristic relaxation times : τ_{th}, and τ_ν. A high Prandtl number implies
that velocity perturbations follow the temperature perturbations, and
the remaining non-linearity is in the heat equation. For stability
analysis, the equations 1-3 are linearized. Elimination of pressure and
projection in the vertical direction gives an equivalent equation of
6th order for the vertical component of the velocity v_z :

$$\left(\dfrac{\partial}{\partial t} - \Delta\right)\left(P_r^{-1}\dfrac{\partial}{\partial t} - \Delta\right)\Delta\ v_z = R_a\Delta_h\ v_z \qquad (4)$$

with Δ_h the horizontal Laplace operator. Velocity is expanded in terms
of Fourier spatial modes in the horizontal direction, of wave-number
$q = 2\pi/\Lambda$, with temporal variation in $e^{\sigma t}$. The solution of eq.(4) depends
on the boundary conditions. In the realistic case of rigid horizontal
plates, impenetrability and adhesive conditions are satisfied when a

non trivial vertical dependency is taken.

Numerical resolution gives, for $Re(\sigma) = 0$ the critical conditions for onset of convection at the threshold parameter $R_{ac} = 1708$ and $q_c = 3.117$ (so the critical wavelength is approximately twice the depth of the layer of fluid).

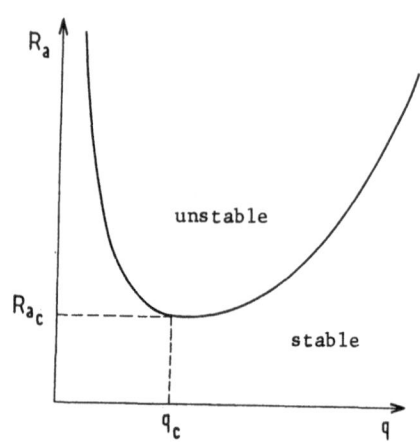

Figure 1 : Linear stability curve.

The non-realistic case, with free-free (no slip) horizontal boundary conditions satisfied by circular functions, is often used. In this case the curve of marginal stability is given by $R_{ao}(q) = (q^2 + \pi^2)^3/q^2$ (Fig.1). Near q_c, the parabolic approximation for this curve is :

$$(R_{ao} - R_{ac})/R_{ac} = \varepsilon_0(q) = \xi_0^2(q - q_c)^2$$

with $\xi_0^2 = 8/3\pi^2$ in the free-free case and $\xi_0^2 = 0.148$ [45] in the rigid-rigid case. The existence of a threshold (critical value of the control parameter) and the cellular structure (critical wave-number) are the main characteristics of the Rayleigh-Bénard convection and of many others instabilities. Analogies were also made with phase transitions [45-47] .

Buoyancy is also responsible for instabilities in the case of a fluid confined in a permeable solid matrix or porous medium, and heated from below [48-50] . The law governing the seeping flow is the Darcy law : $\vec{\nabla}p = -\rho\nu/k\ \vec{U} + \vec{f}$, where $\vec{U}$ is the average velocity and k is the permeability proportional to the square of the "diameter" of the pore. The associated convective instability is called Rayleigh-Darcy convection (see the contribution of Deltour). The Rayleigh number is essentially the same as in the case of a homogeneus medium, but the term in d^3 is modified by d.k, corresponding to the fact that vorticity diffusion occurs in the scale of the dimension of the pores. A similar situation occurs for convection in cells with very small transversal length ℓ ($\ell << d$), namely Hele-Shaw cells [51-52] . In this case $R_a \sim d\ell^2$ for the same reasons as in the porous state.

Convection can also be induced by thermo-mechanical interactions arising through the stress tensor, such as surface-tension-driven convection, provided the surface tension is a function of temperature (Marangoni effect) [53-57] . The contributions of Loulergue and Cerisier

and Pantaloni are devoted to this instability. Interesting thermal instabilities arise in mixtures with both thermal gradients and concentration gradients present, as well as in double diffusive instabilities
[58-59]. Thermal instabilities in nematic liquid crystals in which the
heat diffusivity anisotropy of nematics raises the heat focusing, produced a wide variety of cellular patterns [11] [60-61].

b. Taylor-Couette instability and other hydrodynamical instabilities

The Taylor-Couette instability is one of the best studied hydrodynamical instabilities. G.I. Taylor first calculated the linear instability threshold and investigated experimentally the appearance of
vortices [62]. In this experiment the flow is set in rotation by two
concentric cylinders. Usually, only the inner cylinder is rotating. At
low rotation speeds, one has a cylindrical Couette flow. When the angular speed of the inner cylinder exceeds a critical value, a pattern of
superposed vortices, usually axisymmetric, is observed. This secondary
motion is explained by the existence of an unstable stratification of
angular momentum : when a droplet goes away from the inner cylinder, it
keeps its angular momentum. But the resulting centrifugal force at its
new location is not balanced by the pressure gradient. It then goes
further away. The law of flow conservation implies that the motion is
organized in counter rotating vortices.

Since the pionneering work of Taylor, Taylor-Couette vortex
flow has been subject to intense experimental and theoretical investigation. A recent review can be found for instance in [41].

Görtler instability [63] [64] [65] arises in viscous boundary
layers over concave curved surfaces, on the inner side of airfoils for
instance. It has the same origin : an unstable stratification of angular
momentum. However, as the shape of the boundary layer and the curvature
are spatially varying, the instability involves a strong spatial dependency. This causes considerable theoretical difficulties (see the contribution of M.H. Peerhossaini).

Another instability in rotating flows is the instability of
flow over a rotating [66] [67] disk where a boundary layer of constant
depth exists. This instability has similar properties of spatial
dependency as the previous one. But the physical origin of the instability is the inflexional instability for the boundary layer. It was presented at the colloquium by M.P. Chauve. Many other periodical instabilities
arise in stratified fluids, such as the Kelvin-Helmholtz or Rayleigh -
Taylor instabilities. They are not discussed in this volume : a review
can be found in [68].

c. Other physical and chemical instabilities

A very attractive instability is the electro-hydrodynamical one [69]. It is an instability of dielectric fluids between parallel plates. A description of the driving mechanism has been given by Felici [70], among others:when a roll motion is initiated in such a fluid, it concentrates positive ions near the positive plate and thus creates a larger electrical field at this place. This field in turn enhances the motion. When the fluid is a liquid crystal, an instability can be created with a very large aspect ratio [71] : this is exemplified by the communications of J. Gollub, A. Joets, and R. Ribotta.

There are also very many instabilities related to front propagation [43], [72-74]. Instability of a solidification front was presented by B. Caroli et al. A local bulge in the solidification front can result in a higher concentration gradient in front of it, and enhance the growth of the bulge. The stabilizing effect in this case is the surface tension.

Front instabilities are also observed in combustion fronts [75]-[76] as shown by J. Quinard et al. Spatial structures also arise in some chemical instabilities [77]. Reactive instabilities were presented in the contribution of M.Gimenez and J.C. Micheau. They described two coupled mechanisms:a chemical mechanism involving photo-chemical compounds, and a thermal convection phenomenon due to evaporative cooling.

Our colloquium was not exhaustive in the analysis of instabilities with spatial organization. We mention here, among others the great variety of effects obtained in a ferrofluid by application of a magnetic field [78] [79] .

The results obtained from these studies may be useful for understanding patterns of living organisms [80] or pattern formation inside living organisms [81] [82] [83] .

d. A tentative classification

Many analogies can be made between various large aspect ratio structures of very different origin. A tentative explanation of the different possible behaviours, like transition to turbulence could be based on the possible instability modes in such structures. It is then interesting to estimate,a priori,the possible instability modes by an investigation of the symmetries of the problem.A first category of problems includes one dimensional structures, such as perfectly parallel rolls in Rayleigh-Bénard convection, or axisymmetric flow for the Taylor-Couette instability . Many more instability modes can be found in two dimensional structures. It allows for various planforms, such as rolls, squares or hexagons, with arbitrary orientation and phases. The resulting pattern is often marginally stable with respect to small rotations or small phase motions of the structure. However, in very many systems the rolls have a preferred orientation. In Taylor-Couette flow, they tend to be parallel to the azimuthal direction. In Rayleigh-Bénard convection with a horizontal magnetic field, the rolls tend to align with the field (see the communication of S.Fauve et al.).This is also the case in many electro-hydrodynamical (E.H.D.) instabilities such as those repor-ted in this volume.

The system is fully isotropic for convective structures, such as Rayleigh-Darcy convection in porous media. However the stability of the structure also depends on the more or less non linear character of the instability. In the Rayleigh-Bénard problem, the Prandtl number P_r is low when the inertial terms are important (see part 1.a). Convection in low Prandtl number fluids has characteristics of its own, linked to its highly non linear character, even slightly above onset.

2. THE INFINITE PERIODIC PATTERNS

The first non-linear expansion for the Rayleigh Bénard problem was made by Sorokin [84] and independently by Schluter et al. [85] and Malkus and Veronis [86]. Those authors calculated finite amplitude solu-tions of the Oberbeck-Boussinesq equations, representing convective mo-tion organized in parallel rolls (fig.2). They made an expansion in powers of $\varepsilon = (R_a - R_{a_c})/R_{a_c}$.Gor'kov [87] found by an analogous method a hexagonal pattern.

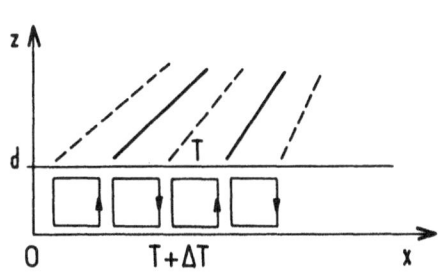

Figure 2 : convection rolls

In all these cases, the bifurcation to the convective state is super-critical i.e. second order or conti-nuous, as shown in [88] (see also [29]). The amplitude of the velocity and temperature perturbations is proportional to $\varepsilon^{1/2}$ near threshold (it was experimentally verified in [89]). This was also found for the Taylor-Couette flow through calcula-tion of the amplitude of the periodic flow (see [41] and for other hydro-dynamical instabilities, ref [35]).

In the case of Rayleigh-Bénard instability , a question can imme-diately be asked : which will be the planform of the convection [90], i.e., will it be organized in rolls, hexagons or squares ?

Stable hexagons are found in experiments only when there is an asymmetry in the vertical direction [91] [92] [93]; corresponding calculations
were made by Palm [40]. Hexagons appear then through a first order transition. Square patterns can be oberved especially as a secondary motion
far above onset, in high-Prandtl convection [94] - [96]. They are also
predicted when horizontal plates are poorly conducting [97] .

   The non-linear theory allows for a family of solutions of different
wavenumber around the critical one q_o (fig.3). The dependency of solutions in the wavenumber cannot be calculated by a small ε expansion
in the region far above threshold, but was investigated in [98].

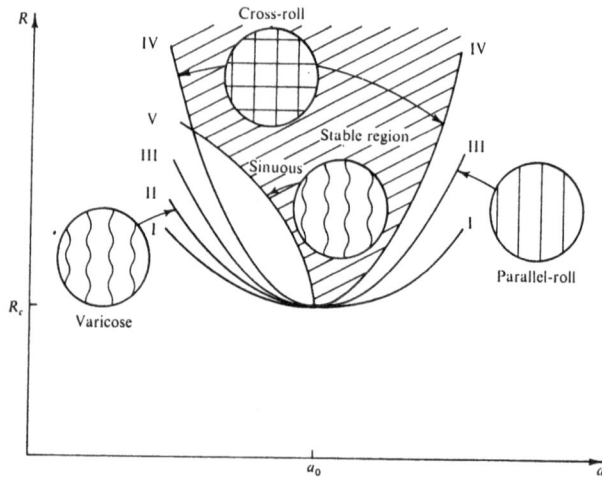

Figure 3 : Sketch of instabilities near threshold for convection at
P_r = ∞ and in porous media : Parallel roll (or Eckhaus), varicose, cross
roll and sinuous (or zig-zag) .[29]

   Once a periodic steady state is known, the next (and indeed much
more difficult) question is : is it stable ? This question leads to
various others, among them, is the obtained pattern stable with respect
to the growth of a small perturbation ? Which will be the final resulting planform   [99] ? Is there a minimization principle ? For a given planform, is there a preferred wavenumber [100] [101] ? Those questions certainly do not have fully analytical answers. The experiments,
wether real life or numerical, are also very difficult in comparison to
similar ones in "small boxes". One of the main difficulties comes from
the very long relaxation time needed to reach an equilibrium state (if
any such state exists).
   We postpone the wavelength selection problem to the next section,
and turn to the stability problem. The first stability calculations

were made by Schlüter et al. in [85] for convective structures near
onset. These results, extended to the fully non linear region are to be
found in [39].

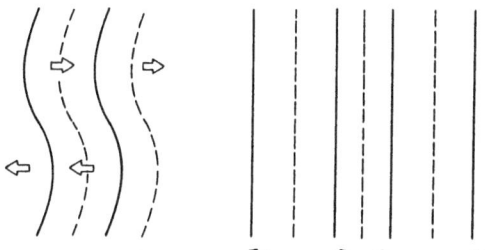

Figure 4 : Two typical instabilities. Lines represent ascending or sinking fluid (see fig.2). Left : zig-zag (sinuous) instability. A large
scale drift or permeation current can be generated (arrows), especially
at low P_r. Right : Eckhaus instability.

Four main types of instabilities near threshold of the roll system
were discussed in [39] : cross roll, skewed varicose, zig-zag and
Eckhaus. The first instability consists in the growth of a roll system of
different wavelength oriented perpendicularly. The skewed varicose is
characterized by a deformation of the nodal lines into a varicose pattern.

The two latter instabilities are long wavelength instabilities :
they appear marginally in the limit of zero wavenumber. The Eckhaus
[102] instability (fig.4) consists in periodic dilation   and compression of the roll system. It is purely one dimensional. As shown in [102],
it appears marginally on a curve in the R_a, q plane defined by near
threshold:

$$R_{a_E}(q) - R_{a_c}(q_0) = 3\left[R_{a_c}(q) - R_{a_c}(q_0)\right]$$

A similar instability has been shown to exist for the Taylor Couette
vortex system in [103], with the same universal law near threshold
[104] [105]. Eckhaus instability is a universal feature in 1D systems
with breaking of translational symmetry, near threshold. The zig-zag
instability consists in a wavy deformation of the nodal lines (fig.4).
For such instability problems which have a potential formulation, q_{zz}
is the optimal wavenumber, i.e. the one which minimizes the "energy" of
the structure, as shown for instance in [106]. This implies that the
corresponding structure is marginally stable with respect to the sinuous,
or zig-zag instability. However, potential formulations are known not

to exist in most problems : they imply that the system monotonically relaxes to a minimum of the potential. This is contradictory with the averted existence of permanent time-dependent states in many instabilities. In the plate buckling problem [14] [107], on the other hand, such a potential exists.

Above threshold, oscillating, or propagating, instabilities may occur, such as the oscillatory instability in Rayleigh Bénard convection at low P_r. In Taylor-Couette flow, in a similar way, a wavy vortex flow appears above T_c [41]. The stability boundaries for Taylor vortex flow have been investigated recently [108], as well as the effects of bottom and top boundaries on the wavy flow [109].

3. ONE DIMENSIONAL STRUCTURES IN FINITE LENGTH : THE WAVENUMBER SELECTION PROBLEM

a. finite size effects

We now turn to a more realistic description of the instability, and wish to account for the influence of finite distance walls. Two kinds of effects have been discussed in this context : the influence of the walls on the onset of the instability, and its influence on the selected wavenumber in the bulk. The presence of boundaries always raises the threshold value R_{a_c} : in a box of length L, $R_{ac} = R_{a_{C_\infty}} (1+\xi_0^2 \pi^2/L^2)$ at first order. When these boundaries exert a destabilizing effect on the system (for instance, heating boundaries in convective problems, or end caps in Taylor-Couette), an imperfect, or smooth, bifurcation occurs [110] [111]. In Taylor-Couette flow, the influence of imperfect boundaries was intensely investigated [112].

The effect of boundaries on wavenumber selection is very different : contrary to the previous ones, this effect is felt in the whole system, however distant are the boundaries. Contrary to the previous ones, it is not a critical effect. All these effects can be accounted for in the framework of the amplitude expansion.

b. Wavelength selection

The determination of the stability of periodic structures with respect to various instabilities, such as Eckhaus or zig-zag, usually leaves a band of stable wavenumbers of width $\epsilon^{1/2}$ near threshold (fig.3). However, experimental study and numerical simulation [114] (on models) of convective roll structures [101] [113] (and others) [100] - [118] usually shows a single wavenumber or a distribution of wavenumbers around a medium one [115] - [116] with a rather small dispersion. For the Taylor-Couette system, however, the wavenumbers are dispersed in a large band, as shown experimentally by Snyder [117].

A first mechanism was proposed [119] - [120] : it shows that the effect of sidewalls, however distant, is to restrict the extent of possible wavenumbers to a band of width $\sigma(\varepsilon)$. This mechanism has undergone several experimental tests [121] [122], and new theoretical analyses [123] were made for the different boundary conditions on the lateral walls. They are reported in the communications of L. Kramer and P.C. Hohenberg, B.Martinet et al., M. Boucif and J.E. Wesfreid, M. Potier-Ferry, and S. Zaleski. For various types of boundary conditions the band of selected wavenumbers is broadened and the lateral walls allow for the whole $\sigma(\varepsilon^{1/2})$ band.

Other wavelength selection mechanisms have been proposed. It was related to marginal stability in the case of Eutectic solidification [124]. It was shown in [125] [126] that for axisymmetric structures, the wavenumber was uniquely determined. Experiments by V. Croquette and A. Pocheau, reported in this book, investigated such structures. Another situation of interest is realized when some external physical parameter of the instability, such as the height of the cell or the temperature difference, has a slow spatial variation [127]. An adiabatic solution of the governing equations can then be constructed. When the variation of the parameters connects a region where the conditions are subcritical to a supercritical region, the wavenumber in the latter region is uniquely determined [128] - [130]. However this is true only if the variation is slow enough : the related non adiabatic effects are discussed in the contribution of Y. Pomeau.

Other mechanisms have been proposed, that involve time dependent effects and can be labelled dynamical : the climb of dislocations in cellular structures (fig.5), as investigated in [131] and [132], is shown to add or remove rolls from the structure, thus letting the mean wavenumber evolve to a final one q_{dis} such that $q_{dis} - q_c = \sigma(\varepsilon)$. The motion of grain boundaries in cellular structures can also result in a tendency to select a given wavenumber [133]. Both these situations must be considered in the general framework of the study of two dimensional structures.

4. SPATIAL ORGANISATION OF TWO DIMENSIONAL STRUCTURES

Whereas much is known presently on one dimensional structures, there is less information about two dimensional problems. In this section we consider only time independent or almost stationary structures. After an introduction to the phenomenology, we briefly review the available theoretical methods.

a. Two dimensional structures

In what follows, we stress the problem of examining which structure will appear in an experiment with perfectly random initial conditions [134] - [139]. It is possible in general to induce a given pattern using for instance a thermal grid [140]-[142]. However, we suppose that no such forcing is present. What can be known about the resulting structure ? Will it be stationary ? A first way to approach the problem is to investigate the various types of defects [16] [113] [141] [143] in the structure : dislocations [142],[144] - [145], grain boundaries [146], umbilics [147], and the matching with lateral walls [148] [149]. What will be then the stationary arrangement of these patterns in a finite box ? In some way, the previous questions lead to a crystallography of the dissipative structures [150] very similar to the problem of layered patterns in smectics for instance [151]. A first very general property of finite structures is the tendency of rolls to be orthogonal to lateral boundaries [135], [140], [146] in the case when the problem has intrinsic rotational invariance. The theoretical argument was given in [119] and developed further in [152] - [153]. In non isotropic instabilities the presence of lateral boundaries also influnces locally on the roll orientation, as shown in the communication presented by E. Guazzelli. In both cases, the merging of rolls with lateral boundaries seems to occur with a preferred orientation, and it implies that no perfectly periodic pattern is possible : there are dislocations and/or grain boundaries [133] [136] present in the box.

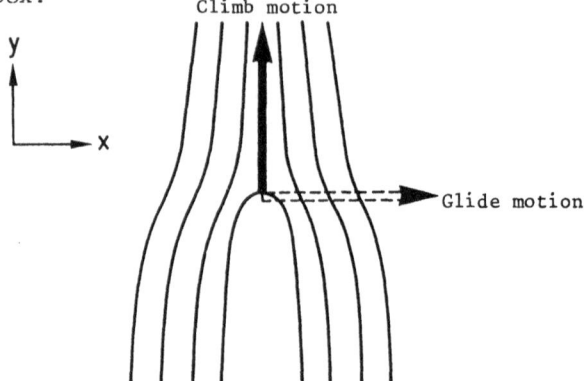

Figure 5 : Climb and glide motion of a dislocation. The sign of the motion may change.

Dislocation motion was first studied experimentally by Whitehead [142]. This author considered an array of regularly spaced dislocations rather far from threshold. Spontaneous occurence of dislocations is frequently observed, and reported in the communications of

Cerisier and Pantaloni, and Ribotta and Joets. Siggia and Zippelius
[131] investigated theoretically a potential approximation of the original equations for convection. They calculated the speed of climb of an
isolated dislocation in a periodic pattern of wavenumber q

$$v_{climb} \sim (q - q_c)^{3/2}$$

A correction to these results was introduced in [132]. Whereas in the
potential case the wavenumber for a stationary dislocation is marginal
with respect to the zig-zag instability, in the more potential case it
is not determined a priori. Whereas climb motion is relevant for wavenumber selection, experimental evidence [154] shows that the displacement of dislocations also consists in glide motion. This glide motion was investigated theoretically in [132].

Perhaps the most difficult question is to determine which are
the stable static structures in a finite box, if they exist. What will
be the final state in a real life or numerical experiment of large
aspect ratio ? How does a given roll structure manage to meet the per-
pendicularity condition on all lateral walls ?

There is no unique answer to this question, despite a tentative
formulation in a potential framework near $\varepsilon = 0$ [152]. Various parameters are involved : geometrical properties of the instability, as mentionned in part 1.d, distance to the threshold, size of the box [155].
In the rather large aspect ratio experiments presented by J.P. Gollub,
there seems to be rather few dislocations, and the structure is made
of sections of curved rolls connected together. This structure might be
the starting point of something like close packing of disks [156] [157]
that would be apparent in much larger aspect ratio systems. In some numerical experiments on model problem [158] a region of parallel rolls
seems to dominate the structure (communication of P. Hohenberg). The
numerical experiments of P. Manneville (this volume) allow one to investigate the effect of the Prandtl number. In all these experiments, the
time of relaxation to the final steady state is very long, and is often
very much larger than the horizontal diffusion time, which is of
order $\frac{L^2}{\nu}$ or $\frac{L^2}{\kappa}$ for a box of size L. This very long transitory time
partly explains why it is often difficult to decide wether a given struc-
ture has reached a "turbulent" regime (non-reproducible states) or still
has a stable steady state.

b. Theoretical and numerical methods

The most widely used theoretical method near onset of convective
motion is the slowly varying amplitude theory, which originated in the

multiple scale expansion widely used in hydrodynamics [35]. Consider
the velocity (u, v, w) and temperature perturbation θ in a convection
problem. Then for rolls parallel to the y axis, a development of the
perturbation can be made in the form

$$\theta(x,y,z) = A(x,y)\, e^{iq_0 x}\, f(z) + c.c. + \mathcal{O}(\varepsilon^{3/2})$$

$$u(x,y,z) = A(x,y)\, e^{iq_0 x}\, g(z) + c.c. + \mathcal{O}(\varepsilon^{3/2})$$

$$w(x,y,z) = A(x,y)\, e^{iq_0 x}\, h(z) + c.c. + \mathcal{O}(\varepsilon^{3/2})$$

where c.c. stands for complex conjugate, f, g, h, carry the vertical
dependency of linear modes (i.e. $e^{iq_0 x} f(z) + c.c.$ is the critical mode
for the linear problem at $R_a = R_{a_c}$). In agreement with previous results,
$A(x,y)$ is $\mathcal{O}(\varepsilon^{1/2})$. Moreover, its variation is assumed to be small, i.e.,
one has for the derivative applied to A :

$$\partial_x = \mathcal{O}(\varepsilon^{1/2}) \qquad\qquad \partial_y = \mathcal{O}(\varepsilon^{1/4})$$

Replacing u, v, w and θ in the original O.B. equations by their expansion
in powers of ε leads to a hierarchy of equations involving A and its
slow derivatives. In most cases, this gives at first order (i.e. $\mathcal{O}(\varepsilon^{3/2})$)
the amplitude equation of Segel [159] and Newell and Whitehead [160].

$$\tau_0\, A_t = \varepsilon A + \xi_0^2 \left(\partial_x + \frac{i}{2q_0}\partial_y^2\right)^2 A - \frac{|A|^2}{A_0^2} A \tag{5}$$

where A_t stands for $\frac{\partial A}{\partial t}$ etc..., $\xi_0,\ \tau_0,\ A_0,$ are the length, time and
amplitude scales. The above equation can be written in a potential form :
$\tau_0\, A_t = -\frac{\partial F}{\partial A}$ where F is the functional:

$$F(A) = -\int \left(\frac{1}{2}\varepsilon A^2 - \frac{1}{2}\xi_0^2\, |(\partial_x + \frac{i}{2q_0}\partial_y)A|^2 - \frac{|A|^4}{A_0^2} \right) dx$$

When the y dependency is dropped, one gets a space dependent Landau-
Ginzburg model. This leads to many analogies with problems in condensed
matter physics. The amplitude equation was calculated for the 1 dimensional case of Taylor-Couette flow in ref [161].

The effects of lateral walls, such as those investigated experimentally in [45] [162], can be explained in the amplitude formalism. The
amplitude equation was verified experimentally in the Taylor-Couette
case (without azimuthal dependency) in [163]. An amplitude equation for
two systems of perpendicular rolls was proposed in [164]. The axisymmetric pattern was investigated using amplitude formalism in [165]

The above amplitude equation in its 2-d form is, however, valid only for systems with rotational invariance. When this invariance is lost the amplitude equation takes a different 2-d form (see communication of P. Tabeling).

In the case of convection in long vertical containers (possibly cylindrical) an expansion in a slowly varying amplitude can also be made, as shown in the communication of C.Normand.The amplitude equation can be written at second order, i.e. $\sigma(\varepsilon^2)$ [120] [123] [129] [132]. The amplitude equation at this order is necessary to get the "thin" effects one expects when $q - q_c = \sigma(\varepsilon)$. As shown above, there is a variational formulation at first order, and it is necessary to push the development to second order to get a non variational problem. Calculations of wavelength selection or of dislocation motion when $q - q_c = \sigma(\varepsilon)$ require a second order amplitude expansion [132].

The above first order amplitude equation (5) is incorrect for Rayleigh-Bénard convection with free-free (no slip) boundary conditions. A coupling must be introduced with the vertical component of the vorticity $\vec{\omega}$, where $\vec{\omega} = \vec{\nabla} \wedge \vec{v}$ [166]. For convection between rigid plates, this coupling arises at next order [168]. The vertical vorticity ω_z is known to be zero for convection in porous media and at $P_r = \infty$ [167]. When it is not zero, it is coupled with the amplitude $A(x,y)$ and has a slow variation near threshold. A non zero ω_z amounts to the existence of a large scale drift velocity field, or permeation current, with non zero components U and V in the horizontal plane only [154] [168]. This drift velocity can arise when rolls are curved as shown on figure 4. Related phenomena are discussed in the communication of P. Manneville and of J. Massaguer and I. Mercader Calvo.

c. Phase diffusion equations

An alternative approach to large aspect ratio structures was introduced by Pomeau and Manneville [106], with the use of diffusion equations for a suitably defined phase variable. Consider a stationary periodic structure above threshold with for instance a temperature perturbation and a velocity field $\theta_0(x,y,z)$, $u_0(x,y,z)$ etc... Translational invariance ensures that $\theta_0(x + \phi, y + \psi, z)$, $u_0(x + \phi, y + \psi, z)$ is also a solution if ϕ, ψ are some constants. Assume now that the phases are slowly varying functions of x and y : $\phi = \phi(x,y)$, $\psi = \psi(x,y)$. Then an expansion can be made in the form

$$\theta = \theta_0(x + \phi, y + \psi, z) + \theta_1 + \theta_2 + \dots$$

$$u = u_1(x + \phi, y + \psi, z) + u_1 + u_2 + \dots$$

etc..., and where θ_1, u_1 are $\sigma(\vec{\nabla}\phi)$, θ_2, u_2 are $\sigma((\nabla\phi)^2)$ etc... When inserted in the original equation, this development leads to a solvability condition that amounts to the following equation for a roll structure :

$$\phi_t = D_{/\!/} \phi_{xx} + D_\perp \phi_{yy} \qquad (6)$$

where $D_{/\!/}$ and $D_\perp$ are diffusion constants (note that for a roll structure ψ is absent). These constants have been measured, investigating space and/or time dependency of the rolls' phase [169]-[171]. Equation (6) allows to recover the zig-zag and Eckhaus instabilities which occur respectively for $D_\perp = 0$ and $D_{/\!/} = 0$ (for stability with respect to zig-zag, see [29] [39] [172]).

The diffusion constants $D_{/\!/}$ and $D_\perp$ can be considered as analogs of Lamé coefficients in elasticity. They can be used to estimate the large deformation field around dislocations [173]. Close to threshold, $D_\perp$ takes the form $D_\perp = C(q - q_c) + E$, where C is a positive constant (it can easily be deduced from the amplitude equation) and E is a positive or null constant. If E is non zero at threshold it tends to align the roll structure with some preferred direction [173] [174]. This occurs in non isotropic instabilities.

The phase diffusion expansion does not assume vicinity to threshold, but only that the gradients $\vec{\nabla}\theta$ and $\vec{\nabla}\psi$ are small, and that the structure is submitted to small rotations only. To allow for large scale drift motions and rotations of the roll axis, more general developments have been proposed recently (see communications of P. Hohenberg and G. Dewel et al). There is no analytical theoretical method presently available to investigate the fully non linear regime with the short wavelength perturbations present in experiments, such as the core region of dislocations, grain boundaries, etc... To get more information we have to rely on numerical models.

d. Numerical methods

Simulations of the original equations governing a given instability have, up to very recently, been made only for two dimensional problems [175]-[178]. Especially for "large box" problems, numerical simulations in three dimension are seldom. They should, however, now become possible with the availability of large vector computers [131]. P. Caltagirone (this volume) presented such simulations for convection in porous media. Some numerical simulations were made on linearized model equations [179] [180].

Other numerical simulations of large scale systems were made recently on non-linear two dimensional model equations. These models display spatial organization with wavenumber q around q_c, and may be obtained by an approximation near threshold of the original Oberbeck - Boussinesq (O.B.) equations for three dimensional convective flow [46]. Such equations are related to models in condensed matter physics 181. They are of the general form:

$$u_t = \varepsilon u - (\Delta + q_c^2)^2 u + NL(u) \tag{7}$$

where NL(u) is a non-linear combination of u and its derivatives. The simplest expression leading to a supercritical (or continuous) bifurcation is $NL(u) = u^3$. Boundary conditions (B.C.) of the form $u = u_{,n} = 0$, where $u_{,n}$ is the derivative normal to the boundary, model lateral no-slip B.C. in convection and are usually taken.

These equations are simpler than the O.B. equations and much easier to integrate numerically. They show spatially disordered patterns reminiscent of those experimentally observed. They allow one to model the main features of large aspect ratio cellular patterns : sensitivity to zig-zag and other instabilities of periodic structures, dislocation motion, etc... However no turbulent or chaotic behaviour has yet been found in such models. Simulations of eq(7) with $NL(u) = u^3$ were reported in [158], showing large straight roll patterns in bulk region. Equation (7) with $NL(u) = \delta u^2 + u^3$ where δ is some constant parameter [182] was also simulated, and it led to hexagons through a first order transition [183].

P. Manneville (this volume) presented simulations of a model system of equations derived from the Oberbeck Boussinesq equations for thermoconvective instability slightly above threshold. Drift velocity terms were included in this model, and appeared to be more and more important as Prandtl number decreases.

5. TIME DEPENDENT AND TURBULENT BEHAVIOUR

a. Transition to chaos

Hydrodynamical and thermoconvective instabilities were largely used to illustrate the concept of deterministic chaos in systems with a small number of degrees of freedom ("small boxes"). It is now conjectured that there are only a few possible stable routes or "scenarios" for the transition to chaos, (except degenerate or "non generic" cases). These routes seem to be well typified presently [184] [185]. For large aspect ratio systems that are considered in this volume, another situation can occur, where a chaotic state was observed just at onset of convection. The experiments of Ahlers and Behringer [155] [186] [187]

show a broad band low frequency noise in the time spectrum of convective
motions. As shown in the communication of S. Fauve et al. (this volume)
and in [188], this behaviour is related to the large number of unstable
modes, or effective degrees of freedom present in low Prandtl Rayleigh-
Bénard convection in large boxes. The frequency spectrum for the chaotic
state in large box instabilities was observed as also reported in [69]
and [155]. The obtained power law dependency might be connected with the
aspect ratio.

The influence of medium range or high Prandtl number for large
aspect ratio systems is evidenced by the experiments reported by
J.P. Gollub (this volume) and V. Croquette and A. Pocheau (this volume).
In these experiments, it seems possible to reach a stationary state above threshold through very long transients. This is to be contrasted with
the finding of low frequency noise for convection in mercury or liquid
helium (i.e. low P_r number fluids). The systematic search for unstable
modes attempted in section 1d. seems to offer a qualitative understanding of the existing variety of transitions to turbulence in cellular
structures. In particular, the shape of the box, circular or rectangular,
influences the existence of more or less unstable modes. There are no
satisfactory theoretical methods up to now to deal with the problem of
weak turbulence in large systems. More quantitative experimental research
of chaotic systems of large aspect ratio is also needed.

Some information on chaos with a large number of degrees of
freedom (a situation expected to occur in hydrodynamical turbulence as
well) can be extracted from numerical simulations of some one dimensional
models of the form (7) [189] [190] with for instance $NL(u) = uu_x$ or u_x^2,
and $\varepsilon = q_c^4$, showed chaotic behaviour [191]. It approximates the problem
of convection between poorly conducting plates [192] [193].

Transition to turbulence in systems with a small number of effective degrees of freedom, but with large aspect ratios, such as the
Taylor-Couette system, is described in the contribution of P. Tabeling
and C. Trakas, and shows a scenario of transition to turbulence typical
of systems with few excited modes. This seems to be related to the strong
tendency of Taylor vortices to align with the azimuthal axis, a phenomenon reminiscent of the alignment of rolls with the magnetic field in the
convective instability of mercury. A feature shared by both systems is
the strictly positive diffusion coefficient $D_\perp$ at threshold [173] [174]
[194].

b. Spatial disorder in cellular structures

Turbulent motion is generally obtained in cellular structures
when the control parameter is raised. A spatial structure is often still

quite recognizable, as in turbulent Taylor-Couette vortex flow [195] [196].
It can undergo a number of transitions at higher values of the control
parameter. A nematic liquid crystal under elliptical shear [138] undergoes a transition from a roll structure to a square lattice of convection cells. The same square pattern is observed in EHD instabilities
[139], where at even higher values of the control parameter, a full disappearance of the structure is observed : it is replaced by dynamic
scattering modes, as reported in the communication by R. Ribotta and
A. Joets. For Rayleigh-Bénard convection, ordered cells may be replaced
by disordered rising thermal droplets (spoke pattern) [39], and large
scale motion [197].

A study of spatial disorganization of structures in Bénard-
Marangoni and stressed nematics was made by R. Ocelli et al. (this volume) through a measurement of correlation functions for position and
orientation of cells. An analogy with phase transitions in bidimensional
systems was made [197] [198] [200].

In an EHD instability experiment [201] presented by J.P. Gollub,
a spatially periodic modulation of the electric potential was present.
Commensurability effects arise from the competition between the natural
wavelength and the external modulation. They are reminiscent of similar
effects in condensed matter physics, and of frequency locking phenomena
in dynamical systems.

6. CONCLUSION

Most of the recent advances in the general understanding of large
aspect ratio cellular structures in instabilities are presented in this
volume. They generally call for further experiments or new theoretical
analyses. Problems related to the space pattern, such as the tiling of
large boxes with almost parallel rolls, are as yet unresolved. An understanding of transition to turbulence in "large boxes" is also lacking. More
generally, cellular structures with a large number of degrees of freedom might help to get closer to a better understanding of classical
"fully developed" turbulence, where coupling between temporal and spatial modes is also present.

REFERENCES

[1] Bénard, H. :"Les tourbillons cellullaires dans une nappe liquide".
Revue générale des sciences pures et appliquées, 11, 1261-1271 and
1309-1328 (1900).

[2] Rayleigh, Lord : "On convection currents in a horizontal layer of
fluid, when the higher temperature is on the under side". Phil.
Mag. 32, 529-546 (1916).

[3] Haken, H. : "Cooperative phenomena in systems far from thermal equilibrium and in non physical systems". Rev. Mod. Phys. 47, 67-121
(1975).

[4] Order and fluctuations in equilibrium and non-equilibrium statistical mechanics. Ed : G. Nicolis, G. Dewel and J.W. Turner, Wiley,
New York, 1981.

[5] Convective transport and instability phenomena. Ed : J. Zierep and
H. Oertel Jr., G. Braun, Karlsruhe, 1982.

[6] Wasiutynski, J. : "Studies in hydrodynamics and structure of stars
and planets". Astrophysica Norvegica, vol.4, J. Dybwad,Oslo, 1946

[7] Problems of stellar convection. Ed. : E.A. Spiegel, and J.P. Zahn,
Springer, Berlin, 1977.

[8] Parsons, B. and McKenzie, D.   : "Mantle convection and the thermal
structure of the plates ".J. Geophys. Res.83,4485-4496 (1978).

[9] Froidevaux, C. and Nataf, H.C. : "Continental drift : What driving
mechanism ?". Geol.Rundschau   70, 166-176 (1981).

[10] Ping Cheng and Lall Teckchandani. : "Numerical solutions for transient heating and fluid withdrawal in a liquid-dominated geothermal reservoir". Geophysical Monograph 20, (p.705-721), The Earth's
crust, by the American Geophysical Union, Washington, 1977 .

[11] Dubois Violette,E., Durand, G., Guyon, E., Manneville, P., and
Pieranski, P. : "Instabilities in nematic liquid crystals". Solid
State Phys., Supp.14, p.147-208 (1978).

[12] Dauzère, M.C. : "Solidification cellulaire". Annales de Physique,
Ser.9, t XII, p.7-106 + plates I-XII, (1919).

[13] Görtler, H. : "On the three-dimensional instability of laminar
boundary layers on concave walls". N.A.C.A. Tech. Memo. 1375 (1954).

[14] Koiter, W.T. : " The stability of elastic equilibrium". Tech. Rep.
AFF DL-TR-70-25, Air Force flight dynamics laboratory, U.S.A. (1970).

[15] Kragerup, J. : "Five notes on plate buckling". Rep. R143 Dep. of
Struct. Eng., Tech. Univ. of Denmark (1982).

[16] The distinction between "small boxes" and "large boxes" was emphasized by P. Bergé : "Experiments on hydrodynamic instabilities and the
transition to turbulence" in Dynamical critical phenomena and related
topics. Ed. : C.P. Enz, p.288-308, Springer, Berlin, 1979 .

[17] Cloupeau, M., Klarsfeld, S. and Grossin, R. : "Visualisation d'isothermes dans un milieu poreux par effet Christiansen". C.R. Acad. Sc. Paris, 269B , 163-166 (1969).

[18] Ahlers, G. : "Heat capacity near the superfluid transition in He⁴ at saturated vapor pressure". Phys. Rev. A 3, 696-716 (1971).

[19] Bergé, P. : "Aspects expérimentaux de l'instabilité thermique de Rayleigh Bénard". J. Physique, Colloque Cl, 37, C1-C23 (1976).

[20] Cooper, T.E., Field, R.J. and Meyer, J.F. : "Liquid crystal thermography and its application to study of convective heat transfer". ASME Journal of Heat Transfer, 97, 442-450 (1975).

[21] Oertel, H. and Kirchartz, K.R. : "Laser-anemointerferometer for simultaneous measurements of velocity and density". Appl. Opt., 17, 3535-3538 (1978).

[22] Meynart, R. : "Equal velocity fringes in a Rayleigh-Bénard flow by a speckle method". Appl. Opt. 19, 1385-1386 (1980).

[23] Moreno, J., Jimenez, J., Córdoba,J., Rojas, E. and Zamora, M. : "New experimental apparatus for the study of the Bénard-Rayleigh problem". Rev. Sci. Instrum. 51, 82-85 (1980).

[24] Ernst, K. and Hoffman, J.J. : "Laser induced convection instability". Phys. Lett. 87A, 133-136 (1981).

[25] Ueda, M., Kagawa, K., Yamada, K., Yamaguchi, C., Harada, Y. : "Flow visualization of Bénard convection using holographic interferometry". Appl. Opt. 21, 3269-3272 (1982).

[26] Koster, J.N. : "Interferometric investigation of convection in plexiglas boxes". Experiments in Fluids 1, 121-128 (1983).

[27] Chandrasekhar, S., Hydrodynamic and Hydromagnetic stability, Dover, 1981.

[28] Gershuni, G.Z. and Zhukhovitskii,E.M., Convective stability of incompressible fluids, Keter Publishing House, Jerusalem, 1976.

[29] Joseph, D.D., Stability of fluid motions, vol. II, Springer, Berlin, 1976.

[30] Platten, J.K. and Legros, J.C., Convection in liquids, Springer, Berlin, 1983.

[31] Lin, C.C., The theory of hydrodynamic stability. Cambridge University Press, Cambridge, 1955.

[32a] Velarde, M. and Normand, C. : "Convection". Scientific American 243, n°1, 92-108 (1980).

[32b] Bergé, P. and Pomeau Y. : "La turbulence", La Recherche 11, 422-432 (1980).

[33] Hopfinger, E.J., Atten, P., Busse, F.H. : "Instability and convection in fluid layers : a report on Euromech 106". J. Fluid. Mech.
92, 217-240 (1979).

[34] Hydrodynamic stability and the transition to turbulence. Ed :
Swinney ,H.L., and Gollub, J.P., Springer, Berlin, 1981.

[35] Stuart, J.T. :" Nonlinear stability theory". Ann. Rev. Fluid Mech.
3, 347-370 (1971).

[36] Koschmieder, E.L. : "Bénard Convection". Adv. Chem. Phys. 26,
177-212 (1974).

[37] Rogers, R.H. : "Convection". Rep. Prog. Phys. 39, 1-63 (1976).

[38] Normand, C., Pomeau, Y. and Velarde, M.G. : "Convective instability
A physicist's approach."Rev. Mod. Phys. 49, 581-624 (1977).

[39] Busse, F.H. : "Non-linear properties of thermal convection". Rep.
Prog. Phys. 41, 1929-1967 (1978).

[40] Palm, E. : "Non linear thermal convection" in Non linear phenomena
at phase transition and instabilities. Ed : T.Riste, p.145-172,
Plenum Press, New-York, 1981.

[41] Di Prima, R.C. and Swinney, H.L. : "Instabilities and transition in
flow between concentric rotating cylinders", in ref.(34) , p.139-
180.

[42] Stability of thermodynamic systems. Ed : J. Casas-Vásquez and
G. Lebon, Springer, Berlin, 1982.

[43] Woodruff, D.P., The solid-liquid interface. Cambridge University
Press, London, 1973.

[44] Langer, J.S. : "Instabilities and pattern formation in crystal
growth". Rev. Mod. Phys. 52, 1-28 (1980).

[45] Wesfreid, J., Pomeau, Y., Dubois, M., Normand, C. and Bergé, P. :
"Critical effects in Rayleigh-Bénard convection". J. Physique 39,
725-731 (1978).

[46] Swift, J. and Hohenberg, P.C. : "Hydrodynamic fluctuations at the
convective instability". Phys. Rev. A 15, 319-328 (1977).

[47] Graham, R. : "Hydrodynamic fluctuations near the convection instability". Phys. Rev. A 10, 1762-1784 (1974).

[48] Lapwood, E.R. : "Convection of a fluid in a porous medium". Proc.
Camb. Phil. Soc. 44, 508-521 (1948).

[49] Combarnous, M.A. and Bories, S.A. : "Hydrothermal convection in saturated porous media". Adv. Hydro. Sci. 10, 231-307 (1975).

[50] Elder, J.W. : "Steday free convection in a porous medium heated from
below". J. Fluid. Mech. 27, 29-48 (1967).

[51] Lyubimov, D.V., Putin, G.F. and Chernatynskii,V.I. : "On convective
motions in a Hele-Shaw cell". Sov. Phys. Dokl. 22, 360-362 (1977).

[52] Müller, U. : "Bénard convection in gaps and cavities" in ref. 5 p.71-100.

[53] Block, M.J. : "Surface tension as the cause of Bénard cells and surface deformation in a liquid film". Nature 178, 650-651 (1956).

[54] Pearson, J.R.A. : "On convection cells induced by surface tension". J. Fluid Mech. 4, 489-500 (1958).

[55] Velarde, M.G. and Castillo, J.L. : "Transport and reactive phenomena leading to interfacial instability". in ref [5] p.235-264.

[56] Linde, H. : "Marangoni instabilities". in ref [5], p.265-296.

[57] Malmejac, Y. Bewersdorff, A., Da Riva, I. and Napolitano, L.G. : "Microgravity research in space". ESA-BR 05, European space agency, Paris, 1981.

[58] Schechter, R.S., Velarde, M.G. and Platten, J.K. : "The two component Bénard problem". Adv. Chem. Phys. 26, 265-301 (1974).

[59] Turner, J.S., Buoyancy effects in fluids. Cambridge University Press, Cambridge, 1973.

[60] De Gennes, P.G., The Physics of liquid crystal. Clarendon Press, Oxford, 1974.

[61] Salan, J. and Guyon, E. : "Homeotropic nematics heated from above under magnetic fields:convective thresholds and geometry". J.Fluid Mech. 126, 13-26 (1983).

[62] Taylor, G.I. : "Stability of a viscous liquid contained between two rotating cylinders". Phil. Trans. R. Soc. Lond. A223, 289-343 (1923).

[63] Bippes, H. : "Experimental study of the laminar-turbulent transition in a concave wall of a parallel flow". NASA Tech. Mem. TM 75243 (1978).

[64] Floryan , J.M. and Saric, W.S. : "Stability of Görtler vortices in boundary layers". AIAA Journal, 20, 316-324 (1982).

[65] Herbert,Th.   : "On the stability of the boundary layer along a concave wall". Archives of Mechanics (Warszawa) 28, 1039-1055 (1976).

[66] Gregory, N., Stuart, J.T. and Walker, W.S. : "On the stability of three dimensional boundary layers with application to the flow due to a rotating disk". Phil. Trans. R. Soc. Lond. A248, 155-199 (1955).

[67] Kobayashi, R., Kohama, Y. and Takamadate, Ch. : "Spiral vortices in boundary layer transition regime on a rotating disk". Acta Mechanica 35, 71-82 (1980).

[68] Drazin, P.G. and Reid, W.H., Hydrodynamic stability, Cambridge University Press, Cambridge, 1981.

[69] Atten, P., Lacroix, J.C. and Malraison, B. : "Chaotic motion in a Coulomb force driven instability: large aspect ratio experiments". Phys. Lett. 79A, 255-258 (1980).

[70] Felici, N. : "Phénomènes hydro et aérodynamiques dans la conduction des diélectriques fluides". Revue Générale de l'Electricité 78, 717-734 (1969).

[71a] Williams, R. : "Domains in liquid crystals". J. Chem. Phys. 39, 384-386 (1963).

[71b] Kapustin, A.P. and Vismin, L.K. : "Ferroelectric properties of liquid crystals". Sov. Phys.-Crystallography 10, 95-97 (1965).

[72] Langer, J.S. and Müller-Krumbhaar, H. : "Theory of dendritic growth I : Elements of a stability analysis". Acta Metall. 26, 1681-1687 (1978).

[73] Langer, J.S. and Müller-Krumbhaar, H. : "Theory of dendritic growth II : Instabilities in the limit of vanishing surface tension". Acta Metall. 26, 1689-1695 (1978).

[74] Müller-Krumbhaar, H. and Langer, J.S. : "Theory of dendritic growth III : Effects of surface tension". Acta Metall. 26, 1697-1708 (1978)

[75] "Propriétés dynamiques des fronts de flamme". Images de la Physique, Supp. au n°39 du Courrier du C.N.R.S., 22-26 (1981).

[76] Clavin, P. : "Dynamical behaviour of premixed flames fronts in laminar and turbulent flows". To appear in Prog. Energ. Comb. Sci. (1984).

[77] Zaikin, A.N. and Zhabotinskii, A.M. : "Concentration wave propagation in two dimensional liquid phase self oscillating system". Nature 225, 535-537 (1970).

[78] Schliomis, M.J. : "Magnetic Fluids". Sov. Phys. Usp. 17, 153-169 (1974).

[79] Rosensweig, R.E. : "Fluid dynamics and science of magnetic liquids". Adv. in Electronics and Electron Physics 48, 103-199 (1978).

[80] Levandowsky, M., Childress, W.S., Spiegel, E.A. and Hutner, S.H. : "A Mathematical Model of Pattern Formation by Swimming Microorganisms" J. Protozool. 22, 296-306 (1975).

[81] Von der Malsburg, C. and Cowan, J.D. : "Outline of a theory for the ontogenesis of Iso-Orientation domains in visual cortex" Biol. Cybern 45, 49-56 (1956).

[82] Bienenstock, E. : "Cooperation and competition in central nervous system development : a unifying approach", in Synergetics of the brain. Ed : E. Baszar et al. p.250-263, Springer, Berlin, 1983.

[83] Swindale, N.V. : "A model for the formation of orientation columns". Proc. R. Soc. Lond. B215, 211-230 (1982).

[84] Sorokin, V.S. : "Stationary motions in a fluid heated from below". (in Russian). Prikl. Mat. Mekh. 18, 197-204 (1954).

[85] Schlüter, A., Lortz, P. and Busse, F.H. : "On the stability of steady finite amplitude convection ". J. Fluid Mech. 23 , 129-144 (1965).

[86] Malkus, W.V.R. and Veronis, G. : "Finite amplitude cellular convection". J. Fluid Mech. 4, 225-260 (1958).

[87] Gork'ov , L.P. : "Stationary convection in a plane liquid layer near the critical heat transfer point". Sov. Phys. JETP 6, 311-315 (1958).

[88] Sorokin, V.S. : "Variational method in the theory of convection". (in Russian). Prikl. Mat. Mekh. 17, 39-48 (1953).

[89] Dubois, M. and Bergé, P. : "Experimental study of the velocity field in Rayleigh-Bénard convection". J. Fluid Mech. 85, 641-653 (1978).

[90a] Ponomarenko,I.B.:"Processes formation of hexagonal convective cells". Prikl Mat. Mekh. 32, 234-245 (1968).

[90b] Buzano, E. and Golubitsky, M. : "Bifurcation on the hexagonal lattice and the planar Bénard problem". Phil. Trans. R. Soc. Lond. A308, 617-667 (1983).

[91] Krishnamurti, R. : "Finite amplitude convection with changing mean temperature Part 1 : Theory. Part 2 : An experimental test of the theory". J. Fluid Mech. 33, 445-455 and 457-463 (1968).

[92] Richter, F.M.:"Experiments on the stability of convection rolls in fluids whose viscosity depends on temperature". J. Fluid Mech. 89, 553-560 (1978).

[93] Dubois, M., Bergé, P. and Wesfreid, J. : "Non-Boussinesq convective structures in water near 4°C". J. Physique 39, 1253-1257 (1978).

[94] Whitehead, J.A., Jr. : "Cellular convection". American Scientist 59, 444-451 (1971).

[95] Whitehead, J.A. : "Dislocations in convection and the onset of chaos". Phys. Fluids 26, 2899-2904 (1983).

[96] Straus, J.M. and Schubert, G. : "On the existence of three dimensional convection in a rectangular box containing fluid saturated porous material" J. Fluid Mech. 87, 385-394 (1978).

[97a] Busse, F.H. and Riahi, N. : "Nonlinear convection in a layer with nearly insulating boundaries". J. Fluid Mech. 96, 243-256 (1980).

[97b] Riahi, N. : "On convection with nearly insulating boundaries in a low Prandtl number fluid". Z.A.M.P. 31, 261-266 (1980).

[98] Dubois, M., Normand, C. and Bergé, P. :"Wavenumber dependence of velocity field amplitude in convection rolls - theory and experiments". Int . J. Heat Mass Transfer 21, 999-1002 (1978).

[99a] Segel,L.A.and Stuart, J.T. : "On the question of the preferred mode in cellular thermal convection". J. Fluid Mech. 13, 289-306 (1962).

[99b] Stuart, J.T. : "On the cellular patterns in thermal convection". J. Fluid Mech. 18, 481-498 (1964).

[100] Stein, M. : "Loads and deformations of buckled rectangular plates". NACA Tech. Rep. R 40 (1959).

[101] Koschmieder, E.L. : "On the wavelength of convective motions". J. Fluid Mech. 35, 527-530 (1969).

[102] Eckhaus, W., Studies in non linear stability theory, Springer, Berlin, 1965.

[103] Kogelman, S. and Di Prima, R.C. : "Stability of spatially periodic supercritical flows in hydrodynamics". Phys. Fluids. 13, 1-11 (1970).

[104] Di Prima, R.C., Eckhaus, W. and Segel, L.A.:"Non-linear wave-number interaction in near-critical two dimensional flows". J. Fluid Mech. 49, 705-744 (1971).

[105] Stuart, J.T. and Di Prima, R.C. : "The Eckhaus and Benjamin-Feir resonance mechanisms". Proc. R. Soc. Lond. A362, 27-41 (1978).

[106] Pomeau, Y. and Manneville, P. : "Stability and fluctuations of a spatially periodic convective flow". J. Physique Lett. 40, L609-L612 (1979).

[107] Pomeau, Y. : "Non linear pattern selection in a problem of elasticity". J. Physique Lett. 42, L1-L4 (1981).

[108] King, G.P. and Swinney, H.L. : "Limits of stability and irregular flow patterns in wavy vortex flow". Phys. Rev. A27, 1240-1243 (1983).

[109] Cole, J.A. : "Taylor vortex instability and annulus length effects". J. Fluid Mech. 75, 1-15 (1976).

[110] Hall, P. and Walton, I.C. : "The smooth transition to a convective regime in a two-dimensional box". Proc. R. Soc. Lond. A 358, 199-221 (1977).

[111] Daniels, P.G. : "The effect of distant sidewalls on the transition to finite amplitude Bénard convection". Proc. R. Soc. Lond. A358, 173-197 (1977).

[112] Benjamin, T.B. and Mullin, T. : "Anomalous modes in the Taylor experiment". Proc. R. Soc. Lond. A377, 221-249 (1981).

[113] Krishnamurti, R. : "On the transition to turbulent convection. Part 1 : The transition from two to three dimensional flow". J. Fluid. Mech. 42, 295-307 (1970).

[114] Pomeau, Y. and Manneville, P. : "Wavelength selection in cellular flows". Phys. Lett. 75A, 296-298 (1980).

[115] Getling, A.V. : "Evolution of two-dimensional disturbances in the Rayleigh-Bénard problem and their preferred wavenumbers". J. Fluid Mech. 130, 165-186 (1983).

[116] Berdnikov, V.S. and Kirdyashkin, A.G. : "On the spatial structure of cellular convection". Isvestiya, Atmospheric and Oceanic Physics 15, 561-565 (1979).

[117] Snyder, H.A.:"Wave-number selection at finite amplitude in rotating Couette flow". J. Fluid Mech. 35, 273-298 (1969).

[118] Schaeffer, D. and Golubitsky, M. : "Boundary conditions and mode jumping in the buckling of a rectangular plate". Comm. Math. Phys. 69, 209-236 (1979).

[119] Pomeau, Y. and Zaleski, S. : "Wavelength selection in one dimensional cellular structures". J. Physique 42, 515-528 (1981).

[120] Cross, M.C., Daniels, P.G., Hohenberg,P.C. and Siggia, E.D. : "Phase-winding solutions in a finite container above the convective threshold". J. Fluid Mech. 127, 155-183 (1983).

[121] Clement, M., Guyon, E. and Wesfreid, J.E. : "Multiplicité de modes de déformation d'une plaque sous compression- Expérience". C.R. Acad. Sci. Paris 293II, 87-89 (1981).

[122] Mullin, T. : "Mutations of steady cellular flows in the Taylor experiment". J. Fluid Mech. 121, 207-218 (1982).

[123] Potier Ferry, M. : "Amplitude modulation, phase modulation and localization of buckling patterns" in Collapse : the buckling of structures in theory and practice. Ed : J.M.T. Thompson and G.W. Hunt, p.149-159, Cambridge University Press, 1983.

[124] Langer, J.S. : "Eutectic solidification and marginal stability". Phys. Rev. Lett. 44, 1023-1026 (1980).

[125] Pomeau, Y. and Manneville, P. : "Wavelength selection in axisymetric cellular structures". J. Physique 42, 1067-1074 (1981).

[126] Manneville, P. and Piquemal, J.M. : "Zig-Zag instability and axisymmetric rolls in Rayleigh-Bénard convection : the effect of curvature". Phys. Rev. A 28, 1774-1790 (1983).

[127] Atten, P. and Wesfreid, J.E. : "Wavenumber variations of spatially damped Rayleigh-Bénard convection". Phys. Fluids 24, 173-174 (1981).

[128] Kramer, L., Ben-Jacob, E., Brand, H. and Cross, M.C. : "Wavelength selection in systems far from equilibrium". Phys. Rev. Lett. 49, 1891-1894 (1982).

[129] Pomeau, Y. and Zaleski, S. : "Pattern selection in a slowly varying environment". J. Physique Lett. 44, L135-L141 (1983).

[130] Cannell,S. Dominguez-Lerma, M.A. and Ahlers, G. : "Experiments on wavenumber selection in rotating Couette-Taylor flow". Phys. Rev. Lett. 50, 1365-1368 (1983).

[131] Siggia, E.D. and Zippelius, A. : "Dynamics of defects in Rayleigh-Bénard convection". Phys. Rev. A 24, 1036-1049 (1981).

[132] Pomeau, Y., Zaleski, S. and Manneville, P. : "Dislocation motion in cellular structures". Phys. Rev. A 27, 2710-2726 (1983).

[133] Manneville, P. and Pomeau, Y. : "A grain boundary in cellular structures near the onset of convection". Phil. Mag. A 48, 607-621 (1983)

[134] Stork, K. and Müller, U. : "Convection in boxes : experiments". J. Fluid Mech. 54, 599-611 (1972).

[135] Stork, K. and Müller, U. : "Convection in boxes : an experimental investigation in vertical cylinders and annuli". J. Fluid Mech. 71, 231-240 (1975).

[136] Bergé, P., Dubois, M. and Croquette, V. : "Approach to Rayleigh-Bénard turbulent convection in different geometries". In [5] p.123-148.

[137] Bergé, P. : "Rayleigh-Bénard convection in high Prandtl number fluid". in : Chaos and Order in Nature. Ed : H. Haken, p.14-24, Springer, Berlin, 1981.

[138] Dreyfus, J.M. and Guyon, E. : "Convective instabilities in nematics caused by an elliptical shear". J. Physique 42, 283-292 (1981).

[139] Kai, S. and Hirakawa, K. : "Succesive transitions in electrohydro-dynamic instabilities of nematics". Supp. Prog. Theor. Phys. 64, 212-243 (1978).

[140] Chen, M.M. and Whitehead, J.A. : "Evolution of two-dimensional periodic Rayleigh convection cells of arbitrary wavenumbers". J. Fluid Mech. 31, 1-15 (1968).

[141] Busse, F.H. and Whitehead, J.A. : "Instabilities of convection rolls in a high Prandtl number fluid".J.Fluid Mech.47,305-320 (1971).

[142] Whitehead, J.A. : "The propagation of dislocations in Rayleigh-Bénard rolls and bimodal flow". J. Fluid Mech. 75, 715-720 (1976).

[143] Donnelly, R.J., Park, K., Shaw, R. and Walden, R.W. : "Early non-periodic transitions in Couette flow". Phys. Rev. Lett. 44, 987-989 (1980).

[144] Guazzelli, E. : "Nucleation homogène d'une paire de défauts dans une structure convective périodique". CR. Acad. Sc. Paris 291B, 9-12 (1980).

[145] Guazzelli, E., Guyon, E. and Wesfreid, J.E. : "Defects in convective structures in a nematic hydrodynamic instability" in : Symmetries and broken symmetries in condensed matter physics. Ed : N. Boccara, p.455-461, IDSET, Paris, 1981.

[146] Croquette, V., Mory, M. and Schosseler, F. : "Rayleigh-Bénard convective structures in a cylindrical container". J. Physique 44, 293-301 (1983).

[147] Dreyfus, J.M. and Pieranski, P. : "Distortion waves and phase slip-page in nematics". J. Physique 42, 459-467 (1981).

[148] Gollub, J.P. and Steinman, J.F. : "Doppler imaging of the onset of turbulent convection". Phys. Rev. Lett. 47, 505-508 (1981).

[149] Gollub, J.P., Mc Carriar, A.R. and Steinman, J.F. : "Convective pattern evolution and secondary instabilities". J. Fluid Mech. 125, 259-281 (1982).

[150] "Morphologie des structures dissipatives". Images de la Physique, Suppl. au n°39 du Courrier du C.N.R.S., 11-16 (1981).

[151] Kleman, M. : Points, lignes et parois, Vol. 1. Les Editions de Physique , Paris, 1977.

[152] Cross, M.C. : "Ingredients of a theory of convective textures close to onset". Phys. Rev. A 25, 1065-1076 (1982).

[153] Zaleski, S., Pomeau, Y. and Pumir, A. : "Optimal merging of rolls near a plane boundary". Phys. Rev. A 29, 366-370 (1984).

[154] Dubois Violette, E., Guazzelli, E. and Prost, J. : "Dislocation motion in layered structures". Phil. Mag. A 48, 727-747 (1983).

[155] Ahlers, G. : "Onset on convection and turbulence in a cylindrical container" in Systems far from equilibrium . Ed: L.Garrido,p.143-161: Springer, Berlin, 1980.

[156] Bidaux, R., Boccara, N., Sarma, G., de Sèze, L., de Gennes, P.G. and Parodi, O. : "Statistical properties of focal conic textures in smectic liquid crystals". J. Physique 34, 661-672 (1973).

[157] Mandelbrot, B., Fractals : form , chance,and dimension. Freeman and Co, p. 185-188, San Francisco, 1977.

[158] Greenside, H.S., Coughran, W.M. Jr. and Schryer, N.L. :"Nonlinear pattern formation near the onset of Rayleigh-Bénard convection". Phys. Rev. Lett. 49, 726-729 (1982).

[159] Segel, L.A. : "Distant side walls cause slow amplitude modulation of cellular convection". J. Fluid Mech. 38, 203-224 (1969).

[160] Newell, A.C. and Whitehead, J.A. : "Finite bandwidth, finite amplitude convection". J. Fluid Mech. 38, 279-303 (1969).

[161] Graham, R. and Domaradzki, J.A. : "The local amplitude equation of Taylor vortices and its boundary condition". Phys. Rev. A 26, 1572-1579 (1982).

[162] Wesfreid, J., Bergé, P. and Dubois, M. : "Induced pretransitional Rayleigh-Bénard convection". Phys. Rev. A 19, 1231-1233 (1979).

[163] Pfister, G. and Rehberg, I. : "Space-dependent order parameter in circular Couette flow transitions". Phys. Lett. 83A, 19-22 (1981).

[164] Brown, S.N. and Stewartson K. : "On thermal convection in a large box". Studies in Appl. Math. 57, 187-204 (1977).

[165a] Brown, S.N. and Stewartson, K. : "On finite amplitude Bénard convection in a cylindrical container". Proc. R. Soc. Lond. A 360, 455-469 (1978).

[165b] Brown, S.N. and Stewartson, K. : "On finite amplitude Bénard convection in a cylindrical container. Part II". SIAM J. Appl.Math. 36,
573-586 (1979).

[166] Siggia, E.D. and Zippelius, A. : "Pattern selection in Rayleigh-
Bénard convection near threshold". Phys. Rev. Lett. 47, 835-838
(1981).

[167] in ref. [29] page 128.

[168] Manneville, P. : "A two dimensional model of three dimensional convective pattern in wide containers". J. Physique 44, 759-765 (1983).

[169] Wesfreid, J.E. and Croquette, V. : "Forced phase diffusion in
Rayleigh-Bénard convection". Phys. Rev. Lett. 45, 634-637 (1980).

[170] Croquette, V. and Wesfreid, J.E. : "Phase diffusion experiment in
Rayleigh-Bénard convection" in : Symmetries and broken symmetries
in condensed matter physics". Ed : N. Boccara, p.399-406, IDSET,
Paris, 1981.

[171] Croquette, V. and Schosseler, F. : "Diffusive modes in Rayleigh-
Bénard structures". J. Physique 43, 1183-1191 (1982).

[172] Manneville, P. and Piquemal, J.M. : "Transverse phase diffusion in
Rayleigh-Bénard convection". J. Physique Lett. 43, L253-L258 (1982).

[173] Guazzelli, E., Guyon, E. and Wesfreid, J.E. : "Dislocations in a
roll hydrodynamic instability in nematics : static limit". Phil.
Mag. A 48, 709-726 (1983).

[174] Dewel, G., Walgraef, D. and Borckmans, P. :"Layered structures in
twodimensional nonequilibrium systems".   J. Physique Lett. 42,
L361-L364 (1981).

[175] Alziary de Roquefort, T. and Grillaud, G. : "Computation of Taylor
vortex flow by a transient implicit method". Comp. Fluids 6,
259-269 (1978).

[176] Davies-Jones, R.P. : "Thermal convection in a infinite channel with
no-slip sidewalls". J. Fluid Mech. 44, 695-704 (1970).

[177] Oertel, H. Jr. : "Thermal instabilities" in [5] p.3-24.

[178] Bauer, L. and Reiss, E.L. : "Non linear buckling of rectangular
plates". J. Soc. Indust. Appl. Math. 13, 603-626 (1965).

[179] Normand, C. : "Convective flow patterns in rectangular boxes of
finite extent". ZAMP 32, 81-96 (1981).

[180] Tabeling, P. : "Convective flow patterns in rectangular boxes of
finite extent under an external magnetic field". J. Physique 43,
1295-1303 (1982).

[181] Brazovskii, S.A. : "Phase transition of an isotropic system to a
nonuniform state". Sov. Phys. - JETP 41, 85-89 (1975).

[182] Sazontov, A.G. : "Concerning the selection of convective structures in a fluid with temperature dependent viscosity". Isvestiya, Atmospheric and Oceanic Physics 16, 319-324 (1980).

[183] Bestehorn, M. and Haken, H. : "A calculation of transient solutions describing roll and hexagon formation in the convection instability". Phys. Lett. 99A, 265-267 (1983).

[184] Eckmann, J.P. : "Roads to turbulence in dissipative dynamical systems". Rev. Mod. Phys. 53, 643-654 (1981).

[185] Ott, E. : "Strange attractors and chaotic motions of dynamical systems". Rev. Mod. Phys. 53, 655-672 (1981).

[186] Ahlers, G. and Behringer, R.P. : "Evolution of turbulence from the Rayleigh-Bénard instability". Phys. Rev. Lett. 40, 712-716 (1978).

[187] Behringer, R.P., Shaumeyer, J.N., Clark, C.A. and Agosta, C.C. : "Turbulent onset in moderately large convecting layers". Phys. Rev. A 26, 3723-3726 (1982).

[188] Fauve, S., Laroche, C. and Libchaber, A. : "Effect of a horizontal magnetic field on convective instabilities in mercury". J. Physique Lett. 42, L455-L457 (1981).

[189] Kuramoto, Y. : "Diffusion-induced chemical turbulence" in : Dynamics of synergetic systems. Ed : H. Haken, p.134-146, Springer, Berlin, 1980.

[190] Sivashinsky, G.I. : "Non linear analysis of hydrodynamic instability in laminar flames. I : Derivation of basic equations". Acta Astronautica 4, 1177-1206 (1977).

[191] Lin, J. and Kahn, P.B. : "Order and turbulence in one dimension" in Systems far from equilibrium. Ed : L. Garrido, p.345-351, Springer, Berlin, 1980.

[192] Gertsberg, V. and Sivashinsky, G. : "Large cells in non linear Rayleigh-Bénard convection". Progr. Theor. Phys. 66, 1219-1229 (1981).

[193] Chapman, C.J. and Proctor, M.R.E. : "Nonlinear Rayleigh-Bénard convection between poorly conducting boundaries". J. Fluid Mech. 101, 759-782 (1980).

[194] Tabeling, P. : "Dynamics of the phase variable in the Taylor vortex system". J. Physique Lett. 44, L665-L672 (1983).

[195] Koschmieder, E.L. : "Turbulent Taylor vortex flow". J. Fluid Mech. 93, 515-527 (1979) and Addendum 93, 801 (1979).

[196] Barcilon, A., Brindley, J., Lessen, M. and Mobbs, F.R. : "Marginal instability in Taylor-Couette flows at a very high Taylor number". J. Fluid Mech. 94, 453-463 (1979).

[197] Krishnamurti, R. and Howard, L.N. : "Large scale flow generation in turbulent convection". Proc. Natl. Acad. Sci. USA 78, 1981-1985 (1981).

[198] Kosterlitz, J.M. and Thouless, D.J. : "Ordering, metastability and phase transitions in two dimensional systems". J. Phys. C $\underline{6}$, 1181-1203 (1973).

[199] Nelson, D.R. and Halperin, B.I. : "Dislocation - mediated melting in two dimensions". Phys. Rev. B $\underline{19}$, 2457-2484 (1979).

[200] Toner, J. and Nelson, D.R. : "Smectic, cholesteric and Rayleigh-Bénard order in two dimensions". Phys. Rev. B $\underline{23}$, 316-334 (1981).

[201] Lowe, M., Gollub, J.P. and Lubensky, T.C. : "Commensurate and incommensurate structures in a non-equilibrium system". Phys. Rev. Lett. $\underline{51}$, 786-789 (1983).

RAYLEIGH-BENARD INSTABILITY : EXPERIMENTAL STUDY OF THE WAVENUMBER
SELECTION

B. Martinet, P. Haldenwang, G. Labrosse, J.C. Payan, R. Payan

Département d'Héliophysique, ERA C.N.R.S. n° 538
Université de Provence, Centre de Saint Jérôme
13397 MARSEILLE CEDEX 13 (FRANCE)

1. INTRODUCTION

The first experiments relating to the cellular pattern selection
and quoted by the Koschmieder's 1974 review article /1/ have pointed out
the presence of a mechanism not yet understood. Indeed, the horizontal
wavelength increase observed with that of Ra was inconsistent with the
behaviour one expects on a naive physical basis, that is to say the
system should increase its energetic dissipation rate by multiplying
the number of the dissipative structures. Among the various reasons
which were put forward to interpret these observations, the sidewall
influence has been mentioned, as testified by the following sentence,
excerpted from page 200 of the above quoted paper /1/ : "It has also
been argued that the increase of λ might be due to effects of the
lateral walls. However, the aspect ratio which Willis, Deardorff and
Somerville use ($\eta = \frac{1}{80}$) makes this argument academic".

Nevertheless, from theoretical models which have been proposed in
the last years /2,3,4/, this argument does prevail today. Moreover,
this mechanism is suggested to understand the observed direct transition
from conduction to turbulence /5/ when the selection mechanism leads to
cellular structures which become unstable with respect to 3-D perturbations /4/.

To our knowledge, the only one model /2/, among the above mentioned,
gives quantitative predictions, as far as the selected structures are
steadily bidimensional. This paper aims to present experimental results
obtained in conditions which allow a confrontation with this model :
an horizontal rectangular geometry, with a large enough aspect ratio.

Finally, why should one discard the Koshmieder's assertion ? We
have tried to interpret our results without making any reference to the
selective influence of the sidewalls : assuming that the wavenumber
selection is associated with the presence of instabilities of a 2-D

layer, we have compared our data with the predictions coming from a
linear stability analysis of an infinite horizontal layer /8/.

2. THE EXPERIMENTAL SET-UP

Our experiments have been performed with two parallelepipedic
cavities whose L_x, L_y, L_z dimensions are respectively :

- Cavity A : 27 x 9 x 1.5 cm³
- Cavity B : 18 x 9 x 1.5 cm³

The cavities are made of two massive copper horizontal plates stuck
to sidewalls in epoxy along the y-direction (short side) and glass
along the x-direction. A constant temperature difference is imposed
between the horizontal plates by using a regulated electrical heating
device for the lower plate and a cold water circuit for the upper one.
Temperatures in the copper plates are homogeneous and stabilized in
time within 0.1°C. Relative thermal conductivities of the sidewall
materials with respect to air thermal conductivity are 10 and 40
approximately for epoxy and glass respectively.

As observed in many other experiments, the horizontal pattern of
the steady convective flows in our cavities is expected to be made of
almost bidimensional rolls, whose axis is parallel to the short side of
the container (y-rolls). To visualize such flows, we have used the
optical interferometry set-up described in detail in /6/.

Giving access to the fluid temperature field integrated along its
optical axis, this device allows a qualitative identification of any
y-independent component of the horizontal flow pattern, if the optical
axis is aligned with the y-direction. But, of course, there is no way
to detect any eventual y-periodic component of the pattern, so that an
interferogram may well come from a 3-D flow made of a y-modulated
(x,z) flow.

3. THE EXPERIMENTAL RESULTS

Figure 1 presents some photographs of typical interferograms
obtained with the optical set-up . Steady 2-D flows, with an odd
(fig. 1.a) or even (fig. 1.b) number of rolls, are observed until a
value ε_{osc} of ε = (Ra - Rac) / Rac, (Rac = 1708), ε_{osc} ≅ 6. Beyond
this value, the flows present a time oscillating behaviour, the
x-periodic component remaining unchanged.

As seen on the interferograms, the transitions between two steady

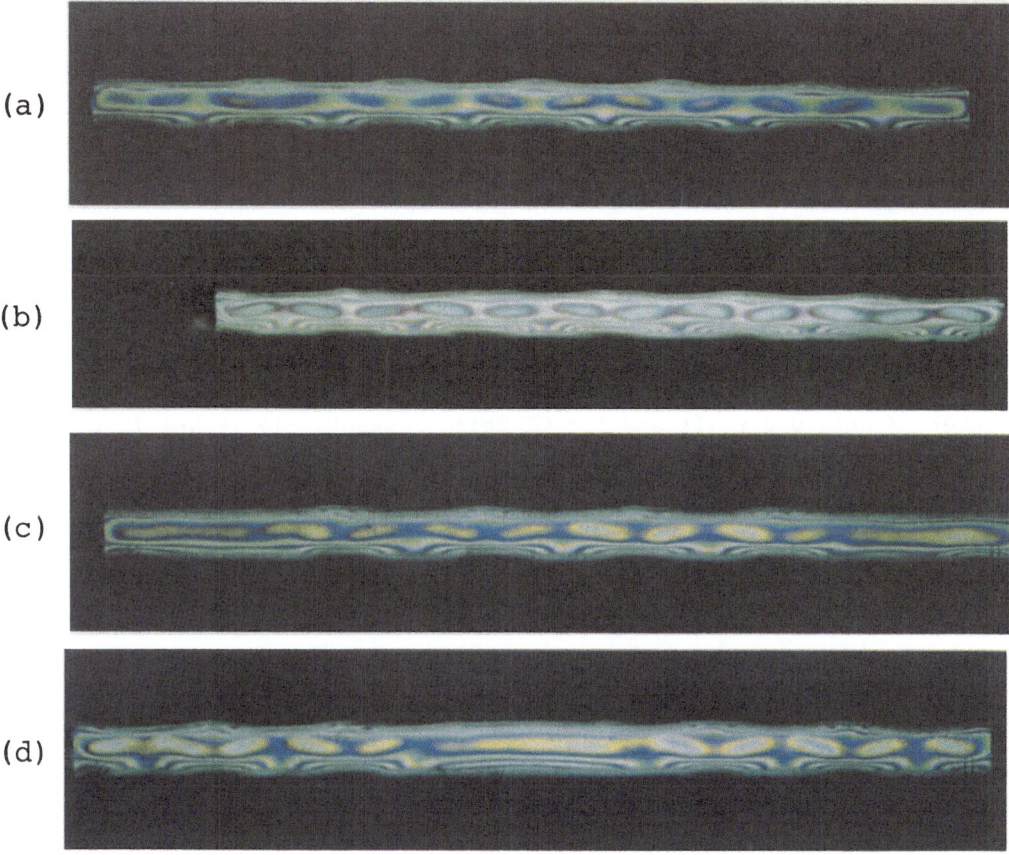

Figure 1

Vertical temperature gradient interferograms corresponding
to the following regimes :
(a) steady 13 rolls
(b) steady 12 rolls
(c) 3-D defects located near both sidewalls
(d) 3-D defect in the central region

2-D flows are different, whether ε is increased or decreased. Transitions
with increasing ε break the bidimensional aspect of the interferometric
patterns, suggesting some kind of 3-D defects, always located near both
sidewalls (fig. 1.c). Sometimes, a defect happens to emerge at the
center of the interferogram (fig. 1.d). It splits up in two defects
which then migrate to the sidewalls, reproducing the previously described
situation. Transitions with decreasing ε occur with a modification of
the x-periodicity localized near only one sidewall, where one roll
expands and splits up.

Figure 2 presents, for each cavity, the experimental behaviour, as
function of ε, of the average x-wavevector q, $q = \pi \frac{N}{\Gamma}$ where N is the
number of observed rolls and $\Gamma = \frac{L_x}{L_z}$ being respectively 18 and 12 for
cavity A and B. As observed elsewhere /1,7/ in different experimental
situations, the wavevector q decreases as ε increases. Furthermore, the
region of allowed x-wavevectors is drastically reduced as compared to
the domain defined by the (ε, q - q_c) marginal stability curve of the
conductive regime. We have sketched on figure 3 the boundaries of the
domain which contains all our data from both cavities, together with the
"conductive" marginal stability curve. The wavevector restriction seems
stronger as ε increases, leading to one well defined wavenumber when
ε > 6. This value turns out to correspond to $ε_{osc}$, the threshold for the
presence of time oscillating flows. Our data are compatible with
published results /6,7/ relative to similar experimental situations and
suggest the presence of a lower limit, of about 2, for the x-wavevector
q in the case of air convective flows in large boxes.

4. ATTEMPTED COMPARISON WITH THEORETICAL PREDICTIONS

Until today no theoretical analysis has been proposed allowing
a realistic modelization of the actual three-dimensional confining and
leading to a physical understanding of the observed wavevector selection
mechanism. Two theoretical approaches can be quoted here to get some
complementary indications.

a) Busse and Clever /8/ (referred as BC) have studied the linear
stability of a 2-D roll pattern with respect to general 3-D perturbations
in an infinite horizontal layer of a fluid confined between two rigid
plates. They predict a stability domain bounded, in our case where
Pr = 0.71, by three marginal stability curves corresponding to the
oscillatory (OS), skewed varicose (SV) and Eckhaus (E) instabilities.
This domain is sketched, on figure 4, together with our experimental

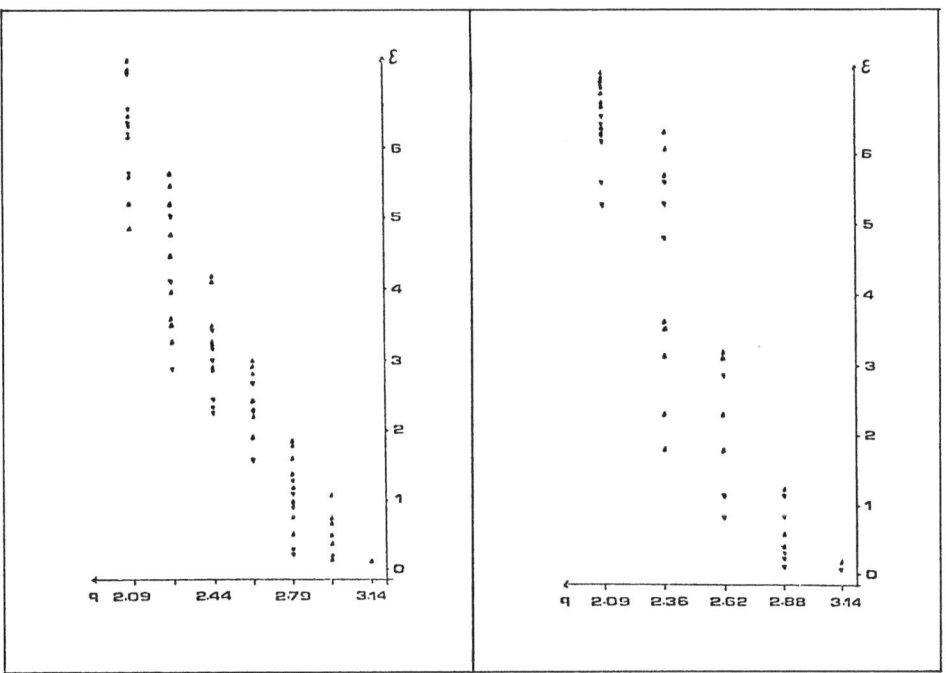

Figure 2.a   Figure 2.b

Experimental wavenumber q plotted as a function of ε for cavities A and B on (2.a) and (2.b) respectively.

data envelope which seems to be related, by some stretching, to the
BC domain. So, the following question can be asked : are the predicted
instabilities effectively observed, and is the stretching of the BC
domain actually due to the presence of the sidewalls ?

The experimental (OS)' boundary is clearly identified and one can
reasonably think of the 3-D confining as being responsible for the
observed significant shift towards higher ε values of the BC (OS)
boundary. This result is confirmed by other published results /6,7,9/.

As it is a 3-D flow, the skewed varicose instability should be
sensitive to the y-confining of the cavities, so that the (B+) boundary
might just be a shifted up (SV) boundary. On the contrary, the Eckhaus
instability could manifest itself more easily, thanks to the translation
invariance breaking which originates near the short sides of the cavities. This would lead to reduce, near the (E) boundary, the stability
domain of the y-roll flows, i.e. to shift up again this boundary to the
experimental (B-) boundary.

For completness sake, we have reported, on figure 4, the $D_1 = 0$
stability curve of the zig-zag instability in air, computed by Picquemal
and Manneville /10/ in the case of an infinite horizontal layer of fluid
submitted to rigid-rigid z-boundary conditions. On the basis of the
previous arguments, the already minor influence of this 3-D instability,
in the case of air, should be still reduced by the y-confining.

Are our experimental observations able to give some indications
about these propositions ?

From what we see on the interferometric patterns, the wavevector
selection mechanims are different, whether ε is increased or decreased.
When we increase ε, we get the (B+) boundary which opens on a region of
unstable 3-D flows with defects, as illustrated on figures 1.c - 1.d.
Undoubtedly, this does not correspond to skewed varicose instabilities.
Integrated on the y-direction by our experimental set-up, these instabilities, as any y-periodic instability, will give a 2-D interferometric
pattern. So, the experimental (B+) boundary must be interpreted as the
marginal stability curve of the farthest (on the ε scale) y-periodic
regime modulating a y-roll system. Nevertheless, the previous discussion
leading to expand the BC stability domain, as far as (B+) boundary is
concerned, cannot explain the observed reduction of this domain at the
$(q - q_c, \varepsilon)$ plane origin vicinity.

Decreasing ε, now, leads to the (B-) boundary corresponding, in our

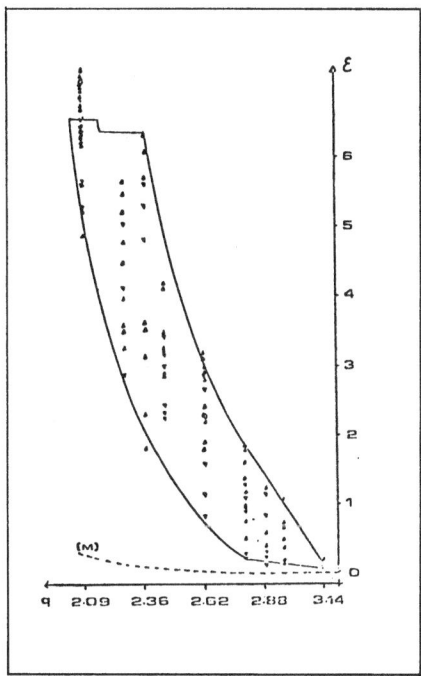

Figure 3

Sketch of the observed selected
domain in the (q, ε) plane
delimited by the marginal
stability curve (M) of the
conductive regime.

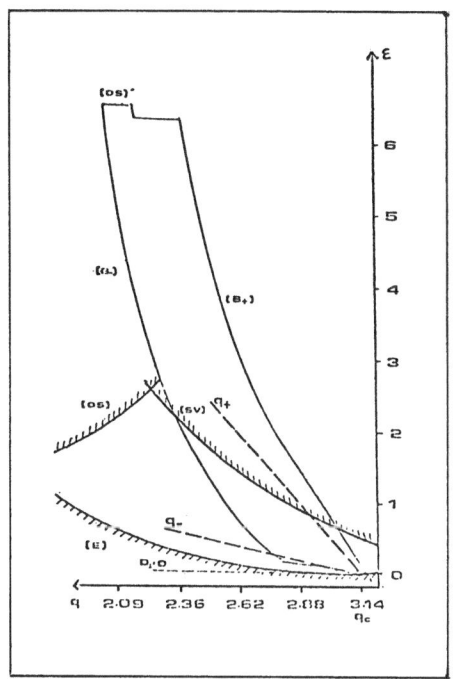

Figure 4

The experimentally selected
domain delimited by (OS)',
(B+), (B-) curves is compared
to Busse and Clever domain
(OS), (E), (SV), to Cross et
al. predictions (q+, q-) and
to $D_\perp = 0$ marginal stability
curve of 2-D roll w.r.t.
zig-zag instabilities.

experiments, to transitions which preserve the y-roll aspect of the interferometric patterns. This seems compatible with Eckhaus instabilities localized near one sidewall.

b) A somewhat different theoretical approach can be now considered. Cross, Daniels, Hohenberg and Siggia /2/ (referred as CDHS) have solved finite amplitude equations for y-roll flows, with free-free z-boundary conditions. Their model takes into account an x-confining, the Prandtl number and the thermal characteristics of the sidewalls being free parameters. Of course, only a small region of our experimental data is concerned with this analysis.

Qualitatively, their predictions are in good agreement with our experimental data, since they propose the presence of a selection mechanism, when $Pr = 0.71$, if their parameter μ, proportional to the sidewall conductance, is $\mu > 2$. We have $\mu = 8\pm2$ which gives to the parameters $|\beta|$ and η of their formula (4.22) the following values :

$$|\beta| = 0.17 \pm .01$$

$$\eta = 1.5 \pm .1$$

and (4.22) reads :

$$(q - q_c)_\pm = - C_\pm \, \epsilon \ , \ C_+ = 0.04 \ , \ C_- = 0.2$$

It is worth mentioning the Pomeau and Zaleski theoretical models /3/ : they do not take explicitly into account the Prandtl number, but one of their models predicts a selection mechanism which is also in qualitative agreement with our data.

Is it meaningful to try any quantitative comparison, based on the CDHS predictions, by taking into account the two main quantitative differences one expects between the finite amplitude treatments of a free-free and a rigid-rigid y-roll system : the marginal stability curves and the finite amplitude expansion parameters ?

Let us define the control parameter ϵ_{FA} of the finite amplitude expansions in both cases, rigid-rigid (RR) and free-free (FF). At first order, we have :

$$\epsilon_{FA} = \frac{R - R_{FF}^{(0)}}{R_{FF}^{(2)}} = \frac{R - R_{RR}^{(0)}}{R_{RR}^{(2)}}$$

where : $R_{FF}^{(0)}$ and $R_{RR}^{(0)}$ are the respective threshold values of Ra for the transition conduction - convection in the free-free and rigid-rigid case

$R_{FF}^{(2)}$ and $R_{RR}^{(2)}$ are the 2nd order (in $\epsilon^{1/2}$) coefficients of the finite amplitude expansions in the FF and RR cases respectively. These coefficients are given, for example, in /10/.

We can superimpose the two osculating parabolae for the marginal stability curves :

$$(q - q_c)^2_{FF} = \frac{R^{(2)}_{FF}}{18\pi^2} \varepsilon_{FA}$$

$$(q - q_c)^2_{RR} = \frac{R^{(2)}_{RR}}{256} \varepsilon_{FA}$$

by a simple matching of the $(q - q_c)_{RR}$ and $(q - q_c)_{FF}$ axis, which is plugged in the CDHS (4.22) relation :

$$(q - q_c)_{\pm FF} = - C_{\pm} \frac{R^{(2)}_{FF}}{18\pi^2} \varepsilon_{FA}$$

to give :

$$(q - q_c)_{\pm RR} = - C_{\pm} \left[\frac{R^{(2)}_{FF} . R^{(2)}_{RR}}{18\pi^2 . 256} \right]^{1/2} \varepsilon_{FA}$$

This relation is presented on figure 4, with :

$$\varepsilon_{FA} = \frac{R^{(0)}_{RR}}{R^{(2)}_{RR}} \varepsilon$$

The surprising and perhaps fortuitous agreement obtained remains despite the above mentioned experimental errors.

Such an agreement of a part of our experimental data with the CDHS model shows an incompatibility with the BC theoretical predictions. Actually, a subdomain of the predicted BC stability region is excluded by the CDHS wavenumber selection mechanism. If it is difficult to give an easy physical interpretation to the CDHS (q±) boundaries, it seems reasonable to think that the confining is the source of some 3-D "defects". Of course, this is beyond the BC analysis.

5. CONCLUSION

We have presented an experiment in which the wavelength selection of parallel rolls pattern is studied for a large range of Rayleigh number. Far from the threshold, the domain of allowed wavenumbers, in our experiment, can be interpreted, to a certain extent, as being a simple distortion of the 2-D rolls linear stability domain /8/. However, when we approach the convection threshold, we must use a selection mechanism which takes into account the sidewalls, in order to interpret our results. Furthermore, our results present, at the vicinity of the threshold, a good quantitative agreement with Cross et al. predictions. The selected domain is reduced in regard to the stability region analysed by Busse and Clever.

The authors present their acknowledgments to S. Zaleski for his helpful suggestions.

This work has been done under CNRS-PIRSEM contract APP 1067.

References

/1/ E.L. KOSCHMIEDER.- Adv. Chem. Phys., 26, p. 177-211 (1974).

/2/ M.C. CROSS, P.G. DANIELS, P.C. HOHENBERG and E.D. SIGGIA.-
   - Phys. Rev. Lett., 45, p. 898-901 (1980).
   - J. Fluid Mech., 127, p. 155-183 (1983).

/3/ Y. POMEAU and S. ZALESKI.- J. Physique, 42, p. 515-528 (1981).

/4/ Y. POMEAU and P. MANNEVILLE.- Phys. Letters, 75A, p. 296-298 1980).

/5/ G. AHLERS and R.P. BEHRINGER.- Phys. Rev. Lett., 40, p. 712-715 (1978).

/6/ B. MARTINET, P. HALDENWANG, G. LABROSSE, J.C. PAYAN, R. PAYAN.- J. Physique-Lettres, 43, p. L161-L169 (1982).

/7/ H. OERTEL, Jr.- "Natural convection in enclosures", presented at the National Heat Transfer Conference, Orlando, Florida. H.T.D., vol. 8 (ASME), p. 11-16 (1980).

/8/ F.H. BUSSE and R.M. CLEVER.- J. Fluid Mech., 91, p. 319-335 (1979).

/9/ J. MAURER and A. LIBCHABER.- J. Physique-Lettres, 41, p. L515-L518 (1980).

/10/ P. MANNEVILLE and J.M. PIQUEMAL.- Phys. Rev. A, 28, p. 1774-1790 (1983).
   J.M. PIQUEMAL.- Thèse de Troisième Cycle, Université Pierre et Marie Curie, Paris VI, (1982), (unpublished).

WAVELENGTH SELECTION AND PATTERN LOCALIZATION IN BUCKLING PROBLEMS

M. Potier-Ferry

Laboratoire de Mécanique Théorique

Université Pierre et Marie Curie

Tour 66 - 4, Place Jussieu

75230 PARIS CEDEX 05

1. INTRODUCTION

In systems with a large aspect ratio, a band of solutions with different wavenumbers q may exist above the instability threshold. Pomeau and Zaleski [13], Cross et al [5] solved the wavelength selection problem by improving the classical amplitude equation method [9] [11] [15]. It was established that the boundary conditions on the short sides restrict the band of admissible wavenumbers. This band has been computed for one-dimensional models [7] [13] [14], for the buckling of long rectangular plates [12] and for Bénard convection with free-free boundary conditions [5].

The buckling of a long rectangular plate has been studied experimentally in the cases of clamped and simply supported short sides (Clément et al [4], Boucif et al [2]). Let λ be the control parameter which is here proportional to the applied force, let λ_c be its critical value and $\varepsilon = \lambda - \lambda_c$. These experiments show that the bandwidth of observed wavenumbers is of order ε if the short sides are clamped. This fact is in agreement with the theory. With simply supported short sides, the observed bandwidth is of order $\varepsilon^{\frac{1}{2}}$, which can be explained by the symmetry $x \longrightarrow -x$ in the boundary conditions [14]. In the clamped case, the experimental band [4] is different from that computed in [12], but it is not surprising and at least four reasons can be proposed to explain this discrepancy.

First, several boundary conditions can be set up within the plate theory and those used in [12] are not in full accordance with the experiment. Next, the theory has an asymptotic character and requires a small ε and a large aspect ratio, which makes doubtful the comparison. Last, it is well known [6] that the imperfections (i.e., mainly, the initial displacement) have a great influence in buckling problems : for instance, in the buckling of a cylindrical shell, the collapse load may be reduced by one half with an initial displacement of the same order as the shell width [18]. Only the important question of boundary conditions will be discussed in this paper.

Wavelength selection in one-dimensional elastic models is studied in Section 2. The method is similar to that of Cross et al [5], but the present solutions are uniformly valid and no matching is needed. This analysis establishes that the band-width of admissible wavenumbers increases if the elastic body is not perfectly clamped.

The influence of boundary conditions on the wavelength selection in the plate buckling problem is discussed in Section 3 (work in collaboration with N. Damil). Our calculations corroborate the result of Pomeau [12] and extend them to some more realistic cases of boundary conditions.

In Section 4, one shows that the amplitude equation method can be used to explain the localization of buckling patterns. This localization has been obser-ved in the experiments by Moxham [10] in many tests carried out on axially compressed rectangular plates of mild steel. After a snap-through, the plate reaches an equi-librium configuration which involves only a few buckles. We establish that there exist localized solutions in systems with a large aspect ratio and such that the bifurcating solutions are subcritical and unstable. This study is complementary to that of the sections 2 and 3, because the wavelength selection problem can be set only in cases of supercritical bifurcations. This approach is compared to that of Tvergaard and Needleman [16] [17].

2. WAVELENGTH SELECTION IN BEAM BUCKLING

2.a. The equation of beam buckling

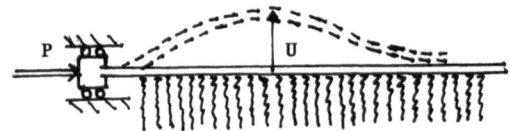

Figure 1

Consider a flexible beam of bending stiffness EI, subjected to a compressive axial force P. The lateral displacements U (s) is restrained by a foundation which provides an elastic force $K_1U + K_3U^3$ per unit length (Figure 1), K_1 being positive. This is the simplest example of wavelength selection in elasticity. As it is usual in elastic stability [6], one can take an eventual small initial displacement $U_o(s)$ into account. The governing equation is [1] [3] :

(1) $\qquad EI\ \partial_s^4 U + P\ \partial_s^2 (U + U_o) + K_1 U + K_3 U^3 = 0$

Let us introduce nondimensional quantities, x, u, λ, τ, u_o :

(2) $\qquad s = (EI/k_1)^{1/4}\, x$

$\qquad U = (K_1/|K_3|)^{\frac{1}{2}}\, u$

$\qquad P = 2\lambda (EIK_1)^{\frac{1}{2}}$

$\qquad U_o = (K_1/|K_3|)^{\frac{1}{2}}\, u_o/2\lambda$

where τ is a small parameter, for instance :

$$\tau = \text{Max}\ \{\ U_o(s)(|K_3|/K_1)^{\frac{1}{2}}\ \}\ .$$

Remark that a natural wavelength is defined by Relation (2). Thus one finds

(3) $\qquad \partial_x^4\, u + \lambda \partial_x^2\, u + u + (\text{sgn}\ K_3)\ u^3 = -\ \tau \partial_x^2\, u_o$

One obtains the marginal stability curve by dropping the nonlinear terms and the imperfection (i.e. $\tau = 0$), next by seeking solutions in the form $\exp(iqx)$. The critical load λ_c and the critical wavenumber q_c correspond to the minimal load :

$$\lambda(q) = q^2 + 1/q^2 \qquad,\qquad \lambda_c = 2\ ,\quad q_c = 1.$$

The length of the beam is assumed to be large as compared to the critical wavelength. Only the case of a half infinite beam will be considered since this appear to be sufficient to study the wavelength selection [5]. At $x = 0$, one assumes boundary conditions in the form

(4) $\qquad u(0) = 0 \qquad\qquad\qquad \partial_x u(0) + k\partial_x^2 u(0) = 0.$

If the coefficient k is zero, the displacement and the slope are prescribed : the beam is said to be clamped. If k is infinite, the slope is free and the bending moment is zero : the beam is said to be simply supported. In the general case, 1/k is positive and is a torsional elastic modulus of the support.

2.b. Uniformly valid solutions

One assumes in this section that the response of the foundation is hardening ($K_3 > 0$) and there is no imperfection ($\tau = 0$). One sets $\epsilon = \lambda - \lambda_c = \lambda - 2$. The problem is to build up solutions of (3) (4) which become periodic for large x. In [5] [14], different forms of solutions are computed in the three regions $x = 0(1)$, $x = 0(\epsilon^{-\frac{1}{2}})$ and $x=0(1/\epsilon)$ and they are matched to one another. To build up a uniformly valid solutions, we seek u(ξ, X) which depends on the two variables

$$X = \epsilon^{\frac{1}{2}} x$$

(5)

$$\xi = q(\epsilon) x \quad , \quad q(\epsilon) = 1 + Q_1\epsilon + Q_2\epsilon^{3/2} + \ldots \, ,$$

and is bounded and 2π -periodic with respect to ξ . As usual, the variable X allows one to take into account an amplitude modulation nearby the boundary. With the so defined variable ξ one can find solutions that are periodic in the large and whose wavenumber $q(\epsilon)$ is not a priori known. One expands u in the form

(6) $$u(\xi ,X) = \epsilon^{\frac{1}{2}} u_1 + \epsilon u_2 + \epsilon^{3/2} u_3 + \epsilon^2 u_4 + \ldots$$

The computations are quite similar to those in [5] [14] and are not reported in detail. One finds

(7) $$u_1 = (A_1(X)\exp i\xi + c.c.) \, 3^{-\frac{1}{2}}$$

(8) $$u_2 = (A_2(X)\exp i\xi + c.c.) \, 3^{-\frac{1}{2}} .$$

The amplitude $A_1(X)$ satisfies the familiar equation

(9) $$4A_1'' + A_1 - A_1(A_1)^2 = 0$$

(10) $$A_1(0) = 0$$

For the sake of boundedness and stability, one selects only the solution

(11) $$A_1(X) = \exp(i\phi) \tanh (X/2^{3/2})$$

ϕ being an arbitrary phase. Taking the derivative of (9) with respect to ϕ, one obtains

(12) $\qquad \mathcal{L}(iA_1) = 0$

where $\mathcal{L}(\cdot)$ is the real linear operator

(13) $\qquad \mathcal{L}(a) = 4a'' + a - 2|A_1(X)|^2 a - A_1^2(X)\bar{a}$.

The second order amplitude $A_2(X)$ satisfies the following equations

(14) $\qquad \mathcal{L}(A_2) = 4iA_1''' + 2iA_1' - 8iQ_1A_1'$

(15) $\qquad A_2(0) = \{(i - 2k)A_1'(0) + (i + 2k)\bar{A}_1'(0)\}/2 = \alpha A_1'(0) + \beta \bar{A}_1'(0)$.

There are generally no bounded solutions of (14) (15), but one can chose Q_1 in order that there exists a bounded A_2. This will define the wavenumber at the order ε. Since the computations in [5], Appendix D, are rather intricate, one uses an alternative method. Let us introduce a bilinear form

(16) $\qquad \langle a,b\rangle = \int_0^\infty \{ a(X)\bar{b}(X) + \bar{a}(X)b(X) \} \, dX.$

Suppose that $A_2(X)$ and the trial function $a(X)$ are bounded and have vanishing derivatives for large X. The following identity results from two integrations by parts :

(17) $\qquad \langle \mathcal{L}(A_2),a\rangle = \langle \mathcal{L}(a),A_2\rangle - 8 \, \mathrm{Re} \{ A_2'(0)\bar{a}(0) - A_2(0)\bar{a}'(0) \}$.

One chooses $a = iA_1$. On account of (12) (14) (15) (17) and of the obvious identities

(18) $\qquad \langle iA_1' , iA_1 \rangle = 1 \qquad , \qquad \langle iA_1''' , iA_1 \rangle = 1/8 ,$

the boundedness of $A_2(X)$ provides the first term Q_1 in the wavenumber expansion :

$$8Q_1 = 5/2 - \mathrm{Im}\,\alpha - \mathrm{Im}\{ \beta \exp(-2i\phi) \}.$$

Because ϕ is arbitrary, there is a band of admissible wavenumbers q, whose extremal values q_- and q_+ are (result given in [14])

$$q_\pm = 1 + \{4 \pm (1 + 4k^2)^{\frac{1}{2}} \} \, \varepsilon /16 + O(\varepsilon^{3/2}).$$

In the clamped case (k = 0), q_+ and q_- are positive. Hence the wavelength must decrease when the load increases. The bandwidth of admissible q increases with the torsional flexibility . When k^2 is greater than 15/4, the wavenumber may remain constant and equal to its critical value (see Figure 2). When k becomes large, the cone of admissible (q,λ) seems to fill a large part of the region above the marginal stability curve, but it is likely that the present expansions no longer hold in this case.

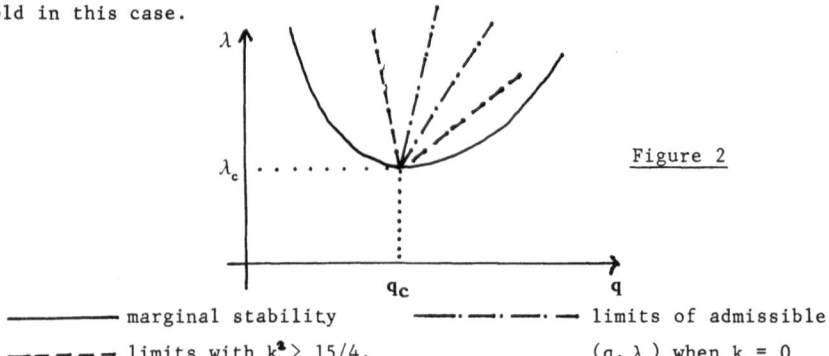

Figure 2

——————— marginal stability   —·——·——·— limits of admissible

— — — — limits with $k^* > 15/4$. (q, λ) when k = 0

2.c. Quadratic nonlinearities

In this section, one assumes that the response of the foundation has, furthermore, a quadratic term $K_2 U^2$. This leads to the equation :

(19) $$(\partial_x^2 + 1)^2 u + \varepsilon \partial_x^2 u + a u^2 + u^3 = 0$$

The computations are performed in the same way as previously. Thus one finds

$$u_1 = \{ A_1(X)\exp i\xi + c.c. \}\gamma$$

$$u_2 = - 2a\gamma^2 |A_1|^2 + \gamma\{A_2(X)\exp i\xi - a\gamma A_1^2 \exp(2i\xi)/9 + c.c.\}.$$

The real number γ can be chosen such that the amplitude A_1 is a solution of the usual equation (9) :

$$\gamma^2 = 9/(27 - 38a^2).$$

This is possible only if

(20) $a^2 < 27/38$.

In the converse case, the nonlinear term in (9) has an opposit sign and the bifurcating solutions exist below the threshold and are generally unstable. The buckling tends to be more localized, as explained in Section 4. If (20) holds, the second order amplitude equation is :

(21) $\qquad \mathfrak{L}(A_2) = 4iA_1''' + 2iA_1' - 8iQ_1 A_1' - \bar{A}_1 (A_1^2)'16ia^2\gamma^2/27.$

On account of the boundary conditions (4) and of the identity

(22) $\qquad \langle i\bar{A}_1 (A_1^2)' \quad , \quad iA_1 \rangle = 1 \quad ,$

one finds the extremal wavenumbers

(23) $\qquad q_{\pm} = 1 + \{ 1/4 - 2a^2\gamma^2/27 \pm\sqrt{1+4k^2}/16\} \varepsilon + 0(\varepsilon^{3/2})$

The quadradic term au^2 does not alter the width of the interval (q_- , q_+), but it shifts the interval to lower values of q. The coefficients of ε in (23) goes to minus infinity if a goes to its limit 27/38.

3. Wavelength selection in plate buckling

We briefly discuss the influence of boundary conditions on the wavelength selection in the buckling of a long rectangular plate subjected to uniaxial compression . A more complete analysis will be published elsewhere with N. Damil.

In Von Karman plate theory [3] [8], the kinematic variables are the three components u,v,w of the displacement. The stresses are modelized by two symmetric tensors : the resultant in-plane stresses N_x, N_{xy}, N_y and the bending moments M_x , M_{xy}, M_y. Here the plate is a half infinite strip, with long sides parallel to the Ox-axis and a short side on the Oy-axis. One applies a compressive load on the short side in such a way that, in the symmetric state, N_x is the only non-zero component of the stress. An Airy function is built up from the resultant stresses arising from a normal displacement w. In terms of non-dimensional quantities, the Von Karman equations can be written as

(24) $\qquad \Delta^2 w + \lambda\partial_x^2 w - [f,w] = 0$

(25) $\qquad \Delta^2 f = - [w,w] /2$

where λ is proportional to the applied load, Δ is the Laplace operator and the bracket is the following bilinear operator

$$[g,h] = (\partial_x^2 g)(\partial_y^2 h) + (\partial_x^2 h)(\partial_y^2 g) - 2(\partial_x\partial_y g) (\partial_x\partial_y h).$$

In the experiment reported in [2] [4] , the normal displacement along the short sides is restrained by rigid vertical blades. On says that the plate is simply supported, which implies that

(26) $\qquad w(x, \pm \pi/2) = \partial_y^2 w(x, \pm\pi/2) = 0.$

With (26), the marginal stability curve and the extremal values λ_c, q_c are

(27) $\qquad \lambda(q) = (q + 1/q)^2 \qquad , \lambda_c = 4 \qquad , q_c = 1.$

Furthermore, there are no in-plane stresses applied on the long sides, which leads to

(28) $\qquad f(x, \pm\pi/2) = \partial_y f(x, \pm\pi/2) = 0.$

Remark that Pomeau [12] considers the alternative condition

(29) $\qquad \partial_y f (x, \pm\pi/2) = \partial_y^3 f(x, \pm\pi/2) = 0 ,$

which makes easier the computations, but it does not correspond to the experimental situation.

The short side is assumed to perfectly or imperfectly clamped, which leads to boundary conditions similar to (4) :

(30) $\qquad w(0,y) = 0, \qquad (\partial_x w + k\partial_x^2 w) (0,y) = 0.$

There the Airy function satisfies conditions such as (28), that are not useful in the present analysis.

With the same notation as in Section 2 and with $\varepsilon = \lambda - 4$,   we find

(31) $\qquad w(x,y) = \varepsilon^{\frac{1}{2}} \gamma(A_1(X)\exp(i\xi) \cos y + c.c.) + 0(\varepsilon)$

and A_1 statisfies the amplitude equation (9) provided that we have

$$\gamma^2 = 2 \qquad\qquad\qquad \text{with (29)}$$

(32)

$$\gamma^2 = \frac{2\pi(\pi + \sinh \pi \ \cosh \pi)}{\pi^2 + \pi \cosh \pi \sinh \pi - \sinh^2\pi} = 2,93 \qquad \text{with (28)}$$

The extremal values q_- and q_+ of the wavenumber are

$$q_{\pm} = 1 + Q_{\pm}\varepsilon + 0(\varepsilon^{3/2})$$

(33)

$$Q_{\pm} = (4 - \gamma^2 \pm (1 + 4k^2)^{\frac{1}{2}})/32$$

This corroborates the result of Pomeau [12] in the case $k = 0$, $\gamma^2 = 2$. As in the one-dimensional model, the bandwidth of admissible q increases when the short side is not perfectly clamped (i.e. $k > 0$). Of course, the interval (Q_-, Q_+) is not the same, with the realistic conditions (28), as what it is with (29), but the two results are not very different. In each case, the wavelength decreases when the load increases, if the short side is perfectly clamped. This does not seem in agreement with the experimental results.

This analysis improves the one of Pomeau, because we consider better boundary conditions on the long sides. But it is still not satisfactory with respect to the short side. Indeed, when a solid body is clamped along a part of its boundary, the displacement of this part is that of a rigid body. Here one shall set

$$u(o,y) = u_o \quad , \quad v(o,y) = 0.$$

Therefore, the prestresses are not uniaxial and Equation (24) does not hold in the region $x \neq 0(1)$. The solution (31) (32) remains valid in the large, but the numerical values of the coefficients α, β in (15) are modified. This study has not yet been carried out.

4. Localization of buckling patterns.

In the buckling of mild steel plate [10], the instability mechanism is a snap-through. This is not unusual, since the bifurcating solutions are generally subcritical in shell instability problems [18]. More surprisingly, it has been observed that the final buckled state involves a localized deformation pattern. Here we use the classical amplitude equation method to show that the localization follows from two points : first a softening nonlinearity which leads to subcritical bifurcation and, secondly, a large aspect ratio.

The model is the beam problem described in Section 2.a, but with K_3 negative. In a first stage, the initial displacement is neglected (i.e., $\tau = 0$ in Equation (3)). One sets

(34) $$\lambda = \lambda_c - m\eta^2$$

where η is a small given parameter and m is of order one and measures the distance to the critical load. The minus sign in (34) has been chosen because one seeks sub-critical solutions. The solution $u(x, X)$ is assumed to depend on x and on the slow variable $X = \eta x$. One expands u into powers of η. Thus one finds

(35) $\qquad u(x,X) = \eta \ \{A(X)\exp(ix) + c.c. \} \ 3^{-\frac{1}{2}} + O(\eta^2)$

(36) $\qquad 4A'' - mA + A|A|^2 = 0.$

In comparison with Equation (9), the main change is the sign of the nonlinear term. If one adds boundary conditions such as (4) at $x = \pm L$, the amplitude $A(X)$ must satisfy

(37) $\qquad A(\pm L\eta) = 0.$

A branch of solutions bifurcates from $A = 0$ at a load slightly greater than λ_c (L being large) :

$$\lambda_{bif} = \lambda_c + \pi^2/L^2.$$

These solutions exist only for λ lower than λ_{bif} and hence are unstable (Figure 3). The shape of the envelope $A(X)$ is modified when one goes away from the threshold. For large m, a good approximation is given by

(38) $\qquad A(X) = (m/2)^{\frac{1}{2}} \ \mathrm{sech}(X/2 \ m^{\frac{1}{2}}) \ ,$

except in the neighbourhood of the boundaries $X = \pm L\eta$. The solution (38) of Equation (36) is small outside an interval $X = O(m^{-\frac{1}{2}})$. When m grows, the maximal amplitude increases and there is a shrinking of the interval where $A(X)$ is not close to zero. This behaviour could explain in many problems the tendency of buckling patterns to localization . Nevertheless, one has to keep on mind that these solutions are uns-table and the final collapse mode can be very different from the latter.

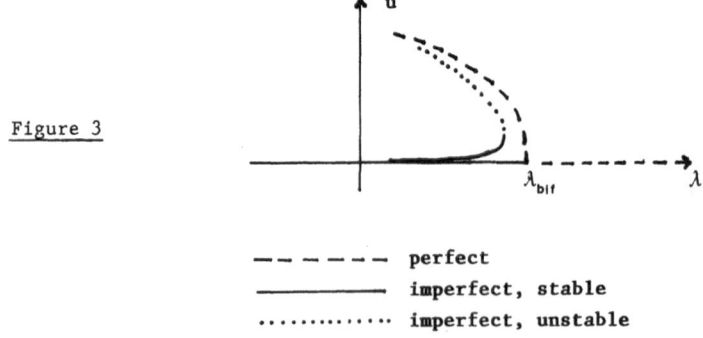

Figure 3

------ perfect
———— imperfect, stable
············ imperfect, unstable

This problem was considered first by Amazigo et al [1], who established Equation (36). But these authors were interested in the effect of localized imperfections and they did not remark that localization appears spontaneously when one follows the bifurcating curve. The solution (38) was given in [14].

Tvergaard and Needleman [16] [17] have proposed an alternative explanation of the localization process. Their analysis relies on several buckling models, whose common properties are softening nonlinearities, rather large aspect ratios and imperfections. They select very special periodic imperfections, very special values of the aspect ratio and they assume that the beam or plate is simply supported in order that the problem have a branch of periodic solutions. Thus this branch has a maximum load point (Figure 3). They established that some branches of non-periodic solutions bifurcate from the periodic branch not too far from the maximum load point. They assert that "the basic mechanism of localization involves a bifurcation subsequent to the maximal load point".

In order to have conditions of the same type as those in [16], let us assume that the initial displacement and the boundary conditions are given by

$$(39) \qquad u_o(x) = 2 \cos x/3^{\frac{1}{2}}$$

$$(40) \qquad u(\pm L) = \partial_x^2 u(\pm L) = 0 \quad , \quad L = (n + 0,5)\pi$$

where n is a positive integer. The parameter η, that remained unspecified in the previous analysis, here is related to the magnitude of the initial displacement :

$$(41) \qquad \tau = \eta^3.$$

A standard computation leads to (35) and to the following amplitude equation

$$(42) \qquad 4A''(X) - mA + A|A|^2 + 1 = 0.$$

One sets $A = r \exp (i\theta)$, which defines the amplitude and the phase of the slow modulation. From (35) (40), one obtains boundary conditions

$$(43) \qquad \theta(\pm L\eta) = k \pm \pi$$

$$(44) \qquad r'(\pm L\eta) = 0 \quad ,$$

where k_+ and k_- are integers. The equations (42) (43) (44) have solutions $A = r_1$

which   are real and independent of X. They satisfy the algebric equation

(45) $r_1^3 - mr_1 + 1 = 0$

which is classical in perturbed bifurcation problems [1]. The response curve is
pictured in Figure 3. Obviously, there is a maximal load point ($m \neq 3.2^{-3/2}$). In
terms of the initial variable u, these solutions are periodic and the wavenumber
is the critical one and also the one assumed for the imperfection (39). It is easy
to show that non-periodic solutions of (3) (40) bifurcate from the fundamental path.
The bifurcation points are characterized by

(46) $2 \, r_1^2 - 1/r_1 = (p\pi/L\eta)^2$

where p is a positive integer. These points always lie on the unstable part of the
response curve. The largest Lη is, the closest to the maximum load point the bifurcation points are. This behaviour is corroborated by the numerical results in [16] [17].
        Since both (38) and the bifurcation study provide unstable solutions, the
analysis remains as much qualitative as the one of Tvergaard and Needleman. Both the
mathematical method and the primary cause of localization are different. Here the
softening nonlinearity and the large aspect ratio give rise to localized solutions.
The bifurcation subsequent to the maximum load point follows from the latter assumptions. Furthermore, it is likely that the tendency to localization persists for
moderate aspect ratios.

REFERENCES

[1] J.C. AMAZIGO, B. BUDIANSKI, G.F. CARRIER : Asymptotic analysis of the buckling of imperfect columns on nonlinear elastic foundations, Int. J. Solids Structures, 6(1970) p 1341-1356.

[2] M. BOUCIF, J.E. WESFREID, E. GUYON, Role of boundary conditions on mode selection in a buckling instability, to be published.

[3] D.O. BRUSH, B.O. ALMROTH, Buckling of bars, plates and shells, Mc Graw-Hill, New-York (1975).

[4] M. CLEMENT, E. GUYON, J.E. WESFREID, Multiplicité des modes de déformation d'une plaque sous compression, C.R. Acad. Sci. Paris, Série II, 293 (1981) p 87-89.

[5] M.C. CROSS, P.G. DANIELS, P.C. HOHENBERG, E.D. SIGGIA, Phase-winding solutions in a finite container above the convective threshold, J. Fluid Mech. 127 (1983) p 155-183.

[6] W.T.KOITER, On the stability of elastic equilibrium, Doctoral Dissertation, Delft (1945). English translation : N.A.S.A. Techn. Transl. F 10, 833 (1967).

[7] L. KRAMER, P.C. HOHENBERG, Effect of boundary conditions on wavenumber selection in spatially varying steady states, to be published or this volume.

[8] L. LANDAU, L. LIFSHITZ, Theory of elasticity, Pergamon Press, New-York (1964).

[9] C.G. LANGE, A.C. NEWELL, The postbuckling problem for thin shells, S.I.A.M.J. Appl. Math. 21(1971) p 605-629.

[10] K.E. MOXHAM, Cambridge Univ. Engnrg Dept Reports, (1971).

[11] A.C. NEWELL, J.A. WHITEHEAD, Finite bandwidth, finite amplitude convection, J. Fluid Mech. 38 (1969) p 279-303.

[12] Y. POMEAU, Nonlinear pattern selection in a problem of elasticity, J. Physique Lett. 42 (1981) L 1.

[13] Y. POMEAU, S. ZALESKI, Wavelength selection in one-dimensional cellular structures, J. Physique 42 (1981) p 515-528.

[14] M. POTIER-FERRY, Amplitude modulation, phase modulation and localization of buckling patters, in "The buckling of structures in theory and practice", Cambridge Univ. Press, Cambridge (1983).

[15] L.A. SEGEL, Distant sidewalls cause slow amplitude modulation of cellular convection, J. Fluid Mech. 38 (1969) p 203-224.

[16] V. TVERGAARD, A. NEEDLEMAN, On the localization of buckling patterns, J. Appl. Mech. 47 (1980) p 613-619.

[17] V. TVERGAARD, A. NEEDLEMAN, On the development of localized buckling patterns, in "The buckling of structures in theory and practice", Cambridge Univ. Press, Cambridge (1983).

[18] N. YAMAKI, Postbuckling and imperfection sensitivity of circular cylindrical shells under compression, Proceedings 14th I.U.T.A.M. Congress, North-Holland, Amsterdam (1977) p 461-476.

WAVENUMBER SELECTION IN BUCKLING EXPERIMENTS

J.E.WESFREJD and M.BOUCIF

Ecole Supérieure de Physique et Chimie de Paris
L.H.M.P.-E.R.A 1000 CNRS
10 rue Vauquelin
75231 PARIS CEDEX 05 - FRANCE-

Elastic buckling of bars,plates and shells are common subjects in mechanics and are classical examples of instabilities.The more well known is the Euler problem of buckling of bars subjected to compressio- nal force along the axes of the bar.Buckling occurs for forces greater than a critical value F_C where F_C is a function of Young modulus E and the geometric parameters of the bar[1]

In the case of compressed thin plates restrained from deflection in the boundaries there is,as in the Euler problem,transition from the undeflected state to the deflected state for forces greater than the critical force,but there is a modulation of the deflection displaying spatially periodic structures.Engineers were initially interested in this case and motivated to know the postbuckling regime in order to im- prove ultimate loads in plates used in steel structures such as brid- ges,airfoils,etc.

Radial and longitudinal distribution of deformation modes are also observed in cylindrical thin walled columns subject to compression. Spherical shells with external radial pressure also presents polyhed- ral distributed deformations and are pleasing examples of elastic bu- ckling.

RECTANGULAR THIN PLATES

The most simple example of periodical structures of deformation is the case of rectangular elastic plates subjected to compression along the two oppositte load edges,with the restriction of deflection in the four sides.This case is often present in compression flanges in thin walled structures and is important in security design when the stress of these flanges lies in the region of critical forces.

When the plates are compressed,elastic shortening occurs and Hooke's linear law gives the relation between shortening and stresses.the in- verse of the slope of this curve is the stiffness and is given by the Young modulus E.Above a certain loading or critical force,elastic shor- tening is not more efficient for this intensity of applied stresses

and the system must choose distributed deflection normally to the plane of the plate.So the shortening-load shows slope increasing and consequently the effective stiffness is diminished.Shortening is a global
parameter and it is an integrated measurement of the full distribution
of deformation modes in the plate.It's analogous to other global measurements in hydrodynamical instabilities such as torque in Taylor-
Couette instability or heat flux in Rayleigh-Bénard convection.

Local measurements of the order parameter are possible by direct
recording of normal deflection by sensors and also by optical methods
as Moire and interferometry techniques.Stress distributions are also
valuables with gauges on the plate.

The deflection profile of the buckling is periodic and the length
of rises and falls are of the order of magnitude of the width b of the
plate.The applied force F may be adimensionalised by the following number $F = 12(1-v^2)bF/Et^3$,where v is the Poisson modulus($v \approx .3$),
t is the thickness of the plate.

Theoretical analyses of stability in the Foppl-Von Karman equations
for thin plates give the diagram for linear marginal stability with
Fourier modes of deflection with wavenumber $q=2\pi/\lambda$ .If the unloaded

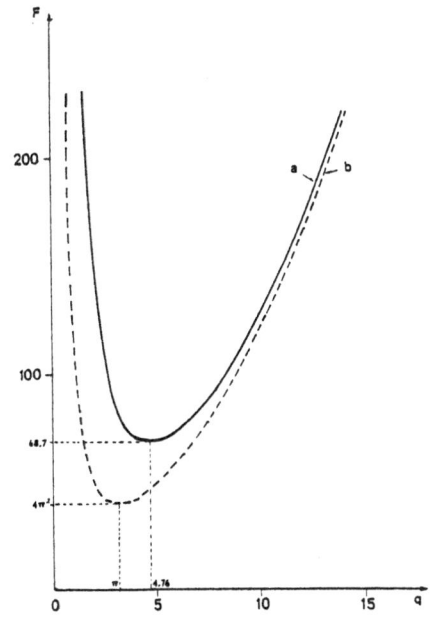

sides are not restricted in the
planeof the plate,the curve of marginal stability is given by:
$$F_0(q) = (\pi^2+q^2)^2/q^2,$$
with critical force $F_c=4\pi^2$ and
critical wavenumber $q_c=\pi$.
When the unloaded boundaries are
clamped the marginal curve (b) is
shifted towards the highest critical force($F_c=68.79$) and wavenumber
($q_c=4.76$).
These boundary conditions on the
unloaded sides are strictly analogous to the boundary conditions
in Rayleigh-Bénard convection(RB).
So the real simple support case is
like the idealised free-free case
in RB,and the clamped support is
like the rigid-rigid case in RB.

In the postbuckling regime there is redistribution of compressional
stresses.Therefore,as in the unloaded sides the normal deflection w is
nil,the localised stresses are greater than the stresses in the middle
of the plate.So a boundary layer for stresses is developed and the wi-

dth δ of this layer is usually related to the effective width of the plate $b_{eff} = 2\,\delta$.

EXPERIMENTS

In our experiments we used rectangular thin plates of brass alloy -"crisocal"-with the following characteristics:

L,long dimension of the plate(unloaded sides)=180 mm.;

b,short dimension of the plate(loaded sides) = 20 mm.;

t,thickness = 0.1 mm.

(aspect ratio Γ = 9,slenderness =200)

The unloaded edges are simple supported.This condition is realised by two pairs of opposite guide-knives-imposing the condition of defle- ction w=0 in these boundaries.These edges are practically hinged and then the edge rotational restraint is zero.The effectivity of this condition is verified by the critical values obtained for the force and the wavenumber.

In the loaded sides(short sides) we used clamped and simple suppor- ted supports.The boundary conditions,for x=0,L-end sides- are:

w $=\partial w/\partial x$ = 0 for clamped boundaries -Figure 2a-

w $=\partial^2 w/\partial x^2$=0 for simple supported boundaries-Figure 2b-

The figure shows schematically the two conditions:

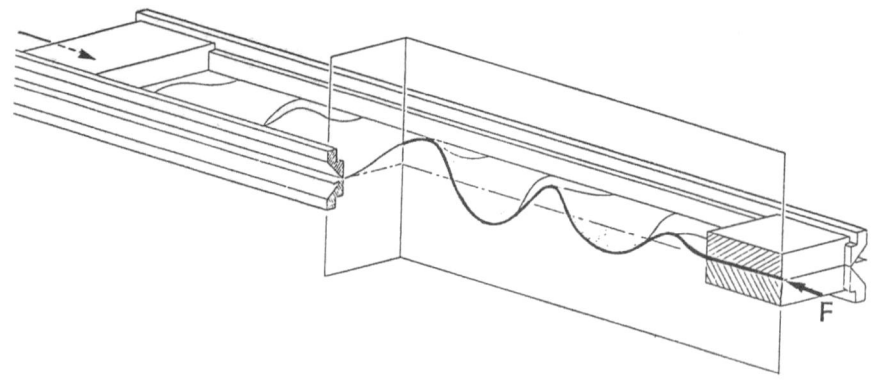

2a

In the clamped case the amplitude modulation shifts the onset of elastic buckling.The critical force is approximately $F_c(L) = F_c(\infty)$. $(1+\frac{2}{\pi}\xi_o^2/L^2)=4\pi^2(1+1/\Gamma^2)$ as may be easily derived by amplitude equations. $\xi_o=b/\pi$ is the length of coherence for spatial non-homogeneities[2]

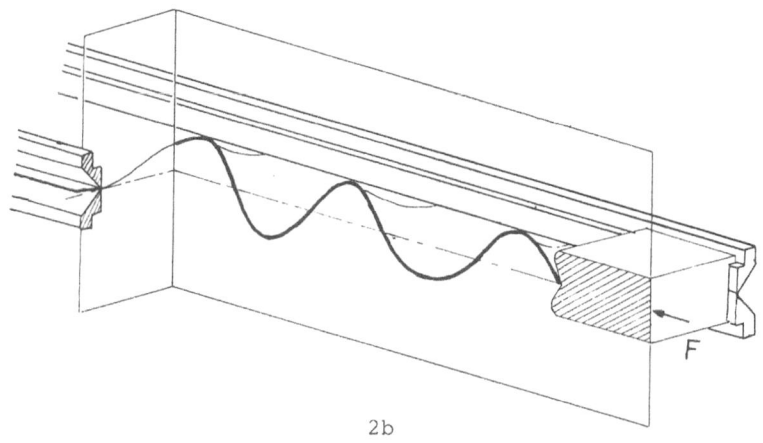

2b

In the simple supported case the measured value of the critical force is F_c=41.6 ± 2.9.The slenderness b/t used in our experiment is sufficiently high for us to work far from the plastic region of permanent
deformation of the plate.

WAVENUMBER SELECTION

Our interest was focused on the problem of wavenumber selection in
the elastic buckling.This problem,known as the jump modes in buckling
was initially studied by Stein[3].In his experiment,realised in plates
of aspect ratio Γ =5.38 he observed jumps between 5 to 6 modes when the
load was increased from the critical value.Additional jumps from 6 to
7 and 7 to 8 modes was observed,but in the region of plastic deforma-
tion.The first time theoretical attention in elastic stability was devoted to calculus of the nonlinear variation of shortening in function
of the wavenumber q,and compared with the experimental results of Ste-
in.The problem of **selection** of wavenumbers was studied theoretically
by Pomeau[4]In the case of amplitude modulation(clamped plates)he found a restrained region of stability,near F_c,bounded by two lines t^-
and t^+(see figure 3).This region allowed a band of wavenumbers $(q-q_c) \sim$
ε ,more restricted than the nonlinear band $(q-q_c) \sim \varepsilon^{\frac{1}{2}} - \varepsilon = F - F_c/F_c.-$
Our experiments was developed to test stability for different wave-
numbers.The procedure used was to force **temporary** deformation with an
imposed quantity of modes(imposed q_1)at sufficiently high forces $(F >> F_c)$
and then the force was slowly decreased.In the simple supported case
-SS-,at a certain value $F_c(q_1)$ jump modes occur from q_1 to q_2 when q_1
$\neq q_c$.We observed suppresion of one wavelength when $q_1,q_2 > q_c$.If $q_1 < q_c$

(large modes of width greater than b) the jumps modes are to $q_2 > q_1$ with nucleation of additional modes,growing initially after a dip in the top of a bulge.In this way we delineated experimentally the boundaries of nonlinear stability,given by full dots in figure 3.These limits lying above those given by the linear theory of stability $F_o(q)=$ $(\pi^2+ q^2)/ q^2$ .The nonlinear mechanism of instability responsible for this mode instability has a bulk origin and is analogous to the Eckhaus instability in fluid dynamics[5].The limit of this instability region,which is originated by parallel phase perturbations of large wavelength,in the lowest nonlinear approximation ($\varepsilon \ll 1$) is given by the parabola (E) : $\varepsilon =3\zeta_o^2(q-q_c)^2$. As far as we know there do not exist calculations for these limits-like the Busse"balloon" in thermal instabilities[6]-in the full nonlinear postbuckling region,where our experiments were performed.

The same kind of experiments were also done in plates clamped along the loaded sides[7].The experimental points(open dots in figure 3) give **higher** limits for $F_c(q_1)$ than the simple supported case.Therefore a more reduced band of allowed modes exist.The mechanism of jump modes is different in this case:addition or substraction of modes are more localised near the clamped edges.

When the force is increased from $F >F_c$ buckling occurs with the critical number of deformation modes(9 in our plate of aspect ratio $\Gamma =9$) and we do not observe changes in the number of modes if the force is continously increased.This observation is valuable in the two cases:

(SS) and (C).Therefore the limits of stability in the clamped case do not agree quantitatively with the cone angle predicted in(4)(t$^-$ and t$^+$).The existence of finite rigidity in the clamped edges may explain this difference.In effect,deviations from the condition of perfect clamping allowed wider bands of accesible states shifting the slope of the t$^-$ line in the direction of our experimental points[8].

However,imperfections are usually present in mechanical experiments The geometrical imperfection[9] or deflection mode with amplitude w_o present without compressional force,are important in the study of the critical region near F_C.

When the imperfections come from residual stresses,competition of modes may arise between the "natural" and the permanent forced modes. This competition is present in the case of compressed plates with regularly distributed stiffner or fibers like orthotropic plates.

On the other hand finite edge rotational rigidity expanded,as we have seen,the band of allowed wavenumbers.

Experiments are also in progress to study the influence of non-homogenous boundary[10]conditions in the loaded edges,sometimes present in the (SS) case as a consequence of out of centre positions of the supports.

ACKNOWLEDGEMENTS

We thank J.C.Charmet,M.Clément and E.Guyon for participation in this project.

REFERENCES

1. Brusch,D.O and Almorth,B.O.**Buckling of bars,plates and shells**(Mc. Graw Hill)1975

2. Wesfreid,J.E.,Pomeau,Y.,Dubois,M.,Normand,C. and Bergé,P.,J.Physique **39**,725(1978)

3. Stein,M.NACA Technical Rep. R-40 (1959)

4. Pomeau,Y.,J.Physique Lett.**42**,L-1(1981)

5. Eckhaus,W.,**Studies in non-linear stability theory**(Springer Verlag) 1965

6. Busse,F.H.,Rep.Prog.Phys.**41**,1929(1978)

7. Clément,M.,Guyon,E.and Wesfreid,J.E. C.R.Acad.Sci.Paris II**293**,87 (1981)

8. Potier-Ferry,M.   in this volume

9. Koiter,W.T.,Technical Rep. AFFDL-TR 70-25(1970)

10. Zaleski,S. in this volume.

EFFECT OF BOUNDARIES ON PERIODIC PATTERNS

Lorenz Kramer

Physikalisches Institut der Universität Bayreuth
D-8580 Bayreuth, West Germany

and

P.C. Hohenberg

Bell Laboratories, Murray Hill, New Jersey 07974, USA

1. INTRODUCTION

The existence and stability of periodic patterns in the semi-infinite region $-\infty < x < 0$ have been studied using one-dimensional model equations with linear boundary conditions at $x=0$. For all boundary conditions there exists at least one static solution whose amplitude near the boundary falls below its bulk value (Type-I solution). A linear stability analysis of the uniform state at threshold reveals that Type-I solutions are often unstable. Then there exists a static solution where the amplitude near the boundary rises above its bulk value ("Type-II" solution), or a limit-cycle solution where the amplitude near the boundary oscillates. Type-I solutions exist in a restricted wavevector band, whereas Type-II solutions typically extend over the whole Eckhaus-stable band of the infinite system. The effect of slow spatial variation of the parameters of a two-component model is studied analytically near threshold, and the nonuniversality of the wavevector selection mechanism explicitly demonstrated.

We primarily wish to study the one-component model[1,2]

$$\partial_t u = \left[\varepsilon - (\partial_x^2 + 1)^2 \right] u - u^3 , \tag{1.1}$$

where $u(x,t)$ is a real function. In the range $0 < \varepsilon < 1$, which we consider here, the homogeneous solution $u \equiv 0$ of Eq.(1.1) is unstable and one has static, linearly stable periodic solutions in the infinite system with wavenumber q in a band plotted as curve 1 in Fig.1. The band is limited by the Eckhaus instability which for small ε is given by

$$|q-1| < q_E = (\varepsilon/12)^{1/2} . \tag{1.2}$$

As can be seen from Fig.1 (dashed curve) this approximation is quite

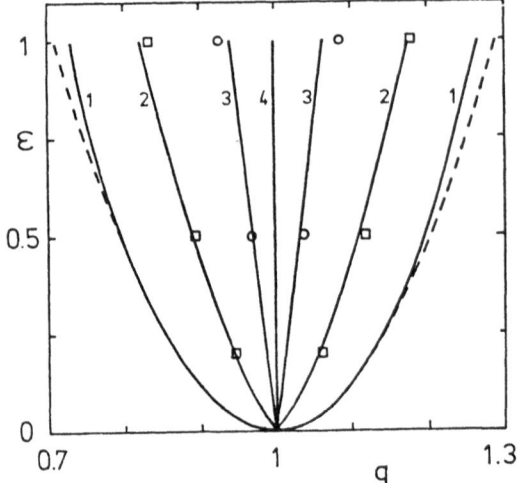

Fig.1: Width of the band of wavevectors q as a function of ε. Curve 1:
full Eckhaus band (dashed curve gives
the small-ε expansion according to
Eq.(1.2)). Curve 2: restricted band
for $\lambda_v = 0$, $|\lambda_p| = 3$ according to Eq.(2.9)
(numerical results for the full equa-
tions (1.1) and (1.3) are given by (□)).
Curve 3: restricted band for $\lambda_{11} = \lambda_{12} = \lambda_{22} = 0$, $1/\lambda_{21} = 0$ according to Eq.(2.14)
(numerical results are given by (o)).
Curve 4: minimizing wavevector.

good up to ε = 1.

For half-space solutions
the boundary conditions at x=0
are written as

$$u + \lambda_{11} \partial_x u + \lambda_{12} \partial_x^3 u = 0 ;$$

$$\partial_x^2 u + \lambda_{21} \partial_x u + \lambda_{22} \partial_x^3 u = 0.$$

(1.3)

For

$$\lambda_{11} - 2\lambda_{12} - \lambda_{22} = 0 \qquad (1.4)$$

there exists a global Lyapunov
functional

$$V = \int_{-\infty}^{0} dx \left[-(\partial_x u)^2 + \frac{1}{2}(\partial_x^2 u)^2 + \right. $$
$$+ \frac{1}{2}(1-\varepsilon)u^2 + \frac{1}{4} u^4 \Big] -$$
$$-\left[\frac{1}{2} \lambda_{12} (2\partial_x u + \partial_x^3 u)^2 - \right.$$
$$\left. -\left(\lambda_{22} - \frac{1}{2} \lambda_{21}\right)\left(\partial_x u\right)^2 \right]_{x=0}$$

(1.5)

which acts as a minimizing potential for the problem. For stationary solutions of (1.1) V attains the value

$$V = -\frac{1}{4} \int_{-\infty}^{0} dx \ u^4 .$$

(1.6)

The wavevector that minimizes V is plotted in curve 4 of Fig.1. It is
very nearly 1 for all ε(<1).

To obtain numerical solutions, Eq.(1.1) was discretized on a finite
(but sufficiently long) strip with "periodic" boundary conditions u=u"=0
on the left and the boundary conditions (1.3) on the right. A Newton iteration scheme proved rather effective for stationary solutions. The method
converges to stable and in some cases also to unstable solutions. Since
for given boundary conditions one usually has a multiplicity of possible
solutions here, the result of an iteration depends on the initial function. Families of solutions may be traced out by changing the values of
λ_{ij}, ε, or the space-increment, in suitable steps using previous solutions as starting points for the next step. To test for stability and to

obtain dynamic solutions of (1.1) an implicit time discretization scheme was used.[2]

In Sec.2 we present results for the existence of half-space solutions using two different analytic approaches and numerical computations. We find that for all boundary conditions there exists in a reduced wavenumber band at least one static solution where the amplitude falls below its bulk value near the boundary ("Type-I" solution). For many boundary conditions there also exist static solutions where the amplitude rises above its bulk value near the boundary ("Type-II" solutions), and sometimes one has a limit-cycle solution where the amplitude near the boundary oscillates. These solutions bifurcate from the homogeneous state below the bulk threshold and therefore remain finite at threshold.

In Sec.3 a linear stability analysis of the uniform state is presented which reveals that Type-I solutions are often unstable and then in fact a stable Type-II or oscillating surface solution exists. In simple cases the multiplicity and types of instabilities (nonoscillatory or oscillatory) correspond qualitatively to the multiplicity and types of solutions (static or oscillatory) they lead too.

In Sec.4 we briefly discuss a different model,[3] consisting of two second-order equations, which do not in general have a minimizing potential in an infinite system. Many of the boundary effects found for Eq. (1.1) also generalize to this model. In addition, we study wavevector selection in situations in which the parameters of the model vary slowly in space.

2. EXISTENCE OF BOUNDARY SOLUTIONS

a. Nearly periodic limit

There exists a subset of boundary conditions of the type (1.3) that can be satisfied by solutions of (1.1) which are periodic in the infinite system. Near threshold this set is characterized by[4]

$$\Delta = \lambda_{11} - \lambda_{12} + \lambda_{21} - \lambda_{22} = 0. \qquad (2.1)$$

When $|\Delta|$ is sufficiently small and $\varepsilon << 1$ there exist "nearly periodic solutions", and the lowest-order amplitude approximation may be used for their description.[5-7] For stationary solutions we then write

$$u = (\varepsilon/3)^{1/2} \left[A(X) \ e^{i(x+\bar{x}_o)} + cc \right] , \qquad (2.2)$$

where $x=\varepsilon^{1/2}X$ ($\bar{x}_o$ is an arbitrary phase). As long as $\varepsilon^{1/2}|A|<<1$ and A varies slowly over the rapid scale 1 this function may be determined from

$$4\partial_x^2 A + A(1-|A|^2) = 0 \ . \tag{2.3}$$

The solutions of (2.3) which become periodic for $x\rightarrow-\infty$ can be put into two classes:

i) Solutions where the amplitude drops below its asymptotic value $(1-Q^2)^{1/2}$ near the boundary are given by

$$A(X) = \sqrt{2}\left[-iQ-\alpha Y(X)\right]\exp\left(iQ[X+x_1]/2\right) \tag{2.4}$$

where

$$\alpha = \left[\tfrac{1}{2}(1-3Q^2)\right]^{1/2} \qquad \text{(real)} \tag{2.5}$$

$$Y(X) = -\tanh\left(\tfrac{\alpha}{2}[X-X_2]\right) . \tag{2.6}$$

The functions u belonging to this class ("Type-I" solutions) are given by

$$u = (8\varepsilon/3)^{1/2}\left(Q\sin[q(x+x_o)]-\alpha Y(X)\cos[q(x+x_o)]\right) \tag{2.7}$$

where $q = 1 + Q\varepsilon^{1/2}/2$.

ii) Solutions where the amplitude near the boundary rises above its asymptotic value $(1-Q^2)^{1/2}$ ("Type-II" solutions) are obtained from (2.4) and (2.7) by replacing (2.6) by

$$Y(X) = -\coth\left(\tfrac{\alpha}{2}[X-X_2]\right) . \tag{2.8}$$

Since Y diverges at X_2, negative X_2 (or, equivalently, $Y(0)<-1$) are not acceptable.

Note that both types of solutions exist only in the Eckhaus-stable band $\alpha>0$. The boundary conditions (1.3) can be used to determine x_o and give for $\varepsilon<<1$ the following condition on $y=Y(0)$

$$Q^2/\alpha^2 + y^2 + [\lambda_v Q+ \lambda_p\alpha y]\varepsilon^{1/2}(1-y^2) = 0 \tag{2.9}$$

with

$$\lambda_v = (\lambda_{11}-2\lambda_{12}-\lambda_{22})/\Delta \ ; \ \lambda_p = -(1+|\underset{\approx}{\lambda}|)/\Delta \qquad (|\underset{\approx}{\lambda}| = \det \underset{\approx}{\lambda}). \tag{2.10}$$

In deriving (2.9) some terms of order $\varepsilon^{1/2}$ were neglected, which is valid as long as $\lambda_v Q + \lambda_p \alpha y \gg 1$. Equation (2.9) must be solved with $y > -1$. For $|y| < 1$ one has Type-I solutions, for $y > 1$ Type-II solutions, and $y = 1$ corresponds to the periodic case.

For $Q = 0$ one has the three roots

$$y_1 = 0, \quad y_{2,3} = \left(\sqrt{2}\, \lambda_p \varepsilon^{1/2}\right)^{-1} \pm \sqrt{1 \pm \left(\sqrt{2}\, \lambda_p \varepsilon^{1/2}\right)^{-2}} \quad . \tag{2.11}$$

Thus for finite λ_p one always has two Type-I solutions corresponding to y_1 and y_2, and for $\lambda_p > 0$ there is a Type-II solution y_3. As $|Q|$ increases the Type-I solutions merge and vanish before the Eckhaus-limit is reached. This band-restriction may be written in the form

$$q_- < q - 1 < q_+ \quad , \quad q_\pm = (q_s \pm q_r) \tag{2.12}$$

where q_s and q_r depend on λ_v, λ_p and ε. For $\lambda_v = 0$ (potential case) the asymmetry q_s is zero, and $q_s \to 0$, $q_r \to q_E$ for $\lambda_p \to \pm \infty$. In curve 2 of Fig.1 the restricted band is shown for $\lambda_v = 0$, $|\lambda_p| = 3$. The squares represent numerical results for the bandwidth with $\lambda_{11} = \lambda_{12} = \lambda_{22} = 0$, $\lambda_{21} = 1/3$. Assuming $y \ll 1$ and $Q \ll 1$ at the band-edge one obtains from (2.9)

$$q_s = -\lambda_v \varepsilon/8 \; ; \quad q_r = \left(\lambda_v^2 + \lambda_p^2\right)^{1/2} \varepsilon/8 \quad . \tag{2.13}$$

This result is valid for $q_r \ll \sqrt{\varepsilon}/12$, or, equivalently, for $3(\lambda_v^2 + \lambda_p^2)\varepsilon/16 \ll 1$.

The Type-II solution exists (for $\lambda_p > 0$) in the whole Eckhaus-stable band. From (2.9) one sees that $\varepsilon^{1/2} y$, i.e. the unreduced amplitude of u at the boundary, remains finite for $\varepsilon \to 0$. In fact a nonzero boundary solution, which now falls off to zero in the interior, persists for $\varepsilon < 0$.

b. The CDHS-approximation

The restriction of the band of wavevectors has been studied previously for particular boundary conditions by Cross, Daniels, Hohenberg and Siggia (CDHS)[7] and others.[8,9] This treatment, which is valid near threshold when the restriction is strong, $q_\pm \ll q_E$, is based on the assumption that the amplitude near the boundary is small compared to its value in the bulk. This assumption is valid for all Type-I solutions in the limit $\varepsilon \to 0$. We have repeated the CDHS-type calculation for the general case (1.3) and find

$$q_s = - \frac{\varepsilon}{8} \lambda_v$$

$$q_r = \frac{\varepsilon}{16} \left[(\lambda_{11} + \lambda_{12} - \lambda_{21} + 3\lambda_{22})^2 + 4(1-|\underset{\approx}{\lambda}|)^2 \right]^{1/2} / \Delta \; .$$

(2.14)

This result exhibits an interesting feature: There may occur perfect selection ($q_r=0$) to lowest nontrivial order in ε. We have determined the bandwidth numerically in several such cases and found that a narrow band does exist with a width that appears to behave like ε^2.

The range of validity of (2.14) is given by q_s, $q_r \ll \sqrt{\varepsilon/12}$ and $\varepsilon \ll 1$. In the nearly periodic limit $|\Delta| \ll 1$, Eqs. (2.14) can be shown to reduce to (2.13) so there is overlap between the two approaches. In curve 3 of Fig.1 the band restriction for $\lambda_{11} = \lambda_{12} = \lambda_{22} = 0$, $\lambda_{21}^{-1} = 0$ is shown, and compared to numerical results (circles).

c. Numerical results

All the above analytic results could be verified numerically in their region of validity. For all boundary conditions we find at least one Type-I solution in a restricted band, and in some regions we find a second (unstable) Type-I solution. It appears likely that there always exists a second Type-I solution. The Type-I solutions may be unstable and then there exists at least one Type-II solution or, for nonpotential boundary conditions, a limit-cycle solution where the amplitude near the boundary oscillates (see Sec.4 below).

In Fig.2 the existence and stability sectors for small ε are shown for the case

$$\lambda_{12} = \lambda_{22} = 0$$

(2.15)

in the plane of λ_{11} vs. $\lambda_{11} + \lambda_{21} = \Delta$ (I_s denotes Type-I stable, I_u denotes Type-I unstable etc.). The coordinate axes and curves 1 and 2 divide the plane into six different regions. The arrows indicate how the static solutions are connected to each other. The horizontal arrows start with the periodic case at $\lambda_{11} + \lambda_{21} = 0$, which separates the Type-I from the Type-II solutions. The wavy arrows that start from the line $\lambda_{11} + \lambda_{21} = \infty$ are to be considered connected to those that start to the left of the line $\lambda_{11} + \lambda_{21} = -\infty$. The dotted arrows indicate that the corresponding I_u-solutions could not be continued numerically, but according to the analysis they probably exist up to $\lambda_{11} + \lambda_{21} = +\infty$ and then again from $-\infty$ to zero.

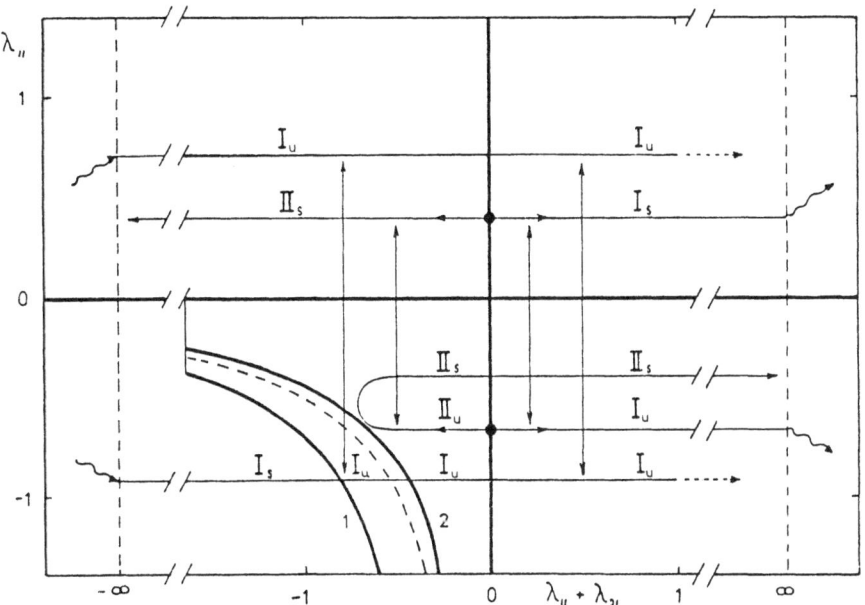

Fig.2:Existence and stability sectors for the case $\lambda_{12}=\lambda_{22}=0, \varepsilon<<1$, in the plane of λ_{11} vs. $\lambda_{11}+\lambda_{21}$. I_u denotes Type-I unstable, I_s denotes Type-I stable etc.. The coordinate axes and curves 1 and 2 give the six sectors. The arrows indicate how the solutions are connected to each other. The oscillating surface solution exists between curves 1 and 2. See text for more details.

The arrangement in Fig.2 is such that solutions with larger amplitude near the boundary are nearer to the abscissa. The vertical arrows indicate how the solutions are connected across the abscissa.

For $\lambda_{11}>0$ in Fig.2 one has everywhere essentially the same situation as in the nearly periodic limit $|\Delta|<<0$: Either there exists a stable Type-I solution $(\lambda_p=-\Delta^{-1}<0)$, or a stable Type-II solution $(\lambda_p>0)$. For $\lambda_{11}<0$ we find two Type-II solutions to the right of curve 2 which merge at curve 2. Between curve 1 and curve 2 we find a (stable) oscillating surface solution. The solution bifurcates from the stable Type-I solution at curve 1 with finite frequency. It merges with the Type-II solution at curve 2 in the following manner: The oscillations become progressively more pulse-like with increasing intervals during which the static Type-II solution is better and better approached. This oscillating surface solution exists down to $\varepsilon=0$, and in general even for negative ε, as does the Type-II solution. Clearly the oscillating case can occur only for non-potential boundary conditions where one does not have a Lyapunov functional.

We also found situations with more than one stable solution (see next section). Under all circumstances the solution with the largest amplitude was stable, which is consistent with Eq.(1.6) for the potential case. Most of the existence and stability properties appear to depend only weakly on ε .

3. STABILITY OF TYPE-I SOLUTIONS

Since Type-I solutions tend to zero for $\varepsilon \to 0$ it is natural to analyze their stability properties near threshold by testing the homogeneous solution $u \equiv 0$ at $\varepsilon = 0$ for stability with respect to surface perturbations. From (1.1) one obtains the stability equation

$$\partial_t u = - \left(\partial_x^2 + 1 \right)^2 u \ . \tag{3.1}$$

Since the problem is linear one may consider complex solutions of (3.1) and (1.3). We look for solutions of the form

$$u = e^{\omega t} \sum_j A_j \exp (iq_j x) \ , \tag{3.2}$$

$$\omega = \omega' + i\omega'', \qquad A_j = A_j' + iA_j'' \ , \qquad q_j = q_j' + iq_j'' \ . \tag{3.3}$$

Then

$$\omega = -(1-q_j^2)^2 \tag{3.4}$$

which has 2 solutions

$$q_1 = -\left(1 + i\omega^{1/2} \right)^{1/2}; \qquad q_2 = -\left(1 - i\omega^{1/2} \right)^{1/2} \tag{3.5}$$

with $q_j'' < 0$ (we adopt the convention $\operatorname{Im} x^{1/2} > 0$ so that a cut is on the positive real axis). In order that u vanish for $x \to - \infty$ one needs $q_j'' < 0$. For instability one needs $\omega' > 0$.

The boundary conditions (1.3) now give

$$e^{\omega t} \left[(1+iq_1\lambda_{11}-iq_1{}^3\lambda_{12}) \ A_1 + (1+iq_2\lambda_{11}-iq_2{}^3\lambda_{12}) A_2 \right] = 0$$

$$e^{\omega t} \left[(-q_1{}^2+iq_1\lambda_{21}-iq_1{}^3\lambda_{22}) A_1 + (-q_2{}^2+iq_2\lambda_{21}-iq_2{}^3\lambda_{22}) A_2 \right] = 0 \ . \tag{3.6}$$

Setting the determinant to zero we find

$$(q_1-q_2) \left\{ (q_1+q_2)(1-q_1q_2|\underset{\approx}{\lambda}|) - i\left[\lambda_{21}-2\lambda_{22}-(\lambda_{11}+\lambda_{22})q_1q_2-\lambda_{12}q_1^2q_2^2 \right] \right\} \quad (3.7)$$

which along with (3.5) are three equations for the complex numers q_1, q_2 and ω. For $\omega \neq 0$ one always has $q_1 \neq q_2$ so the term in curly brackets in (3.8) must be zero. With

$$p = - i(q_1 + q_2)/2 \qquad (3.8)$$

and the relation $q_1{}^2 + q_2{}^2 = 2$, Eq.(3.7), leads to

$$f(p) = 4\lambda_{12}p^4 + 4|\underset{\approx}{\lambda}|p^3 + 2(2\lambda_{12} - \lambda_{11} - \lambda_{22})p^2$$
$$+ 2(1+|\underset{\approx}{\lambda}|)p - \Delta = 0 . \qquad (3.9)$$

All Type-I solutions are expected to be unstable in the limit $\varepsilon \to 0$ when Eq.(3.9) has a root with

$$p' < 0 , \quad p'^2 - p''^2 > - \frac{1}{2} + \left[\frac{1}{4} + 4p'^2 p''^2 \right]^{\frac{1}{2}} . \qquad (3.10)$$

The frequency is

$$\omega = 4p^2(1 + p^2) . \qquad (3.11)$$

The second inequality (3.10) insures that $q_1'' , q_2'' < 0$. When p is real, then ω is also real and automatically positive. If p is complex one has an oscillatory instability and the sign of ω' must be checked.

We now consider some simple cases. In the nearly periodic limit $|\Delta| << 1$, Eq.(3.9) has a real root

$$p = - 2/\lambda_p \qquad (3.12)$$

so that Type-I solutions are unstable when a Type-II solution exists.

In the case (2.15) exhibited in Fig.2, (3.9) reduces to

$$f(p) = - 2\lambda_{11}p^2 + 2p - (\lambda_{11} + \lambda_{21}) = 0 . \qquad (3.13)$$

For $\lambda_{11} > 0$, Eq.(3.13) has a real negative root if $\lambda_{11} + \lambda_{21} < 0$. For $\lambda_{11} < 0$ it has two real negative roots if $\lambda_{11}+\lambda_{21} > 1/\lambda_{21}$ which merge when $\lambda_{11}+\lambda_{21} = 1/\lambda_{21}$ (broken curve in Fig.2). For $\lambda_{11}+\lambda_{21} < 1/\lambda_{21}$ (still $\lambda_{11} < 0$)

Eq.(3.13) has a pair of complex conjugate solutions with q_1', $q_2' < 0$ and

$$\omega' = \lambda_{11}^{-4} \left[\lambda_{11}^2 \lambda_{21}^2 - 4\lambda_{11}\lambda_{21} + 2 - \lambda_{11}^2 - \lambda_{11}^4 \right] \qquad (3.14)$$

which is positive for

$$\lambda_{21} > \lambda_{11}^{-1} \left[2 - \sqrt{2 + 2\lambda_{11}^2 + \lambda_{11}^4} \right] . \qquad (3.15)$$

The stability limit gives curve 1 of Fig.2.

In the general case, Eq.(3.9) can have three or four roots which satisfy (3.10) and in the numerical computations we then found stable coexistence of two or even three solutions, two of which could be of Type-II. The oscillatory instability is then often of the backward type. We have not yet found cases with more than one stable oscillatory so-lution or more complicated dynamical behavior.

4. CONCLUDING REMARKS

We have investigated the influence of general linear boundary. con-ditions of the type (1.3) on the stationary solutions of a simple model for cellular pattern formation in one dimension.[10] It turned out to be useful to divide the static solutions into two classes:

i) Type-I solutions, where the amplitude near the boundary is smaller than in the bulk and which go over into the homogeneous state for $\varepsilon \to 0$, appear to exist for arbitrary boundary conditions in a restricted band of wave-vectors. Analytic treatment is possible for $\varepsilon \ll 1$ when the restriction is either strong or weak. For substantial regions of the parameter space of λ_{ij} the Type-I solutions turn out to be unstable and a simple analy-sis of the homogeneous state at $\varepsilon = 0$ appears to give the essential fea-tures of the stability regions.

ii) Type-II solutions, where the amplitude near the boundary is larger than in the bulk and which remain finite near the boundary for $\varepsilon \to 0$, appear to exist in the full Eckhaus-stable band predominantly for boun-dary conditions where Type-I solutions are unstable. Analytic treatment is possible for $\varepsilon \ll 1$ when the boundary conditions are nearly consistent with periodicity. The Type-II solution with the largest amplitude appears to be stable, but an analytic stability criterion is lacking at this time

In addition we found an oscillating surface solution for nonpotential boundary conditions where Type-I solutions are unstable and Type-II so-lutions cease to exist. The linear stability analysis presented in Sec.3

could easily be generalized to negative ε and would then give the threshold where Type-II solutions or oscillating surface solutions bifurcate from the homogeneous state. A more detailed investigation of dynamic surface states is in progress.

A similar investigation was carried out with a system of reaction-diffusion equations[3,11]

$$\partial_t u_1 = D_1 \partial_x^2 u_1 + a_1 u_1 (1-u_1^2) - b_1 u_2$$

$$\partial_t u_2 = D_2 \partial_x^2 u_2 - a_2 u_2 (1-u_2^2) + b_2 u_1$$

$$(4.1)$$

This system has periodic stationary states if $b_1 b_2$ is smaller than $b_{1c} b_{2c} = (a_1 D_2 + a_2 D_1)^2 / 4 D_1 D_2$ while the inequalities $1 < a_2/a_1 < D_2/D_1$ hold. The wavevector at threshold is $q_o^2 = (a_1 D_2 - a_2 D_1)/2 D_1 D_2$. The influence of boundary conditions is in many ways similar to that found above.

The finding that the band of wavevectors is restricted in situations where the amplitude is reduced at some location in the system appears to be quite general. It also occurs when, instead of applying appropriate nonperiodic boundary conditions, the parameters of the equations vary spatially so that the amplitude is decreased.[11,3] If this change is gradual and the system becomes subcritical somewhere, a unique wavevector is selected. If the system (including the boundary region) has a potential then the minimizing wavevector is selected. Otherwise the selection is nonuniversal and it can be calculated analytically near threshold using the amplitude expansion. For the system (4.1) with slow spatial variations of the parameters $b_{1,2}$ of the form

$$b_1(x) = b_{1c} [1-(1+\varphi)\beta(x)\varepsilon] \qquad b_2(x) = b_{2c}[1-(1-\varphi)\beta(x)\varepsilon] \qquad (4.2)$$

where $\varepsilon \ll 1$, $\beta = O(1)$ and $\varphi = $ const. we find for the selected wavenumber[3]

$$q - q_o = - \varepsilon \frac{(a_1 D_2 + a_2 D_1)^{1/2}}{4 q_o D_1 D_2} \left[\frac{a_1 b_{1c} D_2^2 - a_2 b_{2c} D_1^2}{a_1 b_{1c} D_2^2 + a_2 b_{2c} D_1^2} - \frac{1}{2}\varphi \right] . \qquad (4.3)$$

This result differs from that of Pomeau and Zaleski[12] who used an incorrect amplitude equation, even for the case $\varphi = 0$, where there exists a potential.

5. ACKNOWLEDGEMENT

The authors wish to thank the Institute for Theoretical Physics of the University of California at Santa Barbara, where this work was started, for its hospitality. They have benefitted from discussions with M.C. Cross and H. Riecke. This work was supported in part by the National Science Foundation under Grant No. PHY 77-27084. Travel grants by the Deutsche Forschungsgemeinschaft and the Fulbright Commission (L.K.) and the Emil Warburg Foundation (P.C.H.) are gratefully acknowledged.

REFERENCES

1. This model was introduced by Swift and Hohenberg (Phys.Rev. $\underline{A15}$, 319 (1977)) in the context of Rayleigh-Bénard convection, and its wavelength selection properties were first studied by Pomeau and Manneville (J.Phys.Lett. $\underline{40}$, L-609 (1979); Phys.Lett. $\underline{75A}$, 296 (1980)).
2. L.Kramer and P.C. Hohenberg (to be published).
3. P.C. Hohenberg, L. Kramer and H. Riecke (to be published).
4. We omit all finite - ε corrections.
5. A.C. Newell and J.C. Whitehead, J.Fluid Mech. $\underline{38}$, 279 (1969).
6. L.A. Segel, J.Fluid Mech. $\underline{38}$, 203 (1969).
7. M.C. Cross, P.G. Daniels, P.C. Hohenberg and E.D. Siggia, Phys.Rev. Lett. $\underline{45}$, 898 (1980); J.Fluid Mech. $\underline{127}$, 155 (1983).
8. Y. Pomeau and S. Zaleski, C.R. Acad.Sci. Paris, $\underline{290B}$, 505 (1980); J. Phys.(Paris) $\underline{42}$, 515 (1981).
9. M. Potier Ferry, in "Collapse: The Buckling of Structures in Theory and Practice", J.M.T. Thomson and G.W. Hunt, ed., (Cambridge University Press, 1983).
10. The results are formulated for boundary conditions implemented on the large-x end of the system. If this is reversed the signs of λ_{ij} have to be reversed.
11. L. Kramer, E. Ben-Jacob, H. Brand and M.C. Cross, Phys.Rev.Lett. $\underline{49}$, 1891 (1982).
12. Y. Pomeau and S. Zaleski, J.Phys.Lett. (Paris) $\underline{44}$, L-135 (1983).

CELL NUMBER SELECTION IN TAYLOR-COUETTE FLOW

T. Mullin

Mathematical Institute
24/29 St. Giles
Oxford

1. INTRODUCTION

The Taylor-Couette experiment concerns the behaviour of flow between two concentric cylinders, the outer of which is fixed and the inner rotates. When the speed of the inner cylinder is raised and the length of the apparatus is chosen outside these ranges where cell-number changes occur one finds the following sequence of events in the steady flow regime. At small Reynolds numbers, R, the fluid moves steadily in concentric circles except for some three-dimensional features at the ends. As the rotation rate is increased toroidal counter-rotating steady Taylor vortices are formed, growing smoothly from the fixed ends of the apparatus.

The cell number selection process in Taylor-Couette flow has been investigated experimentally by Benjamin (1978b), Mullin (1982) and Mullin, Pfister and Lorenzen (1982). The results have been interpreted in terms of the theoretical framework of Benjamin (1978a). In particular, information has been obtained concerning mutations of the primary flow as the length of the annulus is changed. For any given proportion of the annulus, the primary flow is uniquely definable as the steady flow realized by very gradual increases in the angular speed of the inner cylinder. The experiments reported here concern the case where the ends of the annulus are fixed and thus we consider the exchange processes for even cell numbers only.

At values of R far above the range discussed above, other steady cellular flows are found. These are the 'anomalous' modes discussed fully in Benjamin and Mullin (1981) and are so called because the direction of spiralling of the cells is such that outward flow is found at one or both fixed end walls. Thus these modes may have either odd or even numbers of cells. They are generally disconnected from any of the primary flows and show strong dependence on finite length effects.

Finally a demonstration of non-uniqueness in steady Taylor-Couette flow will be discussed. Benjamin and Mullin (1982) demonstrated that a surprisingly large number of steady cellular flows exist at a single point in (R, Γ)-plane where Γ is the aspect ratio. The implication of these results will be discussed in relation to the scenario of cell number selection in Taylor-Couette flow.

2. PRIMARY FLOW SELECTION

At small values of R the observed steady motion is mainly in circles, but some three-dimensional features appear near the fixed end walls. As R is increased into a narrow, quasi-critical range, cells grow progressively from the ends, fitting together at the centre to form the primary cellular flow that is uniquely determined over the respective range of aspect ratio, Γ. The cellular mode that will become the primary flow in the next range of Γ is also realizable as a stable steady flow at sufficiently high values of R, and it may be produced in practice when the flow settles down after a sudden start of the rotation to some value substantially above the quasi-critical range. Realized in these circumstances, the flow is called a secondary mode. It evidently cannot survive indefinitely as R is gradually reduced, and it collapses catastrophically at a well defined critical value of R.

The form of the selection process, as represented by a graph of such critical values in the (R, Γ)-plane has been predicted generally in the theoretical work of Benjamin (1978a) and Schaeffer (1980) and has been worked out for a particular case using Schaeffer's model by Hall (1982). In particular, the change-over of the primary mode from one array of cells to another (e.g. four cells giving way to six) is accompanied by a hysteresis, indicated by a cusp pointing obliquely to the R-axis in the (R, Γ)-chart. However, the exact orientation of this cusp, whether pointed up or down relative to the R axis is not predicted by the theory.

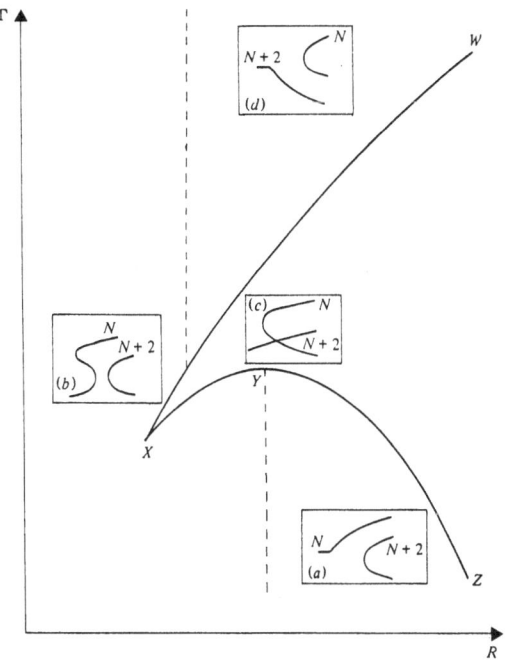

Figure 1

Schematic diagram of exchange procedure in (R,Γ)-plane. Inset figures
show the relevant state diagrams in the (R,f)-plane where f is a linear
functional of the flow. In this example they are ordered (a) → (d) as
Γ increases.

The schematic (R,Γ)-diagram shown in figure 1 illustrates the exchange

process. In the range of Γ below Y the N cell mode is primary and dev-

elops smoothly with gradual increase in R, as is shown in state diagram

(a). The (N + 2)-cell state can exist as a secondary mode to the right

of YZ, collapsing along this line, on reduction of R, to the primary N-

cell mode. An increase in Γ leads to the development of a fold in the

primary locus of the state diagram (b), which is manifested as the cus-

ped region XY in the (R,Γ)-plane. The two state curves then become

linked (diagram (c)) at Y, which is thus the bifurcation point for the

primary flow. In the Γ-range WX the roles are reversed (diagram (d))
and the (N + 2)-cell mode is now primary. Thus one primary locus is
continuously deformed into another by the variation of Γ. A cusp pointing in the opposite direction indicates that the exchange proceeds as
d → c → b → a as Γ is increased.

As a particular example of the exchange procedure we will consider the mutation between the four and six-cell primary modes for various values of Γ. Other exchanges are reported in Benjamin (1978b),
Mullin (1982) and Mullin, Pfister and Lorenzen (1982). In addition,
details of the apparatus used will not be given here and interested
readers are referred to the original reports where full descriptions
are given.

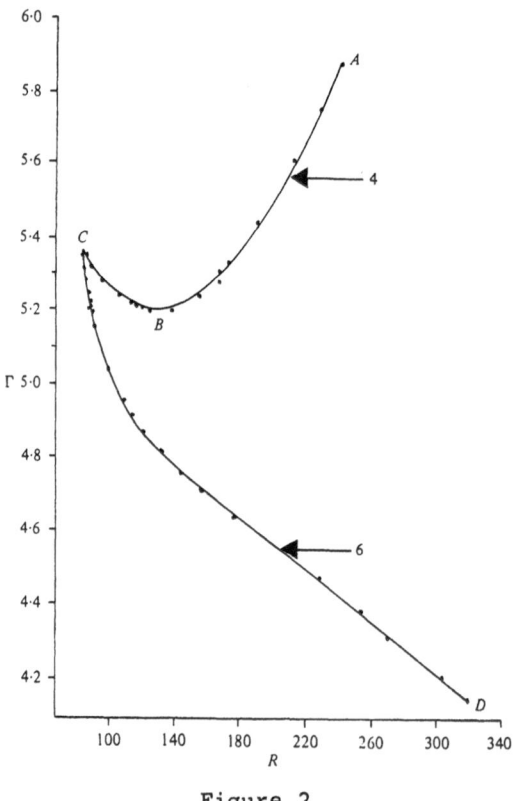

Figure 2

Experimentally determined bifurcation set in the (R, Γ)-plane for four-
cell and six-cell flows. The arrows emphasize that the respective loci
are the lower limits of stability for the secondary modes whose number
of cells are indicated.

The results for this case are presented in figure 2. For parameter values (R, Γ) to the right of line AB in the figure the four-cell flow is realizable as a stable secondary mode by a sudden start of the appratus. On gradual reductions in R, the mode collapses when the line AB is reached, and the system settles into the six-cell mode, which is the primary flow in this range of Γ.

Similarly, the line CD is the locus of terminal states for the six-cell flow occurring as a secondary mode. Here the four-cell mode is the primary flow and is realised upon collapse of the secondary mode. The cusp-shaped region bounded to the right by BC is where the priority of the modes of the primary flow is exchanged. In this range of aspect ratio, a gradual increase of speed from a point to the left of CD will result in the initial four-cell form to jump into the six-cell one as the line BC is reached. If the speed is then gradually reduced, the six-cell mode will collapse back to the four-cell at a lower speed along CD, showing a definite hysteresis. In the ranges where the hysteresis is large the jumps are very distinct, gradually becoming less clear as the aspect ratio is increased.

The results presented above are in good agreement with the theoretical predictions of Benjamin but are nevertheless surprising. The hysteresis is at least an order of magnitude greater than any of the other measured cusps including the two to four cell exchange where it was felt that 'end effects' would be strongest. The results have been confirmed experimentally and, more recently, in a numerical study by Cliffe (1983).

3. ANOMALOUS MODES

Stable steady flows featuring an odd number of cells in symmetric apparatus, therefore having two possible realizations with an abnormally spiralling cell at one end or the other, have been called anomalous modes. This term also includes flows with an even number of cells but with abnormal spiralling in both end cells. Except for the one-cell mode which has other special properties (discussed in detail in Benjamin and Mullin (1981)) these flows apparently have no parametric connection with the primary modes, thus remaining true secondary modes over the whole range of Γ wherein they are realizable.

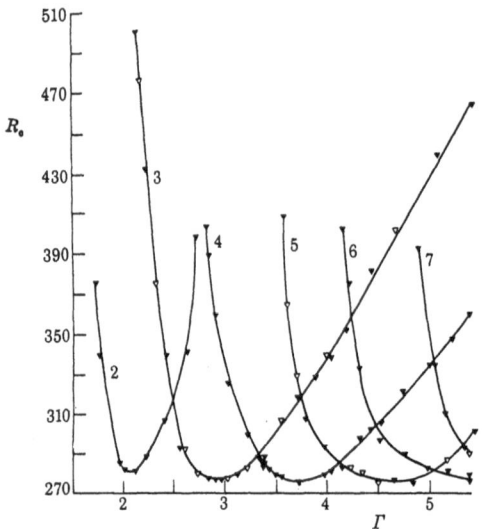

Figure 3

Loci of terminal conditions for anamolous modes with 2 to 7 cells.

The results shown in figure 3 are of a set of measurements of the
limits of stability for anomalous modes with 2 to 7 cells. The mode is
initially formed by a sudden acceleration to a speed well above the
critical range and allowed to equilibriate. The critical speed at
which the respective mode collapses catastrophically is plotted as a
function of Γ , having been estimated as the sequel to very gradual
decreases in speed from much higher values. (The critical values of R
given in this figure must be multiplied by 0.626 to be compared directly with the other figures as the definition of R are slightly different.)
The experimental curves are reminiscent of a result according to the
idealized theory, namely the critical R, for bifurcation from perfect
Couette flow, expressed as a function of axial wavelength. The resemb-
lance is deceptive, however, because the minimum of the curves are at
values of R_c over twice those indicated by the theory. Thus end effects appear to be enormous in separating the anomalous modes from any
counterparts of them that might be identified in the idealized theoret-

ical model.

4. NON-UNIQUENESS

A particularly striking example of multiplicity of the solution set for steady Taylor-Couette flow is given in Benjamin and Mullin (1982). There it was shown that at least twenty different steady flows could be realised at one point in the (R, Γ)-plane. At the chosen value of $\Gamma = 12.61$, the primary flow comprises twelve cells. This flow could be produced by slowly raising the rotation rate, Ω, of the inner cylinder from a small value to its prescribed final value. All the other, secondary, modes were produced by sudden starts of the rotation in respective ranges of Ω found by trial and the speed adjusted to its final value after the desired mode had been established.

Other than the twelve-cell primary flow, the nineteen secondary modes, consisting of between eight and eighteen-cells, all collpase eventually as R is reduced by gradual steps. The critical values R_c of R at which they were found to lose stability and collapse are listed in table 1.

Normally spiralling modes

No. of cells	8	10	14	16	18
R_c	233	81	75	110	197

Anomalous modes

No. of cells	9	10	11	12	13	14	15	16	17
R_c	270	253	244	226	190	201	205	206	212

Table 1. Lower limits of stability for secondary modes $\Gamma = 12.61$.

Table one includes five anomalous modes with an odd number of cells (i.e. from 9 to 17) each of which flows is therefore capable of two different realizations on symmetric boundary conditions. In one realization the abnormally spiralling end cell is at the top of the annulus and in the other it is at the bottom but there is complete parity between the two in the case of symmetric end-conditions. The stability

limits of these two realiations are identical to within experimental
accuracy and therefore the total number of secondary modes is twenty.

The central issues raised by these observations are discussed in
detail in Benjamin and Mullin (1982). Therefore we raise only two of
the questions posed by these observations in relation to steady cell
number selection.

(1) What is the connection between the wavelength of the periodic solutions obtained from an infinite cylinder model and the cell pair sizes
in this experiment? The fourteen cell normal and anomalous modes, for
example, have an identical cellular structure over most of the length
of the cylinders and yet the critical Reynolds numbers for collapse
differ by almost a factor of three. Indeed, the influence of the ends
on all of the anomalous modes is found to be very strong indicating
that the stability of the entire flow may be controlled by effects of
the finite domain size.

(2) What happens when we try to approximate towards the infinite cylinder model by using very long cylinders in an experiment? The results
discussed above indicate that the number of solutions will increase
rapidly as the length of the cylinders is raised and thus delicate effects such as the primary flow exchange process may no longer be amen-
able in an experiment.   External imperfections such as temperature
on speed fluctuations may become an overiding influence on the selection process from a densely packed solution set. Thus the flow may
wander continuously never settling down into any particular solution
near critical points in a manner somewhat analogous to a bubble raft in
a large domain. However, the underlying process is well defined and
it is precisely that which is more readily observable in the experiments at smaller aspect ratio of the kind reported in sections 2 and 3.
Therefore, any theoretical models proposed to explain events at larger
aspect ratios must incorporate these effects.

5. CONCLUSIONS

The importance of end effects in the cellular selection process
in Taylor-Couette flow has been demonstrated. A universal law governing the primary flow exchange process has been demonstrated to hold for

all of the flows studied. In addition, new types of flow viz. the anomalous modes, have been investigated and demonstrated to exhibit regular behaviour. Finally, end effects have been shown to be of the utmost importance in the cell number selection process.

Acknowledgements

This work was carried out in collaboration with Professor T. Brooke Benjamin who initiated much of this research. We acknowledge the support of the S.E.R.C. for our research.

References

Benjamin, T.B. 1978a Bifurcation phenomena in steady flows of a viscous liquid I. Theory. Proc.R.Soc.Lond. A359, 1-26.

Benjamin, T.B. 1978b Bifurcation phenomena in steady flows of a viscous liquid II. Experiments. Proc.R.Soc.Lond. A359, 27-43.

Benjamin, T.B. and Mullin T. 1981 Anomalous modes in the Taylor experiment. Proc.R.Soc.Lond. A377, 221-249.

Benjamin, T.B. and Mullin, T. 1982 Notes on the multiplicity of flows in the Taylor experiment. J. Fluid Mech. 121, 219-230.

Cliffe, K.A. 1983 (In preparation)

Hall, P. 1982 Centrifugal instabilities of circumferential flows in finite cylinders: the wide gap problem. Proc.R.Soc.Lond. A384, 359-379.

Mullin, T. 1982 Mutations of steady cellular flows in the Taylor experiment. J. Fluid Mech. 121, 207-218.

Mullin, T., Pfister G. and Lorenzen A. 1982. New observations on hysteresis effects in Taylor-Couette flow. Phys. Fluids 25, 1134-1136.

Schaeffer D.B. 1980 Analysis of a model in the Taylor problem. Math. Proc.Camb.Phil.Soc. 87, 307-337.

WAVELENGTH SELECTION THROUGH BOUNDARIES IN 1-D CELLULAR STRUCTURES

S. Zaleski

Laboratoire d'Hydrodynamique et de Mécanique Physique
E.S.P.C.I.
10 rue Vauquelin
75231 Paris Cedex 05, France

present address :
Groupe de Physique des Solides de l'Ecole Normale Supérieure
24 rue Lhomond
75231 Paris Cedex 05, France

1. INTRODUCTION

As explained in the general introduction to this volume, wavelength selection is a widely studied experimental and theoretical problem in cellular instabilities. A broad band of wavenumber is theoretically found unstable above the threshold of linear stability, whereas experimental results show various bands of different thicknesses, or even a single selected wavenumber.

The problem has been intensely investigated recently as many mechanisms were proposed, that restrict the allowed wavenumbers to a band of extent $O(\varepsilon)$ near threshold, where ε is the relative distance to threshold. A first such mechanism of wavelength selection, involving the effect of distant lateral boundaries, was proposed in [1a-b] and investigated experimentally in [12]. A situation where the environment is slowly varying was also considered [2a-c]. Such a situation occurs when a physical parameter related to the instability, such as ε, has a slow spatial variation. When the range of variation of the parameters connects a supercritical region to a subcritical one, a unique wavenumber is selected, up to exponentially small terms. Various other selection mechanisms were proposed, in dynamical situations such as dislocation motion [3] and, front propagation [4] or for axisymmetric roll patterns [5].

In the following, we restrict ourselves to one dimensional structures. They correspond to two dimensional Rayleigh-Bénard convection or to a perfectly axisymmetric Taylor-Couette flow. In the plate buckling problem reported in this volume, the structure is also essentially one dimensional. Moreover, we consider only close vicinity to the instability threshold.

We first present a simple derivation of the results of Ref.[1a-b].

We use the slowly varying amplitude expansion, which yields a 2nd order
differential equation for the envelope A(x) of the rapidly varying perturbation. These results are then extented and connected to the case of
imperfect boundary conditions (hereafter labelled B.C.). Imperfect, or
inhomogeneous B.C. arise for instance when some external effect induces
rolls of approximately fixed amplitude near lateral walls. There are
thus subcritical rolls for $\varepsilon < 0$, and the resulting bifurcation is
imperfect (fig.1). Such rolls and/or vortices appear in Taylor Couette
experiments with fixed caps. In Rayleigh-Bénard convection, various
forcing mechanisms were designed. Inhomogeneous B.C. can be modelled
by a condition on the amplitude A which reads at lowest order
$A = \lambda e^{i\phi}$, as first shown in $[6]$, for Rayleigh-Bénard convection with
free-free horizontal B.C.

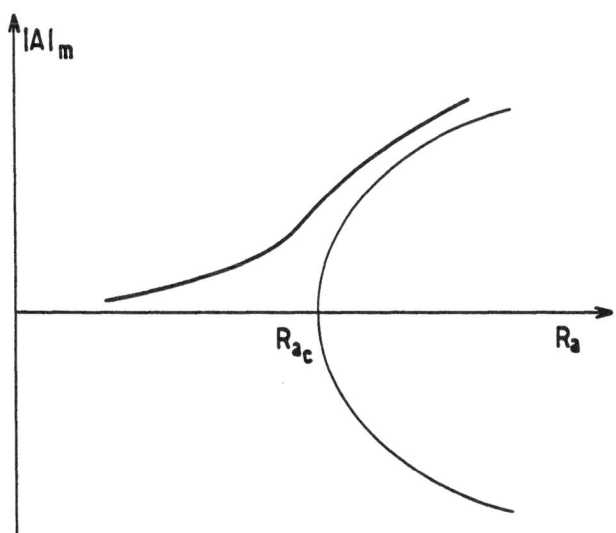

Figure 1 : The imperfect bifurcation. The variation of the modulus of
the slowly varying amplitude in the middle of the cell, $|A|_m$, is plot-
ted w.r. to the control parameter, R. In the perfect case (thin line),
(e.g. A = 0 on boundaries), a sharp bifurcation occurs. It is replaced
by a continuous increase of $|A|_m$ in the imperfect case (thick line).

Wavelength selection with such B.C. has been studied in $[2a]$ in the
limit $\varepsilon \ll \lambda \ll \varepsilon^{1/2}$. We here extend these results to the case $\lambda = (\varepsilon)$
(part 2.) $\lambda = (\varepsilon^{1/2})$ (part 3.). B.C. in the latter case can be called
strongly imperfect, since induced and bulk rolls are of comparable amplitude. When $\lambda \varepsilon^{-1/2}$ exceeds a critical value, B.C. are shown to allow
for all Eckhaus stable wavenumbers. All stationary solutions, stable
or unstable, are within this band. These results are derived in the
limit of half infinite solutions. We also study the case of solutions

of finite size L. B.C. have to be matched on both sides and a "quanti-
zation" of modes occur. Solutions for which the amplitude sinks in the
bulk (Fig.4) are found but are unstable except in the limit $L = \mathcal{O}(\varepsilon^{1/2})$.

In part 4. we recall the mechanism of wavenumber selection in a
slowly varying environment. Such a situation can be realized when a
physical parameter of the instability such as the height of cell in
Rayleigh-Bénard convection, has a slow spatial variation. We review the
previous theoretical results [2-a-b] and the experimental finding of
[7]. The possibility for a coexistence with previously described mechanisms is discussed.

2. WAVENUMBER SELECTION WITH SMALL BOUNDARY FORCING

Our starting point is the amplitude expansion. We simply recall the
main features of this expansion, and we refer the reader to the introduction to this book for further details about the problem of wavenumber selection and the amplitude expansion. The first order amplitude
equation was introduced for the Rayleigh-Bénard instability in [8]-[9].
To introduce it, consider the (rapidly varying) field of temperature
perturbation θ. It is related to the slowly varying amplitude by

$$\theta(x,z) = \frac{1}{2} \left(A(x) \ f(z) \ e^{iqx} + c.c. \right) + \mathcal{O}(\varepsilon) \tag{1}$$

where c.c. stands for complex conjugate, and where $\left(f(z) \ e^{iqx} + c.c. \right)$
is the linear periodic instability mode at threshold, and q_0 is the
critical wavenumber. $A(x)$ is a complex amplitude, and is supposed to
be small near threshold so that $A = \mathcal{O}(\varepsilon^{1/2})$ and $A_x = \mathcal{O}(\varepsilon)$. By substitu-
tion of (1) in the primitive equations for the rapidly varying quanti-
ties, one obtains an equation of the form :

$$\tau_0 \ A_t = A + \xi_0^2 \ A_{xx} - \frac{|A|^2}{A_0^2} \ A \tag{2}$$

where $A_t = \frac{\partial A}{\partial t}$, $A_{xx} = \frac{\partial^2 A}{\partial x^2}$, and where τ_0, ξ_0, A_0 are time, length and
amplitude scales. All terms in (2) are of order $\varepsilon^{1/2}$. Higher order terms
can be obtained as shown for instance in [1a], [2a] and [3]. Since we
are interested in time independant structures, we drop time derivatives
and get at second order :

$$\varepsilon A + \xi_0^2 \ A_{xx} - \frac{|A|^2}{A_0^2} \ A + ig_1 \varepsilon A_x + ig_2 \ |A|^2 \ A_x + ig_3 \ A_x^* \ A^2 = 0 \tag{3}$$

where g_1, g_2, g_3 are real coefficients and A^* denotes the c.c. of A.
This general form obeys a symmetry requirement : it must be invariant
under the simultaneous changes $A \rightarrow A^*$, $x \rightarrow -x$. Actually, these changes
yield the c.c. of (3), which is equivalent to (3). If terms of the form

$|A|^4$ were included there would be no such invariance. Further, there is no A_{xxx} term. This is consistent with B.C. and can be explained as follows. A trivial way to get an amplitude equation at second order is to apply ∂_x on (2). By a suitable combination with the actual second order amplitude equation terms of the form A_{x^3} can be eliminated.

Boundary conditions on eq.(2) and (3) require a special discussion. We propose B.C. of the form [11] :

$$A + \xi_1 A_x + \xi_2 A^*_x = \lambda e^{i\phi_0} \qquad \text{for } x = 0,L \qquad (4)$$

where ξ_1, ξ_2 are some complex, and λ, ϕ_0, some real numbers. In order to derive such B.C., we assume that the perturbation which is responsible for inhomogeneity is small enough, so that in the neighborhood of the boundaries, the solution is reasonably close to that of the linear problem. This is realized if all velocity and perturbated temperature fields remain of order $\varepsilon^{1/2}$ near boundaries. Then the next terms in expansion (1) at the boundary are of order $\varepsilon^{3/2}$ (4) can hence be considered accurate at order $\varepsilon^{3/2}$ (for homogeneous B.C., (4) may be considered as accurate at higher order, as the amplitude is vanishing at the boundaries). On the other hand, if the velocity is not small enough on boundaries, then non-linear terms dominate and the amplitude expansion must be dropped together with (4) near the walls.

We first consider the case where boundary forcing is small, i.e. $\lambda = \mathcal{O}(\varepsilon)$. To find the selected wavenumbers, we match the bulk solutions of (3) with boundaries. Bulk solutions of wavenumber $q = q_0 + \delta$ have the following amplitude :

$$A = A_0 \left(\varepsilon - \xi_0^2 \, \delta^2\right)^{1/2} \exp\left(i\delta_x + i\phi_1\right) \qquad (5)$$

where ϕ_1 is some undeterminaed phase.

Multiply now (3) by $-i\, A^*$ and add the c.c. One gets an expression of the form

$$\frac{dK_1}{dx} = 0$$

where

$$K_1 = \frac{i}{2} \xi_0^2 \left(A\, A^*_x - A^*\, A_x\right) + \frac{1}{4}\left(g_2 + g_3\right) |A|^4 + \frac{1}{2} g_1\, \varepsilon |A|^2 \qquad (6)$$

Multiplying (3) by A^*_x and adding the c.c. leads to

$$\frac{dE_1}{dx} - \frac{i}{2} g_3 A^*_{xx} A|A|^2 - c.c. = 0 \tag{7}$$

where

$$E_1 = \frac{1}{2} \varepsilon |A|^2 - \frac{1}{4} \frac{|A|^4}{A_0^2} + \frac{i}{2} \xi_0^2 |A_x|^2 + \frac{1}{4} g_3 (A A^*_x + A^* A_x) |A|^2 \tag{8}$$

From (2) $A^*_{xx} = \xi_0^{-2} \left[-\varepsilon A^* + \frac{|A|^2 A^*}{A_0^2} \right] + (\varepsilon^2)$

By replacing in (7), one gets

$$\frac{dE_1}{dx} = \mathcal{O}(\varepsilon^{7/2})$$

thus E_1 and K_1 can be considered as invariant at the relevant order. K_1 and E_1 are easily computed away from the boundaries, using (5) and assuming $\delta = \mathcal{O}(\varepsilon)$.

$$K_1 = A_0^2 \varepsilon \left[\xi_0^2 \delta + \frac{(g_2 + g_3) A_0^2 \varepsilon}{4} + \frac{g_1 \varepsilon}{2} \right] \tag{9}$$

$$E_1 = \frac{1}{4} A_0^2 \varepsilon^2 + \mathcal{O}(\varepsilon^3) \tag{10}$$

From B.C. (4), two real parameters remain free at $x = 0$. Define $A_x(0) = C e^{i\Theta}$. Then C and Θ are two such parameters. Since $|A(0)| \ll \varepsilon^{1/2}$, the estimation of E_1 at the boundaries is easy. From (8) one gets

$$E_1\Big|_b = \frac{1}{2} \xi_0^2 C^2$$

Equating with (10) gives

$$A_x(0) = \frac{A_0}{\sqrt{2}} \xi_0^{-1} \varepsilon e^{i\Theta}$$

Replacing in (4) gives

$$A(0) = e^{i\phi} - \frac{A_0}{\sqrt{2}} \xi_0^{-1} (\xi_1 e^{i\Theta} + \xi_2 e^{-i\Theta})$$

From (9)

$$K_1\Big|_b = A_0^2 \varepsilon^2 \left[\frac{\lambda \xi_0}{\sqrt{2}A_0 \varepsilon} \sin(\phi - \Theta) - \frac{1}{2}(\xi_{1_i} - \xi_{2_r} \sin 2\Theta + \xi_{2i} \cos 2\Theta) \right] + \mathcal{O}(\varepsilon^3) \tag{11}$$

where $\xi_j = \xi_{j_r} + i\xi_{j_i}$ for $j = 1,2$

Finally, from (9) and (11)

$$\delta = \frac{\lambda}{\sqrt{2}\,\xi_0\,A_0}\,\sin(\phi-\Theta) - \left(\xi_{1_i} - \xi_{2_r}\,\sin2\Theta + \xi_{2_i}\,\cos2\Theta\right)\frac{\varepsilon}{2\xi_0^2}$$

$$- \left[\frac{g_1}{2} + \frac{(g_2+g_3)}{4}\,A_0^2\right]\frac{\varepsilon}{\xi_0^2} \tag{12}$$

From (12), the precise bounds for δ can be obtained through a 4th order algebraic equation. The resulting diagram in the δ, ε space is plotted on fig.2.

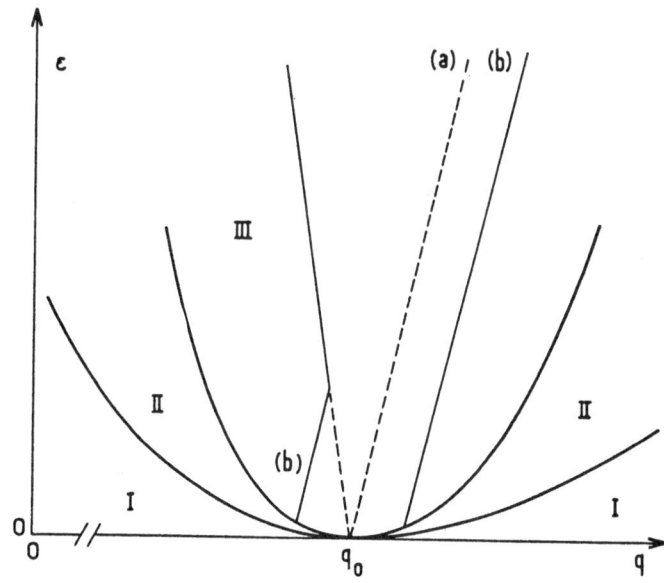

Figure 2 : Various stability and existence analyses are summarized on this diagram. The relative distance to threshold, ε, is represented together with the wavenumber q of the solutions. In region I, the laminar or conducting state is stable. In region II it is unstable, for wavenumbers in a band of extent $\varepsilon^{1/2}$ around the critical one q_0. The roll system itself can be unstable with respect to the Eckhaus instability : stable rolls are restricted to region III. When boundary conditions are accounted for, and when the amplitude is corresponding small on boundaries, the existing solutions (stable or unstable) are restricted to a band of extent ε, between lines (a). For inhomogeneous boundary conditions, i.e. $A(0) = \lambda e^{i\phi}$, this band is broadened (b). Near threshold, its extent is (λ). Line (b) connects with the limit of Eckhaus stable wavenumbers, which is shown to restrict further the band of solutions allowed by the boundaries. The width of the band is $\lambda/\sqrt{2}\xi_0A_0$. On the figure two orders of magnitude are respresented : $\lambda = \mathcal{O}(\varepsilon)$ and $\lambda = \mathcal{O}(\varepsilon^{1/2})$. Hence it only has a schematic character. The limits (b) for allowed wavenumber are computed in Appendix B.

The free parameter, Θ, can be interpreted in the following way : consider the development of A near $x = 0$

$$A(x) = A(0) + x\, C\, e^{i\Theta}$$

for $1 < x < \varepsilon^{-\frac{1}{2}}$, the second term dominates the expression and Θ can be interpreted as the phase of A just outside the boundary layer. For $\lambda = 0$, δ has a periodic variation with respect to Θ, with period π. Thus there are modes with opposite values of A in the bulk and identical wavenumber. For $\lambda \neq 0$, these modes are distinct.

3. CELLULAR PATTERNS WITH STRONG FORCING

In what follows, we consider the case of a strong forcing, i.e. B.C. (4) with $\lambda = \mathcal{O}(\varepsilon^{1/2})$. Thus the amplitude A at the boundaries is of the same order of magnitude as in the bulk region. We first investigate the wavenumber selection in this limit. Then we consider the full problem with B.C. applied at $x = \pm L$. The system is solved by an integration in the limit of large L

a. Wavenumber selection with imperfect boundary conditions

We first restate the stationary amplitude equation (2) in a dimensionless form

$$\overline{A} + \overline{A}_{xx} - |\overline{A}|^2\, \overline{A} = 0 \tag{13}$$

where $\overline{A} = \dfrac{A}{A_0}\, \varepsilon^{-\frac{1}{2}}, \qquad \overline{x} = \dfrac{x}{\xi}$

ξ is the coherence length $\xi = \xi_0\, \varepsilon^{-\frac{1}{2}}$.

The B.C. are now

$$\overline{A} = b\, e^{i\phi_0} \qquad \text{for } x = 0 \tag{14}$$

where $b = \dfrac{\lambda}{A_0}\, \varepsilon^{-\frac{1}{2}}$

The bars are omitted hereafter for simplicity. As in part 1., we connect bulk solutions to boundaries through invariant quantities. Bulk patterns (5) with wavenumber $q_0 + \delta$ correspond to

$$A = (1 - a^2)^{1/2} e^{i(ax + \phi_1)} \tag{15}$$

where ϕ_1 is some arbitrary phase, and $a = \delta \, \xi_0 \, \epsilon^{-1/2}$. Invariant quantities are derived as in 1. and are approximations to E_1 and K_1. Define

$$K = \frac{i}{2} (AA^*_x - A^*A_x)$$

and

$$E = \frac{1}{2} |A|^2 + \frac{1}{2} |A_x|^2 - \frac{|A|^4}{4}$$

then from (2) $\frac{dK}{dx} = \frac{dE}{dx} = 0$

To clarify the physical meaning of E and K consider the modulus r and phase ϕ of A defined by $A = r \, e^{i\phi}$. Then

$$K = r^2 \phi_x \tag{16}$$

and $\quad E = \frac{1}{2} r^2 + \frac{1}{2} r_x^2 + \frac{K^2}{2r^2} - \frac{r^4}{4} \tag{17}$

The variation of A w.r. to x is then identical to a "motion" in central force field with angular momentum K and energy E. An effective potential V_K can be defined :

$$V_K(r) = \frac{r^2}{2} + \frac{K^2}{2r^2} - \frac{r^4}{4}$$

and

$$E = \frac{1}{2} r_x^2 + V_K(r)$$

Thus we must study motion in the potential $V_K(r)$ (fig.(3)). There are two extrema of V_K, S and Q, on the figure (they are function of K) which correspond to periodic solutions. S corresponds to Eckhaus stable and Q to Eckhaus unstable solutions. Eckhaus instability occurs for $a > 1/(\sqrt{3})$ in (15). It is discussed in detail in the introduction to this volume. There are several ways to reach asymptotically point S from any other point, say R or T in Fig.3. They correspond to different types of boundary layers as shown on Figs (4) (5) and (6). Whereas there is no difficulty to reach S from the right on Fig.(3), it is impossible to converge to S for r too small. As b.c. impose $r = \lambda$ at $x = 0$, only a restricted set of values of K, and correspondingly of the wavenumber a, allows for the matching of bulk solutions with boundaries.

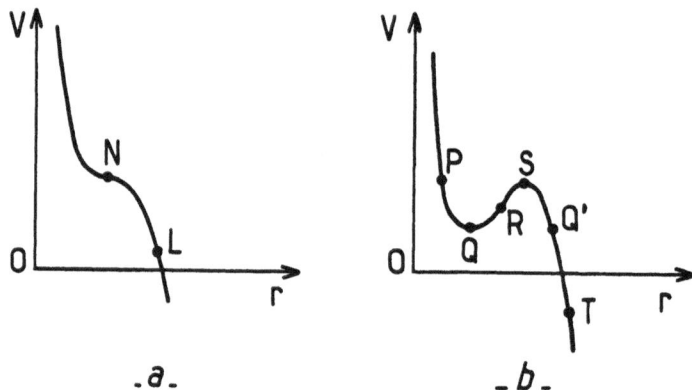

Figure 3 : The potential V defined in text is plotted as a function of the modulus r = |A|. The solutions r(x) for a given value of the invariant K can be represented as motions in this potential.
-a- marginal case : there is only one equilibrium point which corresponds to marginal stability of A(x).
-b- generic case : the equilibrium point S corresponds to stable solutions and point Q to unstable ones. Trajectories in this potential are described in text.

To compute the resulting selection, we first compute E and K. On boundaries, define r_x = c. c is a free parameter. Unless a second B.C. is taken, it is left unspecified.

At the boundary

$$2 E_{|b} = b^2 + c^2 + \frac{K^2}{b^2} - \frac{b^4}{2} \tag{18}$$

In the bulk, (15) gives $r = (1 - a^2)^{1/2}$, ϕ_x = a. Replacing in (16-17) gives

$$K = (1 - a^2)a \tag{19}$$

$$E = \frac{1}{4} (1 - a^2) (1 + 3a^2) \tag{20}$$

Equating (18) and (20), and replacing K by its value (19) gives the following cubic equation for a^2

$$(1 - a^2)^2 a^2 = \frac{b^2}{2} \left[(1 - b^2)^2 - 2c^2 + 2a^2 - 3a^4 \right] \tag{21}$$

Solutions of this equation yield the wavenumber a as a function of the

parameter c. When c is varied the range of possible values of a is
scanned. The number of solutions of (21) corresponds to the various
possible boundary layers discussed previously.

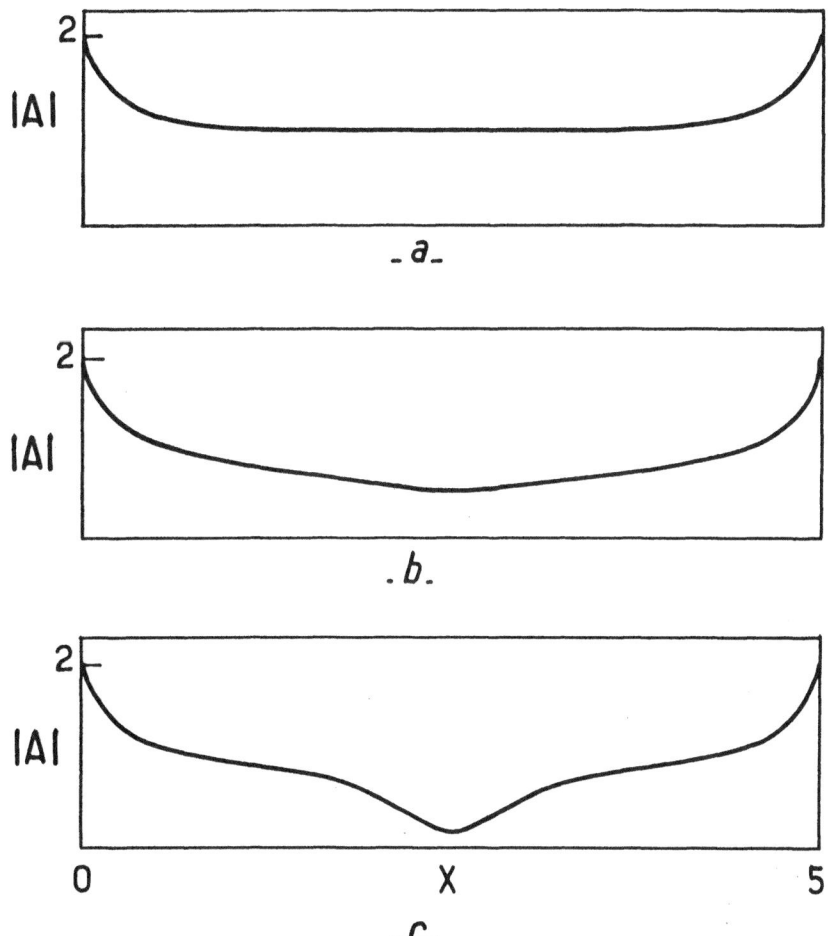

Figure 4 : The modulus r(x) of stationary solutions of eq.(2). The dimensionless amplitude on boundaries is b = 2.
-a- $\phi_+ - \phi_- = 0$.
-b- $\phi_+ - \phi_- = 2$; for large $\phi_+ - \phi_-$, the solution sinks in the bulk
until it reaches zero in the center.
-c- A "kinked" solution. Such a solution is unstable. All solutions
were drawn using numerical integration of (26)-(27).

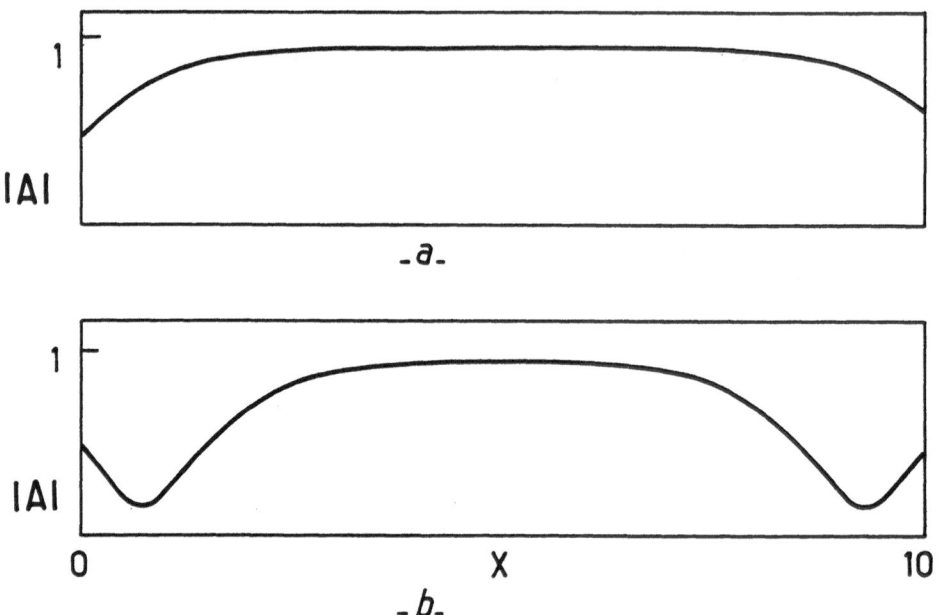

1

|A|

_a_

1

|A|

0   X   10

_b_

Figure 5 : Various symmetric solutions without "kinks" in the bulk are
plotted as obtained, for b < 1 by a numerical shooting method. In all
cases b = 0.5.

To study (21), first notice that its both sides are maximal for
$a = 1/(\sqrt{3})$. For some special values of b and c, double roots appear,
then giving pairs $a = 1/(\sqrt{3}) \pm \eta$, corresponding to pairs of points Q
and S on Fig.(3) (One Eckhaus stable and one Eckhaus unstable). However
solutions with $a > 1/(\sqrt{3})$ are usually spurious. As Eq.(21) merely
consists in equating energies, unwanted trajectories such as TQ', where
Q' is a point with the same potential energy as Q, must be eliminated.

To determine the number of solutions, we compute the discriminant
of (21). Previous remarks on double roots largely help to factorize
it. We find

$$\Delta = - \frac{c^2 b^4}{8} \left[\left(b^2 - \frac{2}{3}\right)^3 - 2c^2 b^2 \right] \tag{22}$$

Three behaviours can now be identified, depending on the value of b.

(i) $\underline{0 < b^2 < 2/3}$

In this range $\Delta < 0$. There is only one solution for each c, but
c can be positive or negative, leading to two kinds of boundary layers
as in Figs (5) and (6). The wavenumber varies monotonically with $|c|$.
For c = 0 it is maximal and (21) is exactly solvable :

$$a = \frac{b}{\sqrt{2}}$$

In physical units, one gets :

$$|\delta| < \frac{\lambda}{\sqrt{2}A_0 \xi_0} \qquad (23)$$

This result was already given in [1a] for the range $\varepsilon \ll \lambda \ll \varepsilon^{1/2}$.

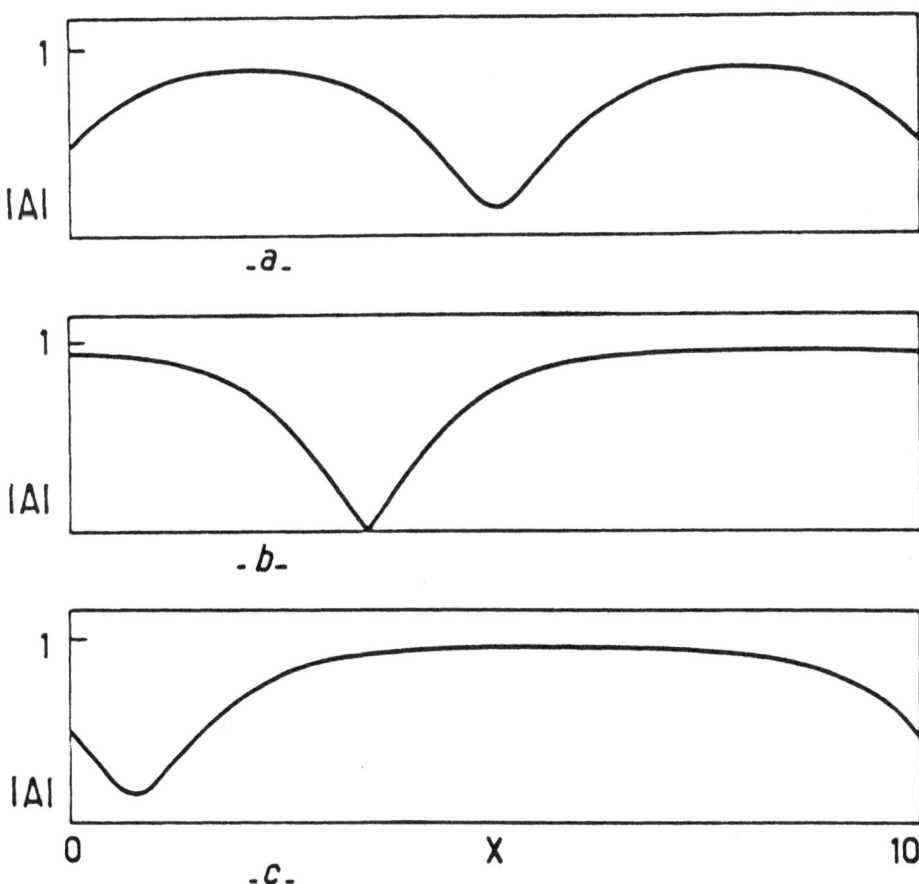

Figure 6 : Other types of solutions for b < 1 :
-a- b = 0.5 : a symmetric "kinked" solution can be found to evolve from 4c.
-b- b = 0.985 : for b ≈ 1, it can also evolve to an asymmetric solution.
-c- b = 0.5 : for lower b, the kink is located at boundaries. All figures are drawn from numerical solutions obtained by the shooting method.

(ii) $\frac{2}{3} < b^2 < 1$

Now $\Delta > 0$ and there are two possible values of a. They cover the full range of wavenumbers a with $|a| < 1/(\sqrt{3})$. Wavenumbers of Eckhaus unstable solutions (E.U.S.) are excluded. This means that E.U.S. are forbidden by imposition of B.C., not because of a stability criterion but because there is no corresponding solution fitting the B.C. However, for c = 0, a particular type of E.U.S. with a = $b/(\sqrt{2})$ can be found.

(iii) $\underline{b^2 > 1}$

There is again a single type of solution with boundary layers as in Fig.(4). The full range of Eckhaus-Stable wavenumbers is allowed.

To summarize these results, the wavenumber selection mechanism through boundaries results in the bound

$$|\delta| < \inf \left(\frac{\lambda}{\sqrt{2}\xi_0 A_0} , \frac{\varepsilon^{\frac{1}{2}}}{\sqrt{3}\xi_0} \right)$$

b. The full boundary value problem

Consider now Eq.(2) with the two B.C.

$$A(\pm \ell) = b\, e^{i\phi_\pm} \tag{24}$$

where $\ell = \frac{L}{\xi}$, and ξ is the coherence length $\xi = \xi_0 \varepsilon^{-\frac{1}{2}}$. We assume B.C. are deduced by translation from (14). The coefficients ϕ_0 can then be related to $\phi_\pm$ in (14).

As Eq.(2) is invariant with respect to a phase change $\phi \rightarrow \phi + \Delta\phi$ it is sufficient to determine the phase difference $\phi_+ - \phi_-$. It can be related to B.C. (14) as follows : consider B.C. at x = L for amplitude A in the region x < L. It can be mapped on B.C. at x = 0 by x $\rightarrow$ L - x. As the rapid variation is related to A by (1), the phase ϕ undergoes the transform $\phi \rightarrow q_0 L - \phi$ and the B.C. at x = L are

$$A + \xi_1 A_x + \xi_2 A^*_x = \lambda\, e^{i(q_0 L - \phi_0)}$$

Then the total phase difference accross the interval (-L, +L) is

$$\phi_+ - \phi_- = 2q_0 L - 2\phi_0 \tag{25}$$

Solutions of (2) can be expressed as simple elliptic integrals. The abscissa x(r) at which $|A(x)| = r$ can be extracted from (17).

Let x be a point in $(-\ell, \ell)$ where r(x) reaches its minimum value

r_m, x_M a point where $r(x)$ has a maximum r_M. In case when $r(x)$ oscillates several times in the potential valley of Fig.(3) let X be the wavelength of these oscillations (i.e. the distance between to such minimum) and N a number such that

$$\left(N - \frac{1}{2}\right) X < x - x_0 < \left(N + \frac{1}{2}\right) X$$

This implies $x_0 + NX$ is the abscissa of the minimum nearest to x. Then

$$X = 2 \int_{r_m}^{r_M} \frac{dr}{(E - V_K(r))^{1/2}}$$

and

$$X(r) - x_m = s \int_{r_m}^{r_M} \frac{dr}{(E - V_K(r))^{1/2}} + NX \qquad (26)$$

where $s = \text{sign}(x - x_0 - NX)$, E and K are the invariant quantities associated to the solution. Let us define the phase change in a period

$$\underline{\Phi} = 2 \int_{r_m}^{r_M} \frac{Kdr}{r^2 (E - V_K(r))^{1/2}}$$

then

$$\phi(x(r)) = s \int_{r_m}^{r} \frac{Kdr}{r^2 (E - V_K(r))^{1/2}} + N \underline{\Phi} \qquad (27)$$

To determine E and K, we apply the b.c. (24) to these solutions. Replacing $x(r)$ by $\pm \ell$ in (26) gives the expression

$$2\ell = s' \int_{r_m}^{b} \frac{dr}{(E - V_K)^{1/2}} + N'X$$

where N' is a positive or nul integer, s' is an integer, with
$|s'| < 2$. N' and s' can be defined by numbering oscillations of $r(x)$.
In the simple case where $r(x)$ increases monotonically in $(0, \ell)$, as in
Fig.(4), one has N' = 0 and s' = 2. In a similar way

$$\phi_+ - \phi_- = s' \int_{r_m}^{b} \frac{Kdr}{r^2 (E - V_K)^{1/2}} + N' \underline{\Phi}$$

The two latter equations relate E and K to ℓ and $\phi_+ - \phi_-$, which are given
by the B.C. (24). Expressions (26)-(27) can be calculated in the limit
where $r(x)$ gets close to its expected bulk value $r = (1 - a^2)^{1/2}$.

E is then close to the energy of a periodic solution of wavenumber a, i.e.

$$E = \frac{1}{4} (1 - a^2) (1 + 3a^2) + \eta$$

for some $\eta << 1$. $E - V_K(r)$ is then small in some region and integrals (26)-(27) are large. In case $N' = 0$ and $s' = 2$, a simple expansion of integrals (26)-(27) in powers of η can be made. In this latter case for $\eta > 0$ trajectories in the potential V_K enter the PQRS region on Fig.(3), and "kinked" solutions result (Fig.4). For $\eta < 0$, the usual pattern (Fig.4a) is obtained. Computation of (26) and (27) gives

$$\ell = O \, (Log \, \eta)$$

and

$$\phi_+ - \phi_- = a\ell + I \, (a,b,\eta) \tag{28}$$

and with :

$$I(a,b,\eta) = m(\eta) \; Arctan \; \frac{1-3a^2}{a\sqrt{2}} - Arctan \; \frac{b^2-2a^2}{a\sqrt{2}} + \quad (\eta) \tag{29}$$

$$m(\eta) = \frac{(sgn/\eta) + 3}{2} \quad , \quad I_\pm(a,b) = \lim_{\eta \to 0_\pm} I(a,b,\eta)$$

The first term in the r.h.s. of (28) is the bulk phase winding of the approximately periodic pattern of wavenumber a. $I_\pm(a,b)$ is the phase difference in the boundary layers of extent ξ near $x = \pm\ell$, and, for $\eta > 0$, near $x = x_m$ where the amplitude sinks to a value $r_m < 1 - a^2$ (see Fig.(6b) and (6c)). Expressions (28)-(29) correspond to the first order of a expansion in powers of $\eta = (e^{-\ell})$. A graphic resolution of Eq.(28) is illustrated on Fig.(7). There is an odd number of solutions with $\eta < 0$ and an even number of solutions with $\eta > 0$. The overall number of solutions is $n \simeq 2\ell/(\pi\sqrt{3})$. Solutions with $\eta > 0$ and $\eta < 0$ appear pairwise   As a single stable solution is expected in the limit $\varepsilon \to 0$, topological degree theory indicates that the first mode should be unkinked, and that "kinked" solutions are unstable (Fig.7).

When $\varepsilon \to 0$, L fixed, one has $\ell \to 0$ and the above approximations are not valid. We performed numerical computations of $I(a,b,\eta)$ in this limit, i.e. for $\varepsilon = (\xi_0^2/(L^2))$. The bifurcation diagram near $\varepsilon = 0$ is shown on Fig.(8). For $\phi_+ - \phi_- \not\to \pi$ the first mode is not kinked, as expected. For $\phi_+ - \phi_- = \pi$, $K = 0$ and r goes to zero in the bulk. This can be understood as follows : $\phi_+ - \phi_-$ increases monotonically with L, and new oscillations of the rapidly varying perturbation θ have to be nucleated. They cannot appear near $x = \pm \ell$ since the phase is fixed in this region. A region of smaller amplitude has then to appear in the bulk. This has been observed in experiments [10a-b]. The bifurcation diagrams can also be compared to those established experimentally in [10a].

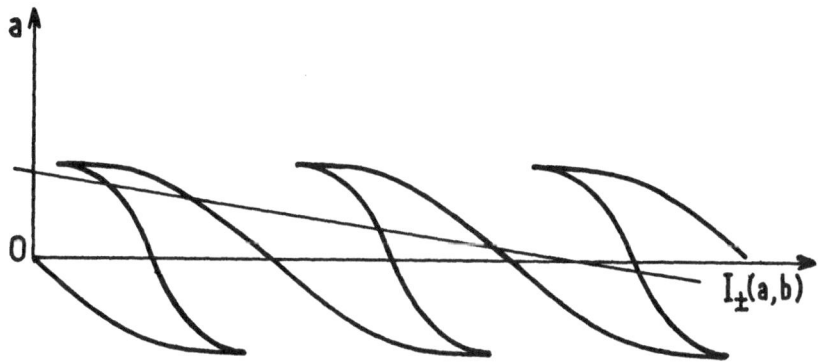

Figure 7 : Resolution of system 13 a-b
For large ℓ an analytic expansion is made. The wavenumber a is related to $I_\pm$ (a,b). I_+ corresponds to short branches in the wavy pattern and I_- to long ones. For b and ℓ fixed, the solutions lie at the crossing of the straight line $I_\pm = \phi_+ - \phi_- - a\ell$ and of the branches $(a,I_\pm)$. "Kinked" solutions, corresponding to $I = I_-$ and unkinked $(I = I_+)$ ones can thus be numbered. Numerical computations of $I(a,b,\eta)$ agree with analytical results very accurately : for $1 > 5$, points calculated numerically and analytically coincide on the figure.

4. WAVELENGTH SELECTION IN A SLOWLY VARYING ENVIRONMENT

We now turn to a rather different situation, where the physical parameters involved in the instability have a slow spatial variation. These physical parameters are for instance the local height of the cell in a Rayleigh Bénard experiment, or thickness in a plate buckling problem, or else the local porosity for convection in porous media. Such a problem has been studied theoretically in Ref.[2a] and [2b], and experimentally in [7]. We review below these results, in view of discussing situations where the two situations - selection through boundaries and through a slowly varying environment - occur simultaneously.

We illustrate the phenomenon of slowly varying environment on amplitude equation (3). However, the resulting selection can be obtained by a non perturbative argument [2a-b], and hence valid outside the range $\varepsilon \ll 1$.

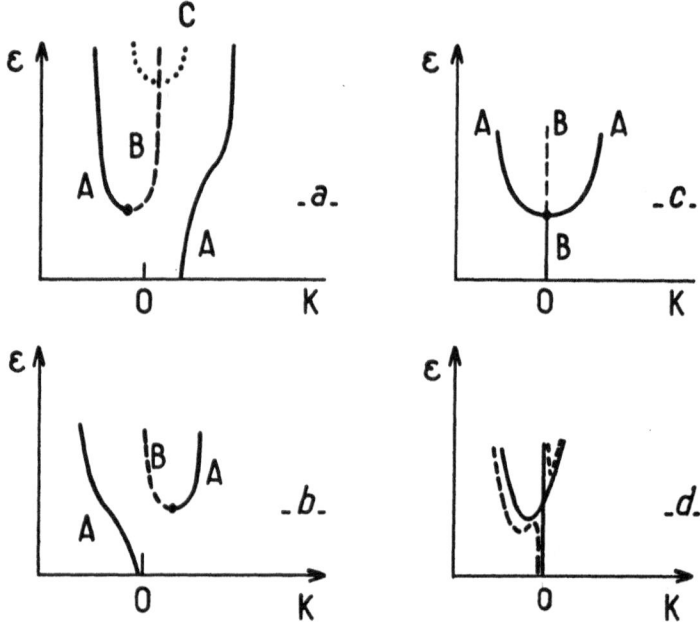

Figure 8 : The bifurcation diagrams near $\varepsilon = \xi_0^2 L^{-2}$ (i.e., $\ell = 1$).
The invariant for K is chosen for parametrization as it measures the
rapidity of phase change (see text). (For lower ℓ, however, the whole
interval is invaded by boundary layers. It hence does not make sense to
consider the bulk "wavenumber" of the solutions).
-a- The first mode has N oscillations (branch A). For larger ε, a solution with N + 2 modes bifurcates (branch B). For even larger ε, i.e.
for b < 1, new solutions appear (Fig.5 and 6) (branch C).
-b- When L is larger, the first mode has N + 2 oscillations.
-c- For a definite value of L, the first mode is "kinked" (branch B).
It bifurcates into solutions of N and N + 2 modes. (branch A).

Our starting point is Eq.(3) where ε is supposed to be x dependent.
Eq.(3) for ε constant has periodic solutions $\hat{A}$ which depend formally on
the three variables ε, ϕ and ϕ_x, where ϕ is the phase and ϕ_x the local
wavenumber

$$\hat{A}(\varepsilon,\phi,\phi_x) = A_0(\varepsilon - \xi_0^2\phi_x^2)^{1/2}\, e^{i\phi} + \mathcal{O}(\varepsilon^{3/2})$$

and where $\phi_x = \delta = \mathcal{O}(\varepsilon^{1/2})$

In the space dependent case, the adiabatic solution is

$$A(x) = \hat{A}\big(\varepsilon(x),\phi_x(x),\phi(x)\big) + A_1 + A_2 + \ldots \tag{30}$$

where $\hat{A}$ depends on $\varepsilon(x)$ and $\phi_x(x)$ as specified above, and $A_1 = \mathcal{O}(\varepsilon_x)$,
$A_2 = \mathcal{O}(\varepsilon_x^2)\ldots$ This adiabatic expansion is valid only if the slow variation of A in (30) is much slower than the scale of variation of A im-

plied by the amplitude expansion i.e. if

$$\frac{\varepsilon_x}{\varepsilon} \ll \xi_0^{-1} \; \varepsilon^{1/2}$$

To put it differently, the coherence length ξ must be much smaller than the length scale for a variation of ε.

Let us put Eq.(3) in the general form

$$\Lambda(\varepsilon, A) = 0 \qquad\qquad (31)$$

then inserting (30) in (31) gives at first order [13]

$$i\left(\varepsilon_x \hat{A}_\varepsilon + \delta_x \hat{A}_\delta\right) \left[2\delta\xi_0^2 + g_1\varepsilon - (g_2 + g_3)|\hat{A}|^2\right] - \frac{1}{2} ig_1 \varepsilon_x \hat{A}$$

$$+ i\xi_0^2 \delta_x \hat{A} + \left.\frac{\delta\Lambda}{\delta A}\right|_{A=\hat{A}} A_1 = 0 \qquad\qquad (32)$$

A solvability condition must be applied to (32), multiplying by an element of the Kernel of the linearized operator $\delta\Lambda/\delta A$. From translational invariance, if A is a solution of (31), A $e^{i\phi_1}$, ϕ_1 constant is also a solution. For $\phi \ll 1$, A $\rightarrow$ A $- i\phi A$ so that the kernel of the linearized operator is iA, multiplying (32) by iA is equivalent to collecting the two first terms. Collecting terms in the domain $\delta \sim \varepsilon$, one has

$$\varepsilon_x\left[\delta \; \xi_0^2 - \frac{g_1}{2}\varepsilon - \frac{(g_2 + g_3)}{2} A_0^2 \; \varepsilon\right] + \delta_x \; \varepsilon \; \xi_0^2 = 0$$

Or else $\frac{dK_1}{dx} = 0$

where K_1 is defined as in part 2., but ε and δ are now space dependent. The curves K_1 = constant draw a system of hyperbolae in the plane (ε, δ) with the common asymptotes $\varepsilon = 0$ and

$$\delta = -\frac{2g_1 + (g_2 + g_3) A_0^2}{4\xi_0^2} \varepsilon \qquad\qquad (33)$$

When the domain of instability includes a subcritical region, the local values of δ and ε lie on the degenerate hyperbolae made of the two asymptotes. The expected selected wavenumber is then given by (33) and is identical to the center of the band of wavenumber selected through boundaries for homogeneous B.C.

However, as mentionned above, the adiabatic expansion breaks for $\varepsilon \rightarrow 0$ when $\varepsilon \simeq \varepsilon_c$ with

$$\xi_0^{-1} \, \varepsilon_c^{3/2} = \varepsilon_x$$

Thus all values of K_1 in the band of Eckhaus stable wavenumbers for $\varepsilon = \varepsilon_c$ are allowed. The band of selected wavenumbers is hence bound by two hyperbolae (Fig.9).

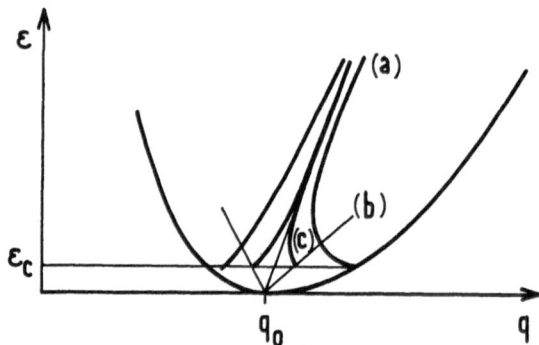

Figure 9 : The cooperation of wavelength selection through boundaries and by a slowly varying environment is shown. In a slowly varying environment all wavenumbers between hyperbolae (a) are allowed for $\varepsilon > \varepsilon_c$. Boundaries select wavenumbers between lines (b). When the two mechanisms are present, the allowed wavenumbers at $\varepsilon = \varepsilon_c$ are restricted to a narrower band by the effect of boundaries. For $\varepsilon > \varepsilon_c$, allowed wavenumbers are between hyperbolae (c).

This fits the experiments of [7]. This broadening must not be confused with the broadening due to exponentially small terms in the amplitude expansion mentionned in [4]. These terms are of order $\exp\left[- \varepsilon^{3/2}/(\xi_0 \, \varepsilon_x)\right]$. Up to such small terms, the adiabatic invariant K_1 is not a constant, and unicity of δ is not ensured for $\varepsilon \to \infty$.

The difference occuring in numerical results of [2b] and experimental results of [7] can possibly be explained : when a selection through boundaries is superimposed to the previously described mechanism, a much thinner band near threshold is recovered. As boundary forcing prevents wavenumber selection in the Taylor Couette case, the selected band in this latter case is large near threshold.

5. CONCLUSION

We have restated different results about wavelength selection in one dimensional cellular structures. The wavelength selection mechanism through boundaries, when boundary forcing is small or zero, was reviewed and the resulting bounds for the allowed wavenumbers were calculated again using the framework of amplitude theory at second order. Amplitude expansion at first order is sufficient to compute the selection effect when boundary forcing is strong. Wavenumber selection through boundaries is shown to disappear when end rolls have a larger amplitude than center ones. As inhomogeneous boundary conditions fix the phase of the amplitude at the boundaries, a periodic solution in the bulk must obey a phase matching condition. In some cases, this results in the appearance of "kinks" is the slowly varying amplitude.

The case of a slowly varying environment was also discussed using second order amplitude expansion. When a subcritical region is related to a supercritical one by a slow variation of the parameters, selection of a unique wavenumber results from an adiabatic theory. If some non adiabatic effects are considered a band of wavenumbers is expected. The mean wavenumber selected through this process and the central one in the band of wavenumbers allowed by boundaries coincide, thus the two mechanisms can cooperate.

The above calculation exemplify simple methods based on the invariants of amplitude equations. The derivation of such an equation is not always possible in the two dimensional case. When possible, however, the use of adiabatic invariants again can lead to selection mechanisms as happens for instance in axisymmetric structures.

The author acknowledges for many fruitful discussions with Y. Pomeau on the subject of part 3. Parts 2. and 4. review previous work with Y. Pomeau (Ref. [1b] and [2a]).

REFERENCES

1a M.C. Cross, P.G. Daniels, P.C. Hohenberg and E.D. Siggia, J. Fluid Mech. 127, 155 (1983).
1b Y. Pomeau and S. Zaleski, J. Phys., Paris, 42, 515 (1981).
2a Y. Pomeau and S. Zaleski, J. Phys. Lett. Paris, 44, L-135 (1983).
2b C. Kramer, E. Ben Jacob, H. Brand and M.C. Cross, Phys. Rev. Lett. 49, 1981 (1982).
2c L. Kramer and P.C. Hohenberg, this volume.
3   Y. Pomeau, S. Zaleski and P. Manneville, Phys. Rev. A27, 2710 (1983).
4   Y. Pomeau, this volume.
5   Y. Pomeau and P. Manneville, Phil. Mag. A, to appear.
6   P.G. Daniels, Proc. R. Soc. Lond. A358, 173 (1977).
7   D.S. Cannell, M.A. Dominguez-Lerma and G. Ahlers, Phys. Rev. Lett. 50, 1365 (1983).
8   A.C. Newell and M.A. Whitehead, J. Fluid Mech. 38, 279 (1969).
9   L.A. Segel, J. Fluid Mech. 38, 203 (1969).
10a T. Mullin, J. Fluid Mech. 121, 207 (1982) and this volume.
10b M. Boucif and J.E. Wesfreid, private communication.
11 S. Zaleski, to appear.
12 B. Martinet et al., this volume, and M. Boucif and J.E. Wesfreid, this volume.
13 Careful examination of Eq.(3) shows that the additional term $\frac{1}{2} ig_1 \varepsilon_x A$ is necessary to preserve the potential character of the linearized equation, in case of a spatially varying parameter ε.

WAVENUMBER SELECTION IN RAYLEIGH-BENARD CONVECTIVE STRUCTURE

V. CROQUETTE , A. POCHEAU

CEN Saclay - Orme des Merisiers
91 191 GIF-SUR-YVETTE cedex France

We present a study of wavenumber selection mechanisms occurring in convective
structures: In the first part, we analyse structural defect motions which allow a
wavenumber selection to be made. We found experimentally that, when the Prandtl number is large (Pr = 70), dislocation and grain-boundary motions lead to the same preferred wavenumber, which corresponds to the zig-zag instability threshold $D_\perp$ = 0. In
the second part, we consider the case of axisymmetrical structure, and we show that
a symmetry-breaking occurs in such structures. We analyse the possible origins of
this symmetry breaking and its consequences on the wavenumber selection produced by
the roll curvature, when Pr = 70 and 14.

INTRODUCTION : In the approach to turbulence in Rayleigh-Bénard convection, experiments made in liquid Helium[1] have indicated that in some particular cases, turbulence may appear as soon as convection sets in. Such an observation has motivated
various studies in order to understand how such a surprising situation is possible.
These studies, both theoretical and experimental, led to the conclusion that the convection patterns may be disordered. This disorder can be described by structural defects like dislocations or grain-boundaries, and the turbulence in [1] is likely to
be associated with defect motions in these structures. In this spirit, theoretical
calculations have been developed about dislocation motion topic[2,3,4], and we have
performed experiments which try to reproduce as well as possible the motion of a single dislocation in an infinite bidimensional roll structure. These studies have empha
sized the fundamental role played by the so-called "wavenumber selection": the convective patterns during their evolutions select a defined wavelength; we will see
further how dislocations enable such a selection mechanism. To complete the dislocation study we have performed a second kind of experiments concerning the wavenumber
selection produced by another mechanism: the curvature in axisymmetric structures.
In this last series of experiments a symmetry-breaking occurs; we discuss its possible origin and its consequences on the selection mechanism.

1. NATURAL STRUCTURES

In this article we shall describe convection experiments performed using silicon oils with moderate and high Prandtl number Pr = ν/κ (the kinetic viscosity divided by the thermal diffusivity).

This convective fluid is confined in a container with large horizontal dimensions compared with its depth. The container may be of rectangular or cylindrical shape, and is sandwiched between a copper plate at the bottom and a sapphire monocrystal plate at the top. The temperature of both plates is controlled by water circulations, so that it is possible to establish a well defined temperature gradient on the fluid layer. The sapphire plate offers the interesting property of being transparent and of having a very good thermal conductivity (220 times greater than the oil); furthermore the copper plate is polished to mirror finish. These features enable convenient optical measurements to be made. The pictures that we present in this article have been obtained by the focalization technique[5] more accurate measurements, obtained by the deflexion of a laser beam (detected by a photodiode which gives locally the direction and magnitude of horizontal thermal gradients) are presented on the recordings of figures 13-15-16.

If we increase very slowly the temperature of the bottom plate relative to that of the top plate, convection motions appear at a well defined temperature difference ΔTc ; these convective motions are organized as rolls with relatively well defined width, but with disordered orientations. The convection patterns then observed are typically like the one of Picture 1 obtained in a cylindrical container.

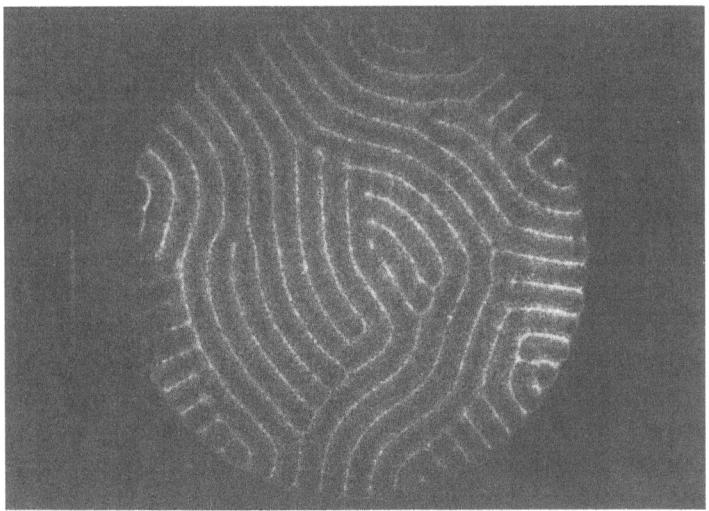

Figure 1: "Natural structure" observed in a cylindrical container, this structure corresponds typically to what we observe when we increase slowly the Rayleigh number Ra above Rac without any induction process (ε ≃ 1, $\Gamma_r = R/d = 20$).

We observed disordered structures of the same nature in rectangular containers as well. These structures may be considered as bidimensional crystal-like roll patterns with a great amount of defects such as: dislocations, disinclinations and grain-boundaries. These disordered patterns present slow time evolutions: just at the beginning of the experiment, the convective pattern "tries" so to speak, to adapt itself as well as possible to the container geometry; during this phase dislocations glide or climb, grain boundaries move in order to favor a definite set of rolls, etc. In all our experiments (P = 14 or 70, cylindrical container with Γ_r = R/d = 20 or rectangular one with Lx = 38d, Ly = 24d) this phase appears as a transient towards a stationary structure (as far as we know); nevertheless, this transient phase may have lasted as long as one or two weeks, which is 10 or 20 times longer than the largest thermal diffusion time of our experimental cell. At the end of this phase, the stationary structure presents a few remarkable features: the width of the rolls is fairly constant in all the structure, the rolls near the container boundaries come perpendicular to them and grain-boundaries seem to prefer meeting at right angles. The influence of the Rayleigh number Ra (which measures the normalized temperature difference Ra = $\dfrac{\alpha\, g\, d^3\, \Delta\, T}{\kappa\, \cdot\, \nu}$) on these structures does not seem very important. We may notice however that, when Ra comes closer to Rac (the convection threshold), the number of defects present in these structures decreases[6]. On the other hand the width of the rolls depends on the Rayleigh number Ra, as we shall see in detail later; consequently if we increase the Rayleigh number Ra when a pattern has reached its stationary state, a new transient phase will occur until the pattern finds a state compatible with the new rolls width and with the container geometry.

These results concerning natural structures are in agreement with the experiments performed by Gollub et al [7] using water at 70°C (Pr ≃ 2.5).

2. CLIMB OF DISLOCATIONS

A dislocation may be built by inserting a pair of rolls of limited length in a perfect bidimensional roll arrangement as drawn on Figure 2. This defect is very simple and, as in solid state physics, is very common in Rayleigh-Bénard patterns. (see for instance, Picture 1). In order to understand the dynamics of the disordered structures, we have investigated first the dynamics of an isolated dislocation. Obviously this requires the use of a somehow ideal structure compared with natural ones. Such structures may be induced using the thermal printing technique of Chen and Whitehead[8] : this technique uses the very high thermal sensitivity of the fluid layer when Ra is just smaller than Rac ; if we illuminate the oil layer by a powerful light beam passing through a grid, we produce small thermal gradients which trigger the convection when we increase Ra above Rac. The pattern obtained by this method is just an image of the grid so that we may obtain an isolated dislocation in an otherwise perfect roll structure,by drawing the corresponding grid. The container used for this experiment was rectangular (Lx = 38×d, Ly=24×d, d=2,56 mm).

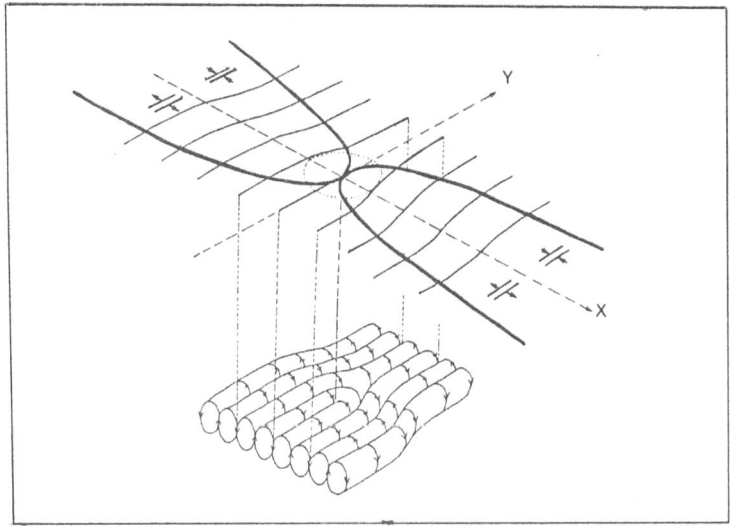

Figure 2: Sketch of a dislocation in a roll structure

This imposes some considerations as to the possible patterns to be made. The number of rolls in the cell must be an integer, as a dislocation consists in inserting a pair of extra rolls on a finite length of the structure; the dislocation grid is made up of two parts: one producing "n" parallel rolls and the other one "n + 2".The dislocation appears at the boundary between these two sets of rolls. The rolls are induced parallel to Ly and the dislocation position is determined along the y-axis by the length of the extra roll pair and along the x-axis by the point where the two sets of rolls are just of opposite phase. In these experiments the two sets of rolls were chosen to be just in phase at x = 0 and x = Lx so that the dislocation was right at Lx/2. An induction grid is presented in Fig. 3.

Such a structure was relatively easy to induce but an instability occurred very soon for the rolls located at x = 0 and x = Lx. These rolls are parallel to the cell sidewall and they break into small rows of rolls perpendicular to the sidewalls producing a structure different from that required. To prevent such an instability we induced a small thermal gradient at the two sidewalls at x = 0 and x = Lx (details may be found in Ref. 9). The effect of this thermal gradient is to stabilize the rolls parallel to the sidewall by triggering an uprising flow at both sidewalls With the combination of the thermal gradients and the induction technique we have been able to stabilize a single dislocation in an otherwise regular roll pattern as that presented on Fig. 4. Now, since the cell has two uprising flows at both ends (x = 0 and Lx) the number of rolls is forced to be an even integer n = 2 × N, where N stands for the number of wavelengths in the cell.

Figure 3: Induction grid for an isolated disloca-tion between two structures with the wavevectors 18 × (k$_c$/19) and 19 × (k$_c$/19).

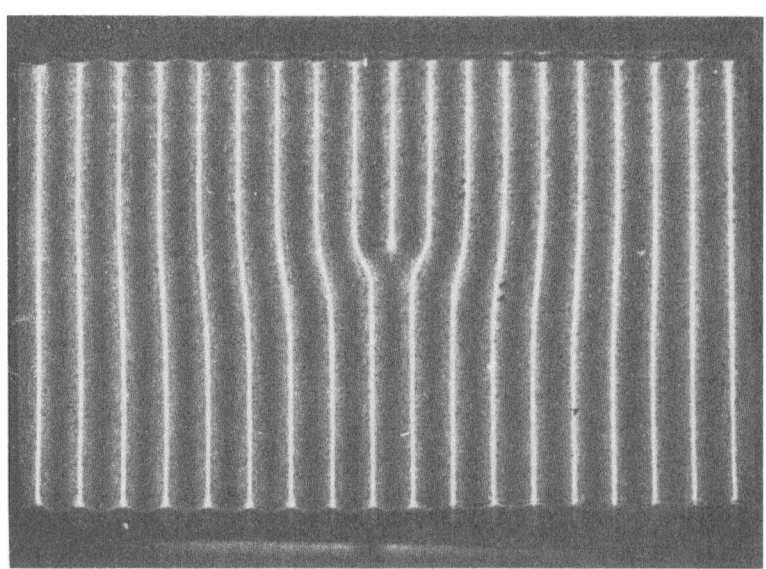

Figure 4: Focalization image of an isolated dislocation. P = 70, ε = .5, K$_1$ = 18 × (k$_c$/19) and K$_2$ = 19 × (k$_c$/19).

A dislocation pattern may be characterized by two parameters: the Rayleigh number Ra or more conveniently the reduced Rayleigh number: $\varepsilon=(Ra-Rac)/Rac$ and a couple of consecutive integers N and N+1 signifying that the dislocation associates a structure with N wavelengths to one with N+1, the average wavevector of the structure being $K=(2N+1).\pi/Lx$. The pattern of Fig. 4 is defined by $\varepsilon=.5$ and the couple 18/19.

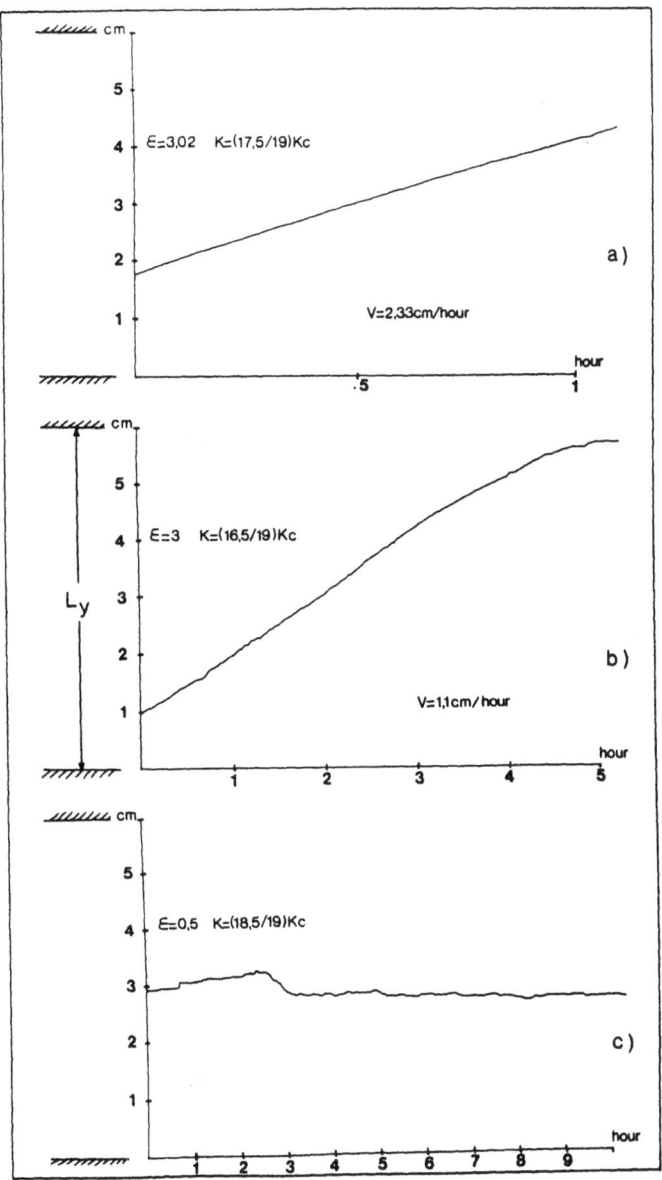

Figure 5:Recording of the dislocation position versus time.
A)Uniform behaviour of the velocity, commonly observed; p=70, K=17.5 (k_c/19), ε=3.02.
B)Influence of a sidewall: the dislocation is trapped at a short distance from the sidewall; p=70, k=16.5 (k_c/19), ε=3.00.
C)Very low velocity subjected to noisy variations; p=70, k=18.5 (k_c/19), ε=0.5.

Now we may wonder what will happen to a dislocation $(\epsilon, K=(2N+1)\pi/Lx)$ after the induction process: usually the dislocation climbs; we have followed the dislocation position on the y-axis as a function of time using an automatic tracking technics, typical recordings are presented on Fig. 5. On Fig. 5a we see the most common case: the dislocation velocity is quite uniform, that is the structure with, for instance, 2N rolls, progressively extends over the whole cell. When the dislocation comes near the sidewall two situations are possible: if the dislocation moves with a large velocity, the dislocation will slightly slow down near the sidewall but it will disappear leaving a perfect structure with 2N rolls. But if the dislocation has a small velocity it stops near the sidewall and stays there, this corresponds to the velocity recording of Fig. 5b. The structure then looks like in Fig. 6.

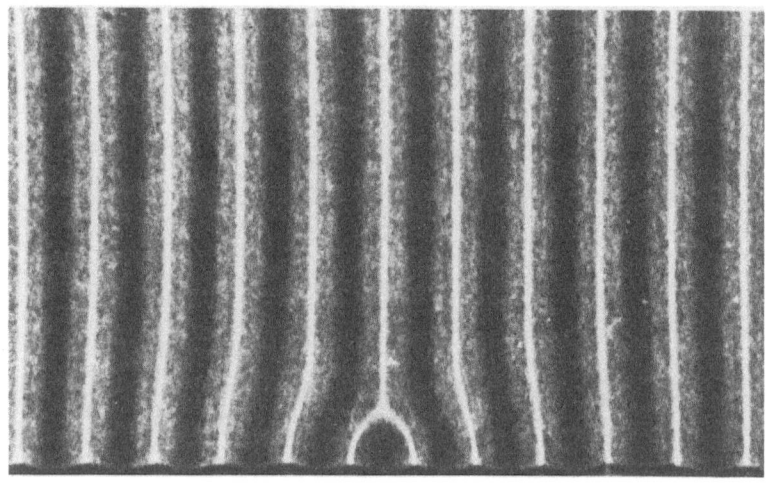

Figure 6: Trapped dislocation at the sidewall $\epsilon = 1.8$, $K_1 = 17 \times (k_c/19)$, $K_2 = 18 \times (k_c/19)$.

It is also possible that the dislocation does not move in a noticeable manner, which corresponds to the two structures with 2N and 2N+2 rolls being equivalent see the recording of Fig. 5c. Finally the entire structure may become unstable relative to zig-zag instability which destroys the entire structure. If we gather these different cases in the (ϵ, K) plane, we find a line where the dislocation velocity vanishes, see Fig. 7.

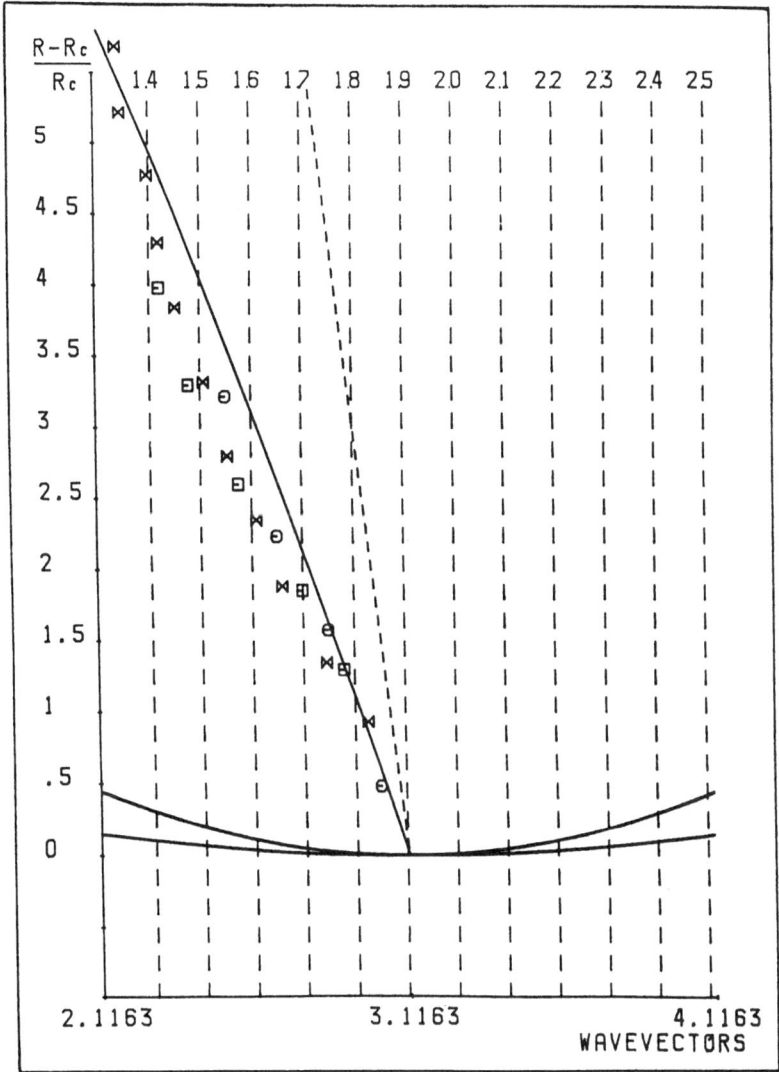

Figure 7:Experimental wavenumber selection compared to the $D_\perp = 0$ crite-rion. The circles correspond to the zero dislocation velocity criterion. The wavenumber reported is an half-integer which is the arithmetic value of the two wavenumbers of the structure. These points were determined by va-rying the Rayleigh number until the dislocation stops, or by extrapola-tion of results obtained at a fixed wavenumber. The squares are the velo-city zeroes deduced at a constant Rayleigh number: from the fit of the various velocities obtained at the same Rayleigh number using the 1.5 power law for the wavevector variation we determine the wavenumber at which this velocity should cancel. The crosses correspond to the mean wave number of the central set of rolls of Fig. 8, and thus give the wa-ve number selection by grain-boundaries. The full-line is the computed $D_\perp = 0$ curve at $P = 70$ (marginal stability against zig-zag) and the dotted line is that for $D'_\perp = 0$ at the same Prandtl number.

On the right of this line the dislocation has a positive velocity that is the structure with N wavelengths expands while the one with N+1 shrinks; the velocity rapidly increases as the distance to the line increases. On the left of this line the
patterns are unstable with respect to the zig-zag instability. Nevertheless it is
possible to observe a negative velocity of the dislocation as far as the distance
from the zero velocity line is small: the growth rate of the zig-zag instability
increases with the distance to the zero velocity line so that when we are just slight
ly at the left of this line the zig-zag takes a long time to develop, and during
this period it is possible to observe a negative dislocation velocity. As the line
of zero velocity for the dislocation is not vertical, it is possible for a given
pattern to change the sign of the dislocation velocity just by changing ε ; for
example for the dislocation with K = 17.5 (K_c/19) it is possible to have it move
back and forward by changing ε from 1.3 to 1.8. This is the case of Fig. 8, on
Fig. 8a the velocity is positive. Notice that the rolls are straigth. On Fig. 8b
the velocity is negative, notice that the rolls tend to zig-zag especially in the
part of the structure where the wavelength is the larger. This clearly demonstrates
that the zero velocity line is quite close if not identical to the zig-zag instability threshod in this case.

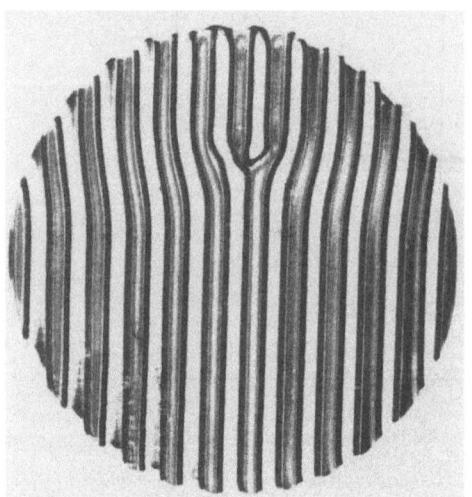 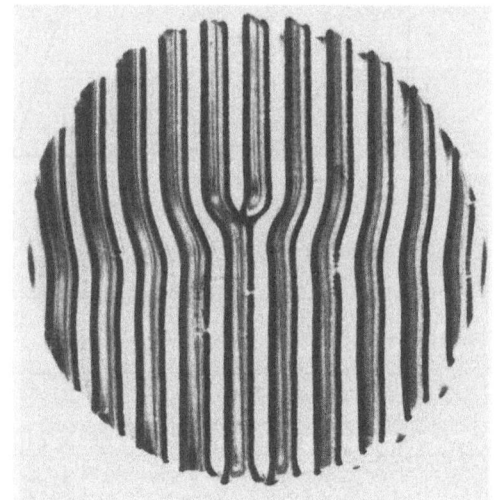

Figure 8: A dislocation moving back and forward:
a) ε = 1.8 the pair of extra rolls shrinks corresponding to a positive climb velocity
b) ε = 1.3 the dislocation velocity is now negative; notice the zig-zag shape of the
rolls.(These picture represent the positive horizontal isogradient lines of tempera-
ture, their round shape is due to the dimension of the lens which are necessary to
obtain such pictures; the dislocation structure was placed in a rectangular con-
tainer as usual).

The motion of a dislocation in a dissipative structure has motivated several theoretical approaches[2-3-4]; the Rayleigh-Bénard case at finite Prandtl number is not completely understood but these approaches have given interesting insight of the subject. We shall not describe them in detail but we shall try to describe their main results using very crude approximations: we shall assume that the convection may be derived by the amplitude equation of Newell and Whitehead[10]; this equation can be derived from a functional form which may be considered as the free "energy" of the system. If we consider that when the dislocation moves of dy, one thin layer of roll with wavevector $2(N+1) \times \pi/Lx$ has been converted to rolls with wavevector $2N\pi/Lx$, the free-energy has changed to an amount $dF = \frac{dF}{dK} \times \frac{2\pi}{Lx} \times Lx \times dy$. This change of energy corresponds to the work of the Peack-Köhler force acting on the dislocation: $f_{P.K} = \frac{dF}{dy} = 2\pi \frac{dF}{dK}$. When such a functional exists, one can show that dF/dK is proportional to the Diffusion coefficient $D_\perp$[3]. This coefficient measures the stiffness of a Rayleigh-Benard roll structure against a zig-zag distortion[11] The peack-Köhler force governs the motion of the dislocation, but the velocity is limited by some friction process, and when the velocity is steady we have $f_{P.K} = f_D$ where f_D is a damping force. This friction force may be written:

$$f_D \propto \left(\frac{2D_{//}}{D_\perp}\right)^{1/2} \cdot v$$

Where $D_{//}$ is the diffusion coefficient corresponding to the stiffness of the structure against compression or dilatation[11]. This implies that the dislocation velocity is proportional to $D_\perp^{3/2}$ so that the zero velocity line coincides with $D_\perp = 0$ at this approximation level.

At this point we would like to compare the result with the experiments, but this is still too early: the description that we have given is based on Newell-Whitehead amplitude equation and Siggia and Zippelins have shown[12] that this amplitude equation neglects large scale flows which develop when a roll structure is bent . These flows have non-trivial contribution, for instance, to the skewed-varicose instability at low Prandtl number. Their magnitude is related to the amount of bending and proportional to ϵ and P_r^{-1}. Unfortunately an account of these large scale flows involves a second equation in addition to the Newell and Whitehead equation, which becomes insufficient to describe roll structure.

The fundamental assumption that convection may be formulated within a functional formalism does not apply any longer so that there is no reason now why the zero velocity line of dislocation should coincide with $D_\perp = 0$ line. The presence of these large scale flows increases considerably the difficulty of the theoretical study of the dislocation motion and actually no complete theory has been proposed yet. In spite of this lack we can try to compare our results with earlier theories. Indeed ,large scale flows certainly act on the dislocation since they induce rolls bending, but these flows are proportional to P_r^{-1}, so that when $P_r = \infty$ the theoretical models would be valid. Moreover when P_r is large the large scale flows will not have a dramatic effect and the simplified analysis sketched above should apply. This is what we have done but we have used the expression for $D_\perp$ which includes the contribution of the large scale flows calculated by Manneville and Picquemal[13]. The secondary flow generated by the roll curvature associated with zig-zag distortions acts in a such way to reduce the bending, this stabilizing effect corresponding to an increase of $D_\perp$ as can be seen from the comparison with the diffusion coefficient $D_\perp'$ associated with secondary flow forced to zero. On fig. 7 we have drawn the condition $D_\perp = 0$ in full line and $D_\perp' = 0$ in dashed line; the zero velocity condition for dislocation motion turns out to be in agreement with $D_\perp = 0$ condition, though this might be a coincidence. As a matter of fact our experimental results do not contradict the crude theory developed above so far as $D_\perp$ has the correct (ε, K) dependence. We have found also that $v \propto D_\perp^{1.26 \pm 0.2}$, which is not far from the prediction $v \propto D_\perp^{1.5}$. Further we have found that the zig-zag instability threshold coincides with in experimental accuracy, with the zero velocity condition for the dislocation as predicted by the simplified theory.

However if we look in detail at the motion of a dislocation around the line $D_\perp = 0$, we may notice that the law $v = D_\perp^{3/2}$ is no longer valid there. This means that the interpretation given above is only approximative and that in this limited region the dislocation behaviour may not correspond to what we have described, and that perhaps its velocity may be governed by the large scale flows around it. Such effects will become larger at smaller Prandtl numbers, but in the case of large Prandtl numbers we may say that the selection criterion is $D_\perp = 0$ to a sufficient accuracy level. We have used the word selection since the dislocation motion allows clearly a choice between two structures with different wavevector to be made. We found it interesting to measure what selection criterion we obtain when using another kind of defect. On Fig. 9 we see the structure we used for this purpose: a set of rolls parallel to the y-axis in the middle of the cell is ended by two grain boundaries with rolls parallel to x-axis. Clearly the rolls in the middle of the cell are free to expand or to contract since the length of the rolls along x axis can change continuously. We can follow this expansion when we increase Ra on Fig. 9b and c. The wavevector at equilibrium for various ε is reported in Fig. 7 which shows without possible doubt that the selection criterion by grain-boundaries is the same as for dislocations.

A remarkable result of this study is that dislocation motion provides a well defined wavenumber selection mechanism, which cannot be distinguished from that provided by the grain-boundaries defects one. However the criterion can be derived from simplified theory of dislocation motion which simply leads to $D_\perp = 0$. However this agreement seems possible only because the Prandtl number is high (70) and it is worth to wonder what happens when Pr decreases and also if this criterion is truly unique. These are the questions which have motivated us to consider an other experiment in a cylindrical container.

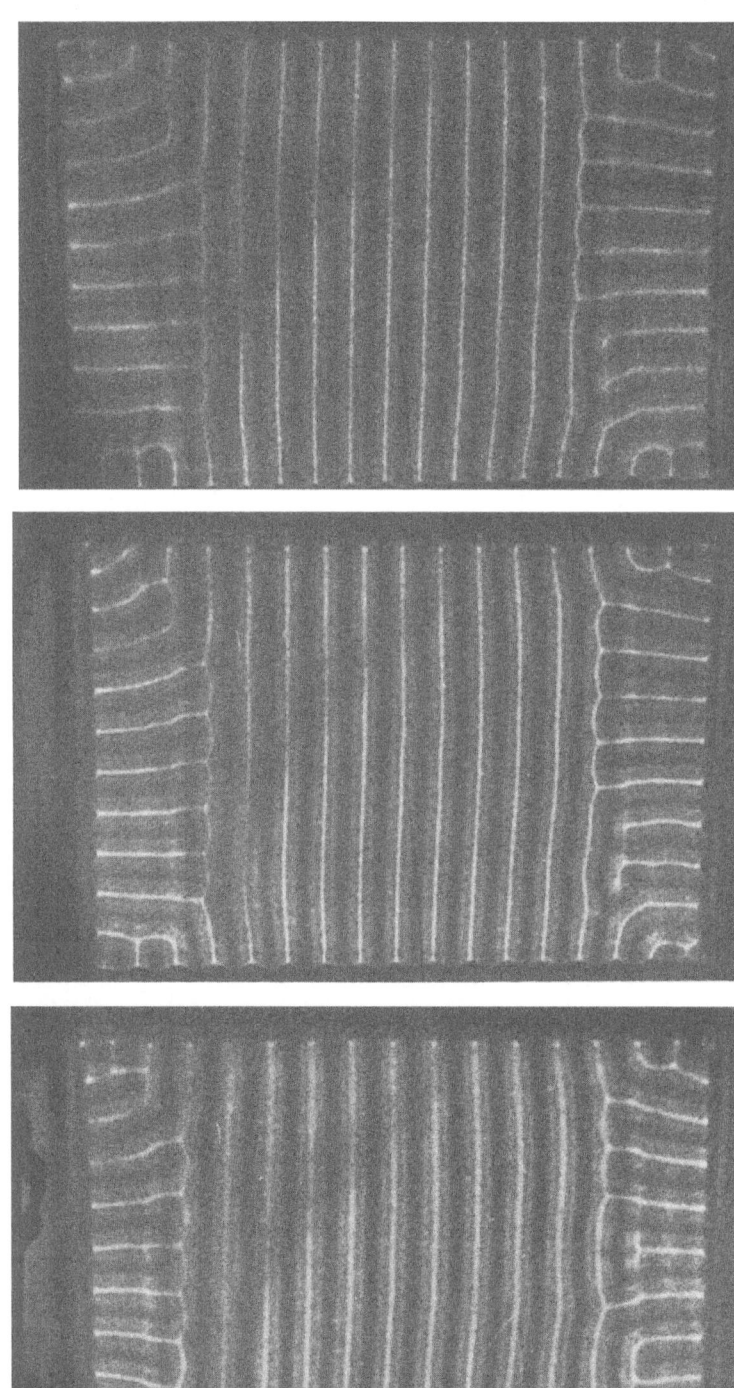

a

b

c

Figure 9: Grain-boundary pattern used to determine the wavenumber selection curve. The two rows of lateral rolls enables the principal set of rolls to expand or to contract freely. Picture a) corresponds to ε = 1.30, b) to ε = 1.90, c) to ε = 3.35.

3. WAVENUMBER SELECTION IN A CYLINDRICAL CONTAINER

In 1974 Koschmieder and Pallas[14] reported their observations of axisymmetric patterns in a cylindrical container with a corresponding well defined wavenumber selection. The symmetry of their pattern raises interesting problems[13]: in an axi-symmetric structure, rolls are bent   and this should induce a large scale radial flow which obviously cannot exist since it would bring some fluid continuously to-wards the center of the cylinder (think at the continuity equation), so that a ra-dial pressure gradient builds up and inhibits this flow. Such a structure would be really peculiar since it forbids the occurrence of large scale flows. In fact one can show that the wavenumber selection in these structures cam be understood as the loss of stiffness of the structure against zig-zag deformations subjected to the condition of no induced large scale flow[13] i.e. to $D'_\perp = 0$. This selection criterion is different from $D_\perp = 0$ when Pr is finite, and already at Pr = 70 the difference is really noticeable: consider the full line ($D_\perp = 0$) and dashed line ($D'_\perp = 0$) on Fig. 7. Koschmeider's experiment used a very high Prandtl number fluid for which $D'_\perp = 0$ could not be distinguished from $D_\perp = 0$, so that we decided to perform a similar experiment with a smaller Prandtl number in order to test these predictions.

Considering what we have said about natural structures, Koschmeider's axisymme-tric structures are surprising since the roll all along the cylinder sidewall should break into small rolls perpendicular to the boundary.

Thus we infer that there must be a small radial gradient at the sidewall to sta-bilize the roll perpendicular to it. In the experiment that we have performed we have placed a resistive wire at mid-height of the sidewall, all along it, very near to the oil boundary[15]. An ajustable current flows in the wire producing a controlled thermal

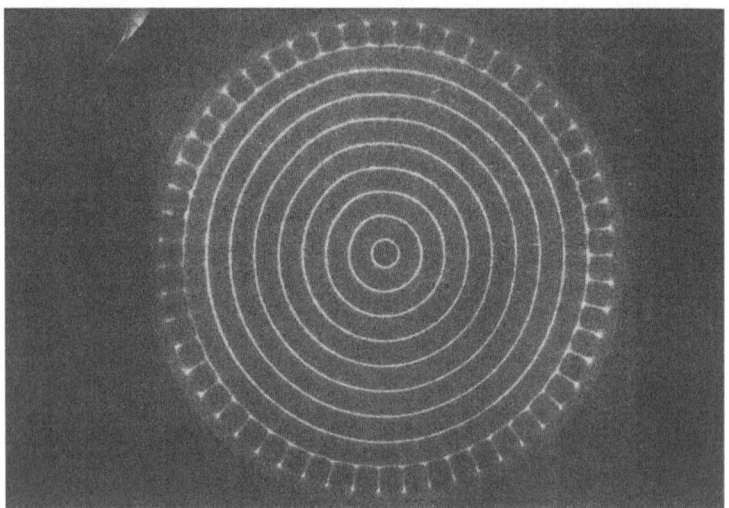

Figure 10: Instability of the roll at the sidewall when the cur-rent producing the radial thermal gradient is switch off. (Pr = 70, ε = 0.21).

gradient. In the absence of current and even if we induce an axisymmetric structure, the roll at the sidewall breaks as in Fig. 10; when the current is sufficient, the rolls are stable and axisymmetric structures can be observed see Fig. 11.

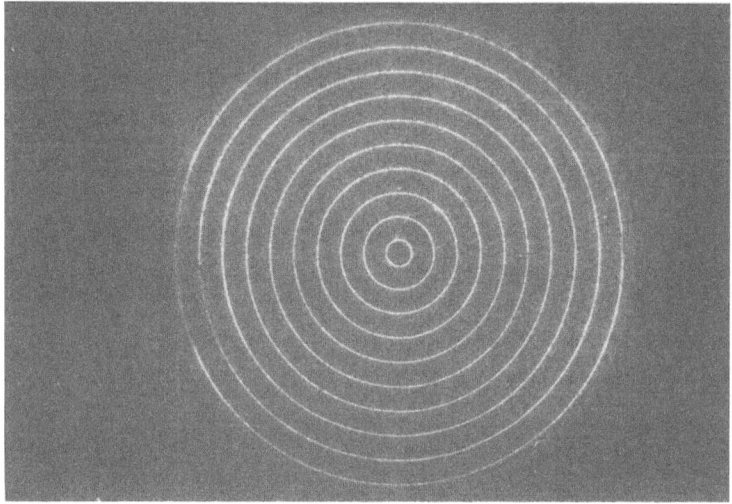

Figure 11: Stable annulus roll pattern (Pr = 70, ε = 0.315)

How is selection possible in these axisymmetric structures. At first glance, the number of rings must be an integer and there is no obvious reason why this integer may change. In fact the umbilicus that is the central ring, is a weak point in the structure, One ring may disappear or appear at the umbilicus as in Fig.12 decreasing or increasing the number of rolls which allows the wavelength adjustement with Ra.

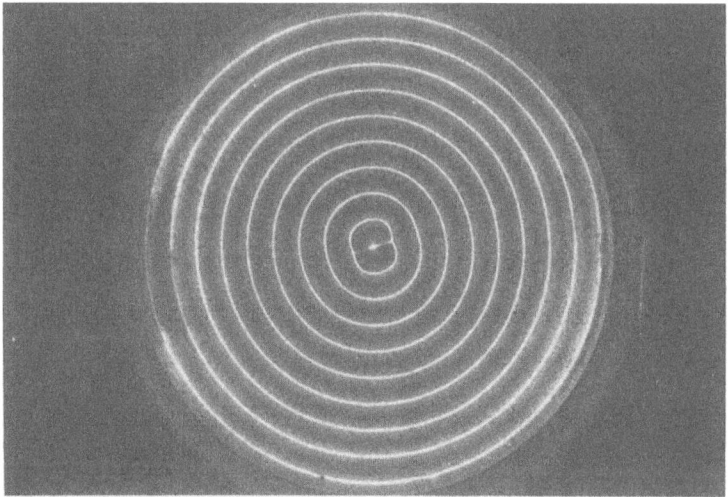

Figure 12: Transition from 18 rings to 19 rings (Pr = 70, ε = 0.926).

In this experiment, we measure the thermal gradients induced by the convection using the deflexion of a laser beam passing vertically through the convecting layer. This measurement is local and may be done everywhere in the container by translating the convection cell with step by step motors. Measurements have been performed along one diameter of the cylindrical container   800 points (1 every tenth of mm) enable us to get an accurate knowledge of the convective structure since the aspect ratio Γ_r = R/d = 20 limits the possible roll number to 40 along a diameter.

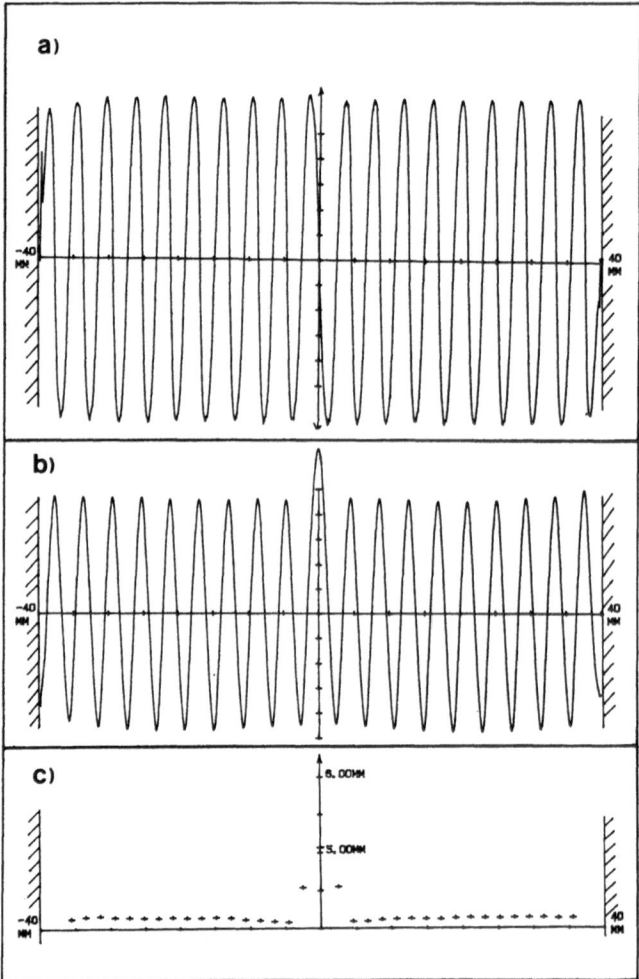

Figure 13: *Recordings along a cell diameter with Pr = 70 and ε = 0.376*

a) Radial component of horizontal thermal gradient plotted along one diameter of the cell (arbitrary units).

b) Numerical integration of the data of a). This plot corresponds to the temperature profile along a diameter.

c) Roll width in millimeter, measured between the zero of the temperature gradients.

On Fig. 13a, we present a typical recording of the thermal gradients along the scanning direction Fig. 13b is just the numerical integration of the thermal gradients and so represents the temperature variations along the diameter. Notice its maximum value at the umbilicus[6]. Finally we have calculated the roll width as a function of their position along the diameter, see Fig. 13c. Once more this width is very uniform, except in the umbilicus.Such measurements were repeated along different diameter angles to check the symmetry of the structure. And here we come to the striking point of this study : we have found that the convective structures,like that of Fig. 11 were not exactly axisymmetric! We first thought that our cell was not perfectly round or was not really horizontal but all these experimental imperfections were really too small to explain the disymmetry of our structures and, furthermore the dependence of this disymmetry on the physical parameters, indicates that it has its origin in the convection itself rather than in experimental imperfections.

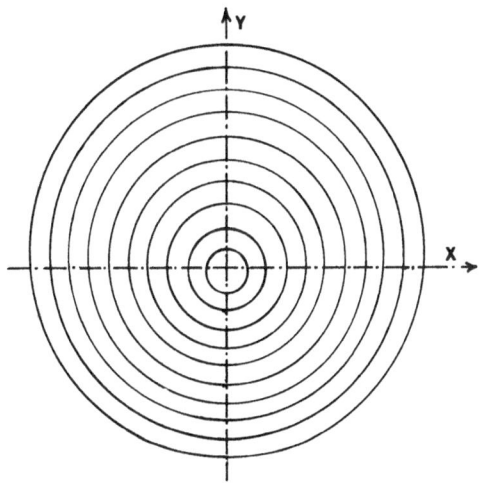

Figure 14: Sketch of the pattern,
the disymmetry has been exaggerated
on purpose

What is this disymmetry?. In Fig. 14 we have exaggeratedit to make it visible . The umbilicus is shifted radially so that when we measure along the X diameter we obtain symmetrical recordings about the center of the cell, like those of Fig. 15a,but when we measure along the Y diameter, the recordings are no longer symmetrical, see Fig. 15b. This effect is not very large, in our experimental conditions, but it is really noticeable: the umbilicus typically is shifted by one millimeter (R = 40mm).

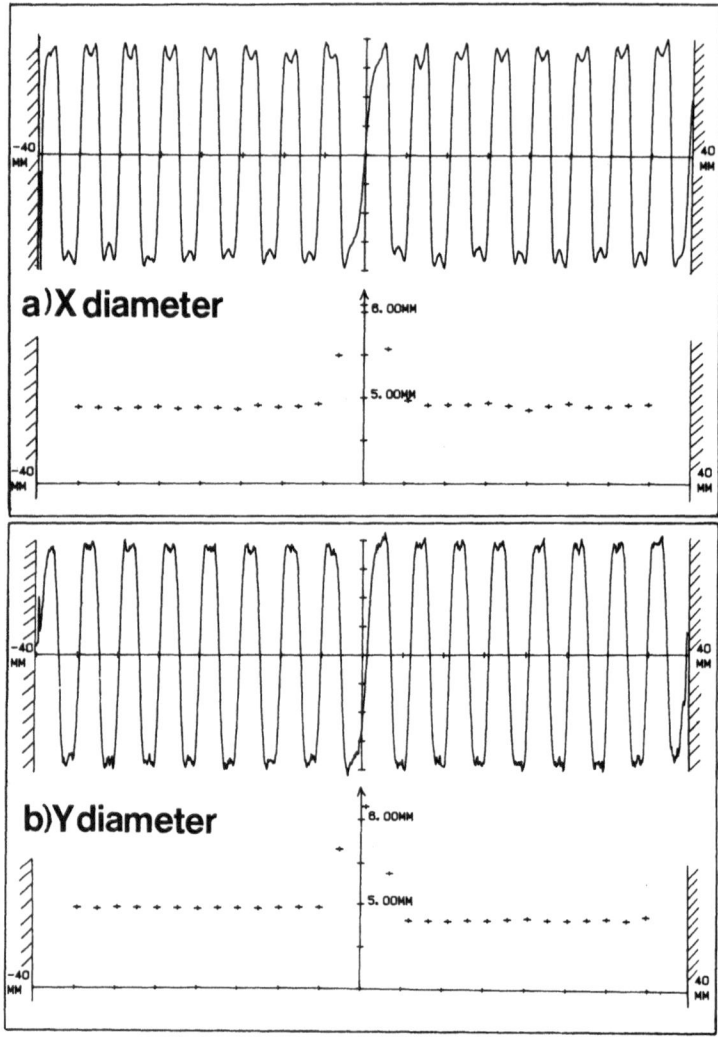

Figure 15: a) *Thermal gradients and roll widths measured along an*
X diameter; notice that the recordings are symmetrical about the
center of the cell (Pr = 14, ε = 2.58).

b) *Thermal gradients and roll widths measured along a Y diameter*
(perpendicular to X) on the same structure (Pr = 14, ε = 2.58).
Notice that the umbilicus does not coincide anymore with the center
of the cylinder and that the roll widths are different from left to
right.

We shall return to the effect of this dissymmetry later but let us now come back
to the experiment itself: after the induction of the axisymmetric structure we have
proceeded to a series of measurements while   increasing slowly the Rayleigh number and then decreasing it. We have performed the experiment for two Prandtl numbers
P ≐ 70 like for the dislocation and a lower value P=14. When we increase the Rayleigh

number the roll width tends to increase , and we were surprised to notice that the evolution of the structure was rather fast about 30 minutes, compared with the thermal diffusion time ($R^2/\kappa \approx$ 300 minutes), and that the width of the rolls was very uniform along the diameter, even for rolls which are higly bended. As the roll width increases, the umbilicus diameter decreases drastically.

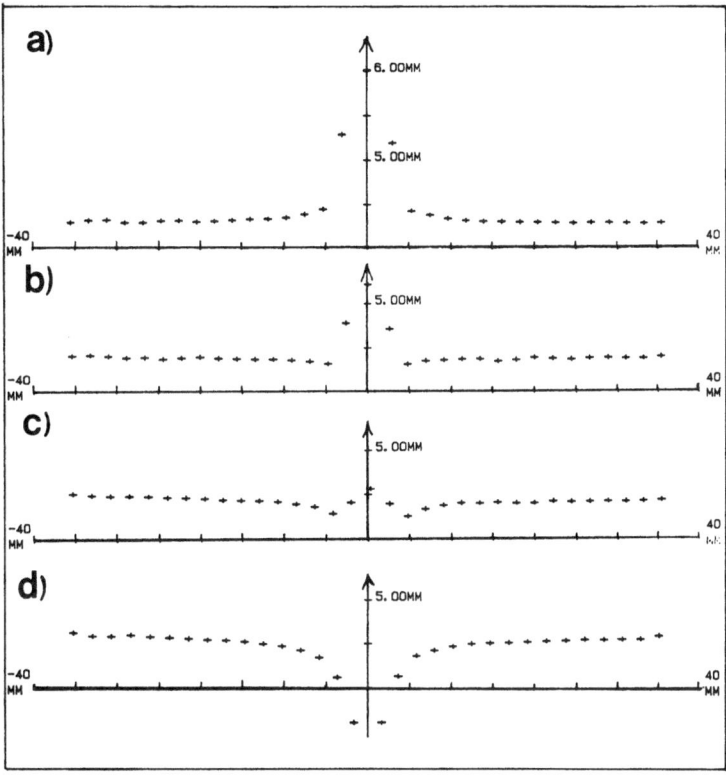

Figure 16: Variations of the width of the umbilicus when the Rayleigh number is increased (Pr = 70).

a)ε = 1.11,umbilicus diameter: 3.04 × d

b)ε = 1.38,umbilicus diameter: 2.64 × d

c)ε = 1.55,umbilicus diameter: 2.31 × d

d)ε = 1.83,umbilicus diameter: 1.66 × d

The dimension of the umbilicus may vary in a large manner at least by a factor of two; on Fig. 16 we reproduced its variations when increasing the Rayleigh number. Clearly we may distinguish two cases: in Fig. 16a and b the umbilicus is in depression so that the neighbour rolls have a width somewhat larger than the others, in Fig. 16c and d the umbilicus is in compression. When the compression is too large a roll disappears,at the center and the same process will repeat. This mechanism gives rise to hysteresis since the disappearance of a roll when increasing the Rayleigh

number does not correspond to the birth of a new roll when decreasing Ra. However
at a given Ra it is possible to observe two structures but never three, so that the
wavenumber selection does not give rise to a band but to a curve with hysteresis
loops. This evolution suggests that the structure is usually submitted to stresses
so that one could have thought that the symmetry breaking of the structure was the
result of a kind of buckling of the pattern. But in this case we should observe this
symmetry breaking only when the structure is under compression, however the symmetry
breaking does not depend significantly on the stress state of the structure.

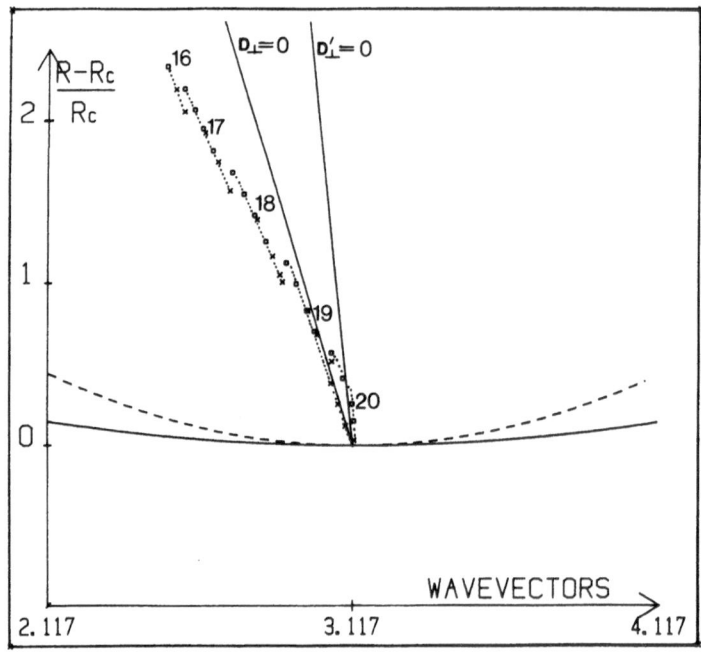

*Figure 17: Wavenumber selection curve when Pr = 70. square points
were obtained while increasing ε as cross point while decreasing ε . The numbers indicate how many rings the structure contains.*

We have measured the roll width in the region of the cell where it is uniform and
on a diameter along which the structure appears symmetrical. In the case of Pr = 70
we present the curve obtained on Fig. 17. The selection is well defined but does not
correspond to $D'_\perp=0$ at all and rather fits with what we have observed with dislocations
at the same Prandtl number ($D_\perp = 0$).Whether or not large scale flows are still present and play a defined role in the selection cannot be answered yet. Let us consider the case of the lowest Prandtl number; at Pr = 14 the selection curve Fig. 18
is very similar to the one with Pr = 70, except that the hysteresis loops are larger.
What is more astonishing is that the slopes of the two curves are nearly the same.
Now this slope corresponds neither to the curve $D_\perp = 0$ or to $D'_\perp = 0$ (calculated for

P_r = 14). We cannot explain this last result but we think that it would be interesting to measure the selection obtained with the dislocations at P = 14 before trying
to compare with any theory.

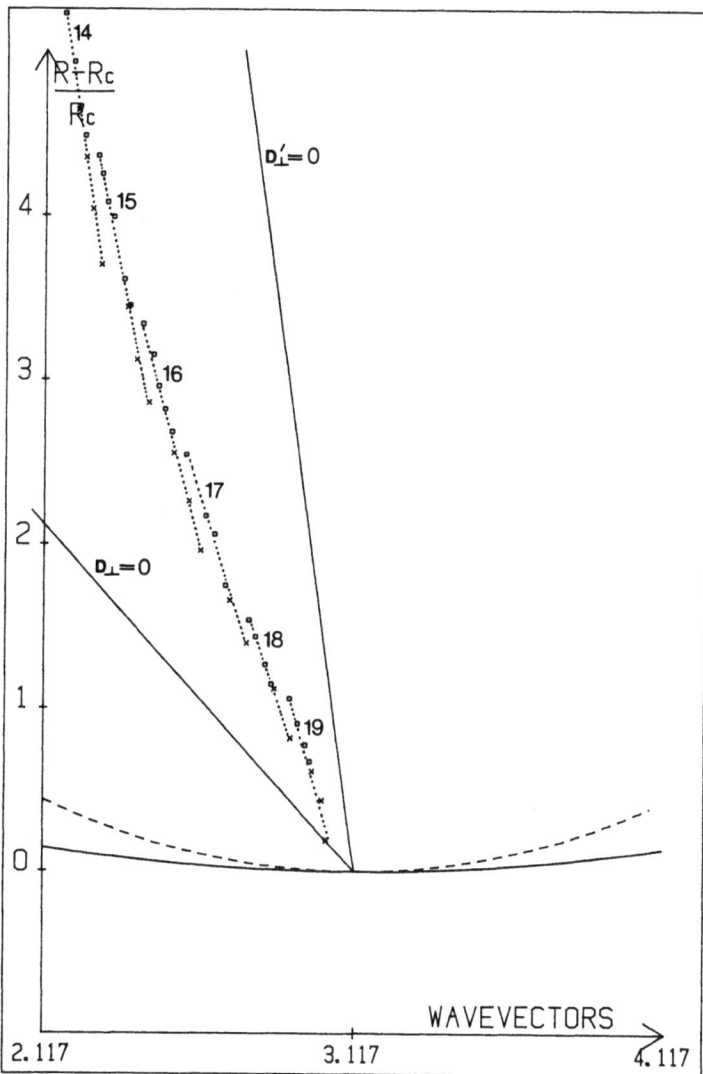

Figure 18: Wavenumber selection curve when Pr = 14.

This experiment which was expected to bring simple results has failed to enligh-
ten the problem. The main reason for that is the symmetry breaking of the structure
this phenomenon has already been observed in another instability by J.P Gollub [16].
This effect rules out previous theoretical considerations on wavelength selection in
strictly axisymmetric structures [13] since this last property is not verified. But

unexpected results are more stimulating than results which merely confirm one's view: in our case it is worth trying to understand what is at the origin of the symmetry breaking of the structure. We have a conjecture that cannot be checked using our experimental results, but which is compatible with them and would be really interesting to test. The selection criterion in strictly axisymmetric structures is based on the fact that to be compatible with the continuity equation the radial large scale flow must cancel . Now if we assume that the symmetry is broken, the argument no longer holds: large scale flows may be present and we think that the situation where these flows exist, even if they have a special pattern, is more favorable for convection than the situation where they are absent.

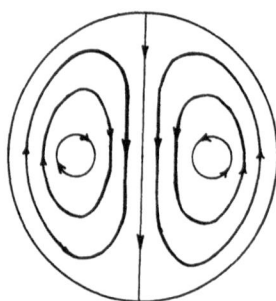

Figure 19:Streamline of the large scale flows suspected to be at the origin of the symmetry-breaking of the pattern.

We conjecture that the large scale flows have a dipolar like pattern, as sketched in Fig. 19, and that they induce the symmetry-breaking of the structure. This conjecture is supported by some experimental observations. First we may measure the symmetry-breaking by two methods: the displacement "d_0" of the umbilicus relative to the center of the cell, and the difference between the roll widths "$\Delta\lambda$" on the left of the umbilicus and that on the right . These two measurements give consistent results:"d_0" and "$\Delta\lambda$" are found to increase when we increase the Rayleigh-number; in the case of P_r = 70 the relationship is clearly proportional , so that very near to the convection threshold, the pattern is axisymmetric within our experimental accuracy. In the case of P_r = 14,"d_0" and "$\Delta\lambda$" are quite larger than for P_r = 70, at the same ε, some preliminary experiments in methyl alcohol have shown that the effect was even directly visible at P_r = 7. This indicates that "d_0" and "$\Delta\lambda$" increase as a function of Pr^{-1} this is clearly in agreement with the conjecture on the role of the large scale flows since they are expected to vary as $\varepsilon.P_r^{-1}$. Another proof that the symmetry breaking is of hydrodynamic origin is the following: for P_r = 14, the structure becomes threedimensional around $\varepsilon \simeq 10$, see Fig. 20. When the Rayleigh number is increased beyond this threshold the threedimensionality invades the structure progressively from the boundary to the umbilicus and "d_0" and "$\Delta\lambda$", which were very large just before, rapidly decrease and vanish at least for sufficiently large ε.

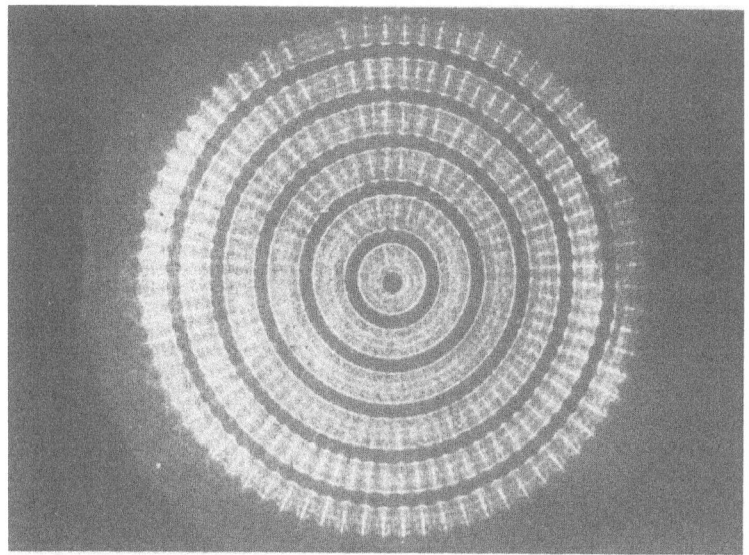

Figure 20:P_n *= 14,ε = 10.8. The threedimensionality invades progressively the annulus pattern from the boundary towards the center of the cell.*

This suggests that threedimensionality destroys the large scale flows. Obvious-ly the simplest way to check this conjecture would be to actually measure the hori-zontal velocity profile but this is not very easy to achieve experimentally.

CONCLUSION

The dislocation motion experiments we have performed, illustrate how the convecti-ve pattern, containing a single defect, finds a stationary state after a well defined evolution. This simple behaviour confirms that the transient evolution of textures at high Prandtl number may be understood as a phase of minimization of energy. The motions of dislocations also indicate that in our experiments a well defined prefer-red wavenumber does exist at each Rayleigh number, this selection criterion is the same as that offered by grain-boundaries. This agreement between selection crite-rions has motivated us to consider that offered by the curvature in an axisymmetric strucutre. This last experiment has shown the occurrence of an unexpected symmetry-breaking of the structure, we conjecture that the large scale flows   are at the origin of this symmetry-breaking. This is supported by our experimental observations. This symmetry-breaking destroys a part of the selection argument concerning axisymmetric structures; however our experimental selection curves reported here, agree with the zero velocity dislocation curve. This agreement is astonishing since it does not seems to depend on the Prandtl number as expected.

REFERENCES

[1] Ahlers, G. and Behringer, R.P. (1978) Phys. Rev. Lett. 40 712
    Ahlers, G. and Walden, R.W. (1980) Phys. Rev. Lett. 44 445
    Libchaber, A. and Maurer, J. J. de Physique Lett. 39 1 369
[2] Siggia, E.D., and Zippelius, A. (1981) Phys. Rev. A 24 1036
[3] Pomeau, Y., Zaleski, S. and Manneville, P. (1983) Phys. Rev. A 27 2710
[4] Dubois Violette, E., Guazzelli, E., Prost, J., J. Phis. Mag. A (to appear)
[5] Berge, P. and Dubois, M. p 493 of "Scattering techniques applied to supra-
    molecular and nonequilibrium systems" Ed. Sow-Hsin Chen, Benjamin Chu and Ralph
    Nossal 1981
[6] Croquette, V., Mory, M. and Schosseler, F., (1983) J. de Physique 44 293
    Berge, P."Chaos and order Nature" (Elmou 1981) Synergetics (Springer Verlag)
[7] Gollub, J.P., Mc Carriar, A. and Steinman, J.F. (1982) J. Fluid Mech. 125 259
    Gollub, J.P., and Steinman, J.F. (1981) Phys. Rev. Lett. 47 505
    Gollub, J.P. and Mc Carriar, A.R. (1982) Phys. Rev. A 26 3470
[8] Chen, M.M. and Whitehead, J.A. (1968) J. Fluid Mech. 31 1
[9] Pocheau, A. and Croquette, V., J. of Physique (to appear) Jan. 84
[10] Newel, A.C. and Whitehead, J.A., (1969) J. Fluid. Mech. 38 279
[11] Pomeau, Y. and Manneville, P. (1979) J. Physique Lett. 40 L 609
    Croquette, V and Schosseler, F. (1982) J. Physique 43 1183
[12] Zippelins, A. and Siggia, E.D Phys: Fluids 26 (83) 2905
    Siggia, E.O. and Zippelius, A. (1981) Phys. Rev. Lett. 47 835
[13] Manneville, P. and Piquemal, J.M. (1983) Phys. Rev. A 28 1774
    Cross M.C., Phys. Rev. A 27 (1983) 490
[14] Koschmieder, E.L. and Pallas, S.C. (1974) J. Heat: Mass. Trans. 17 991
[15] Croquette, V. :future publication
[16] Gollub, J.P. and Mayer, C.W. (1983) Physica D 60 337

CONVECTION PATTERNS IN LARGE ASPECT RATIO SYSTEMS

P. C. Hohenberg

Bell Laboratories
Murray Hill, New Jersey 07974

This talk summarizes recent theoretical work /1,2,3/ done at Bell
Laboratories and the University of Arizona on convection patterns in
Rayleigh-Benard systems containing many rolls. The work is both
analytic and numerical, and focuses primarily on simple model equations, whose pattern forming behavior is thought to be similar to that
of the Boussinesq equations of hydrodynamics, at least near threshold.
The analytic work of Cross and Newell /1/ treats the statics and slow
dynamics of convective rolls via an expansion in the roll curvature
about a state consisting locally of parallel rolls. The theory is
quite general and applies in principle to any system with periodic
steady states, including the Boussinesq equations. It is not restricted to the vicinity of the threshold, nor to small deviations
from a state consisting globally of parallel rolls. Wavenumber
selection, the shape of patterns, their stability, and the time dependence resulting from long-wavelength instabilities, are discussed.

The theory includes the notion of the Busse stability balloon, it
reduces near threshold to the Newell-Whitehead-Segel amplitude equations /4,5/, and it contains the Pomeau-Manneville phase equations /6/
as a special case.

The models which have been investigated include the real or complex Swift-Hohenberg equation /7,8/, as well as various other models
obtained by modifying the nonlinear term, or by introducing a coupling
to the vertical vorticity field. The stability analysis of parallel-
roll states in these models has been carried out by Greenside and
Cross /3/ using a standard Galerkin procedure. For those instabilities which occur at long wavelengths the results have been checked by
applying the Cross-Newell scheme /1/.

Numerical simulations of the Swift-Hohenberg model have been performed by Greenside and Coughran /2/, on rectangular cells with aspect
ratios ranging from $L_x = L_y \backsim 5$ to $L_x = 29$, $L_y = 20$. The integration
times were sufficient to reach stationary conditions, even for states
containing significant numbers of defects and sizable roll curvature.
In the latter cases the integration times went up to 25,000 τ_v

for large systems, where τ_v corresponds to the vertical diffusion time of the Rayleigh-Bénard system. Effects studied include the influence on pattern formation of initial conditions (regular vs. random), of lateral boundaries, and of the various instabilities (zig-zag, cross-roll, Eckhaus) occurring in regular roll patterns. Wave-number selection is investigated using spatial Fourier analysis, and the time evolution of the wavevector distribution is displayed. Many of the results obtained agree with observed experimental features of convective pattern formation, but some new as yet unobserved features are also found.

References

/1/ M. C. Cross and A. C. Newell, "Convection Patterns in Large Aspect Ration Systems", Physica D (1983).

/2/ H. S. Greenside and W. M. Coughran, Jr., "Nonlinear Pattern Formation Near the Onset of Rayleigh-Bénard Convection", Phys.Rev.A(1983).

/3/ H. S. Greenside and M. C. Cross, "Stability Analysis for Equations Modeling Convection", Phys. Rev. A (1984).

/4/ A. C. Newell and J. A. Whitehead, J. Fluid Mech. 38, 279 (1969).

/5/ L. A. Segel, J. Fluid Mech. 38, 203 (1969).

/6/ Y. Pomeau and P. Manneville, J. Phys. (Paris) 42, 1067 (1981).

/7/ J. Swift and P. C. Hohenberg, Phys. Rev. A15, 319 (1977).

/8/ Y. Pomeau and P. Manneville, Phys. Lett. 75A, 296 (1980).

THREE DIMENSIONAL CONVECTIVE STRUCTURES IN A HORIZONTAL OR TILTED POROUS LAYER

J.P. CALTAGIRONE

Laboratoire Energétique et Phénomènes de Transfert
E.R.A. CNRS N° 1026 - Esplanade des Arts et Métiers
33405 TALENCE CEDEX

SUMMARY

The natural convection placed in a horizontal or tilted porous layer of large shape ratio factor subject to a temperature gradient, depends not only on imposed stresses in the geometry of the system but also on initial conditions.

A three-dimensional modelization of spectral type helps to understand the induced flows from arbitrary initial conditions. The main convective structures thus obtained correspond to those studied experimentally.

1. INTRODUCTION

When a horizontal or tilted porous layer of great lateral extensions is heated differentially, the natural convection develops for a Rayleigh number higher than the critical value Ra_c^*, such as $Ra_c^* \cos \phi = 4\pi^2$.
The linear theory of stability enable us to determine this critical condition of transition between the unicellular flow and a three-dimensional regime but not to foresee either the structure of the induced flow or the heat transfer due to the convection. However J.E. Weber (1974) shows that a flow in transversal rolls is less stable than a three-dimensional flow.
Previous experiments (S. Bories and M. Combarnous 1973) carried out in a cell of great transversal sizes show a wide diversity of stationary flows obtained by varying the Rayleigh number and the inclination of the experimental cell. For lower Rayleigh numbers, it's the unicellular flow which takes place very quickly in the cell, this regime remaining in all the other flows with which it combines. The criterion $Ra^* \cos \phi = 4 \pi^2$ is rather well confirmed experimentally ; over this value of $Ra^* \cos \phi$, two types of flow appear naturally : a three-dimensional flow in cells of polyhedric type, of right section roughly hexagonal for small values of ϕ, and for angles superior to about 15° the stream corresponds to the superposition of the unicellular regime and rolls of axes parallel to the largest slope of the cell. Finally when the Rayleigh number is superior to about 250 the fluctuating regim appears to give longitudinal rolls which oscillate around their axis.
A few two-dimensional numerical studies (Walch and Dulieu 1979) and three-dimensional (J.P. Caltagirone and S. Bories 1980) for limited configurations (weak aspect ratios) however, reveal the importance of initial conditions over calculated struc-

tures.

We shall only give here results owing to a three-dimensional numerical modelization, yet trying to place them back in the context of theoretical and experimental researches actually conducted in this configuration.

2. FORMULATION OF THE PROBLEM AND NUMERICAL PROCESSING

Let us consider a porous layer of heigh H and of aspect ratios $A = L/H$ and $B = M/H$ tilted with an angle ϕ in relation to the horizontal. The porous medium is constituted by a solid matrix with a porosity ε and a permeability K, and a saturating fluid defined by kinematic viscosity ν thermal expansion coefficient β and heat capacity $(\rho c)^* = \varepsilon(\rho c)_F + (1-\varepsilon)(\rho c)_S$.

Using dimensionless equations of conservation reveal a parameter of similarity, the Rayleigh number $Ra^* = g\,\beta(\rho c)_F\,\Delta TKH/\nu\lambda^*$ ratio of convective and dissipating effects representative of the amplitude of convective movements in porous media. Dimensionless equations are written as (J.P. Caltagirone 1982) :

(1) $\nabla . \underline{V} = 0$

(2) $- \nabla p + Ra^*\,\underline{k}T - \underline{V} = 0$

(3) $-\dfrac{\partial T}{\partial t} = \nabla^2 T - \underline{V} . \nabla T$

where $\underline{k} = \sin\phi\,\underline{e}_1 + \cos\phi\,\underline{e}_3$ is the unit vector gravitational acceleration, $\underline{V} = U\,\underline{e}_1 + V\,\underline{e}_2 + W\,\underline{e}_3$ is the filtration velocity, p the pressure and T the temperature. Horizontal surfaces for $\phi = 0$ are supposed to be isothermal (T = 1 for z = 0 and T = 0 for z = 1) and impermeable, and the other lateral surfaces adiabatic and impermeable.

The system of equations (1 - 3) is solved by a spectral method of Galerkin type based on Fourier's functions :

(4) $T = (1-z) + \displaystyle\sum_{l=0}^{L}\sum_{m=0}^{M}\sum_{n=1}^{N} a_{lmn}(t)\,\cos l\pi x\,\cos m\pi y\,\sin n\pi z$

(5) $U = -A^2 \displaystyle\sum_{l=1}^{L}\sum_{m=0}^{M}\sum_{n=1}^{N} b_{lmn}\,\sin l\pi x\,\cos m\pi y\,\cos n\pi z\;(ln\pi^2)$

(6) $V = -B^2 \displaystyle\sum_{l=0}^{L}\sum_{m=1}^{M}\sum_{n=1}^{N} b_{lmn}\,\cos l\pi x\,\sin m\pi y\,\cos n\pi z\;(mn\pi^2)$

(7) $W = \displaystyle\sum_{l=0}^{L}\sum_{m=0}^{M}\sum_{n=1}^{N} b_{lmn}\,\cos l\pi x\,\cos m\pi y\,\sin n\pi z\;(1^2\pi^2 + m^2\pi^2)$

The experimental functions of these developments respect the determined boundary conditions as well as the continuity equation of incompressible fluid (1). Spectral coefficients of temperature $a_{lmn}(t)$ and of velocity $b_{lmn}(t)$ allow one to come back at any time to the real fields.

The introduction of these developments in the equation of impulse (Darcy's law) after a few elementary processes leads to a direct relation $b_{ijk} = F(a_{ijk})$. Similar

operations on the energy equation lead to a differential system which is not linear
on spectral coefficients of the temperature a_{ijk} which can be written as :

$$\frac{da_{ijk}}{dt} = L\, a_{ijk} + N$$

where L and N are linear and non-linear operators of the system. It is to be noted
that the linear operator defines the system stability through the research of its
own values. As for the non-linear operator it represents the coupling between the
spectral coefficients and the physical system damping.

The non-linear terms of the energy equation of the type $\underline{V}.\ \Delta T$ are determined thanks
the Fast Fourier transforms by carrying out products in the physical space and
coming back into the spectral space through an inverse transformed. The differential system is then solved by a special scheme (J.P. Caltagirone) considering the
non-linear terms represented by the e_{ijk} in the spectral space, as constant : the
system becomes linear and allows a solution of exponential type, with arguments
f_{ijk}, representative of the linear operator L. The integration is clearly carried
on by consultating again the non-linear terms e_{ijk}.

The discretisation scheme can be written as :

$$(9) \qquad a_{ijk}^{n+1} = (a_{ijk}^{n} + e_{ijk}^{n}/f_{ijk})\ \exp\ (f_{ijk}\ \Delta t) - e_{ijk}^{n}/f_{ijk}$$

The terms e_{ijk} are calculated directly at the step n or estimated with a diagram more worked out.

Initial conditions are laid down through the temperature field $To(x,y,z)$ defined
by the initial spectral coefficients $a_{ijk}(o)$. Besides these, coefficients present
an obvious physical reality since they square with the convective modes. In some
cases a white noise is chosen as initial conditions $(a_{ijk}(o) = 10^{-5})$ in others,
some modes are privileged.

Calculations are made on CRAY 1 with approximations of 32 x 16 x 8 components,
spectral coefficients are stocked for a later use, the purpose being a graphic
representativity of the flows.

3. DISCUSSION OF THE RESULTS

Numerical calculations are made with approximations L = 32, M = 16 et N = 8
and a choice of aspect ratios A = 6 and B = 4. A compromise between the aspect ratios in all directions and the number of development terms (4-7) necessary for the
good representatitivy of flows is to be found.

- For values of the Rayleigh number and of angle ϕ such as $Ra^{*}\ \cos \phi = 4\pi^{2}$ the calculation converges to the only solution of an unicellular flow whatever the imposed
initial conditions may be. The fluid heated by the lower surface goes up to the
lower half and then down to the other side. In the center of the layer, far from

the boundaries, the velocity distribution is almost linear, the temperature distribution corresponds in actual fact to pure conduction ; this is a temperature gradient, constant along the layer, in the way of the steepest slope in the center of that one for z = 0.5.

It snould be noted that the criterion indicated above is good only for a tilted porous layer of infinite extension ; a study of relative stability for a layer with finite dimensions (J.P. Moron 1981) reveal that the flow there is more stable This difference can also be checked numerically.

- In the instability zone of the unicellular stream or which is the same thing in that case, in the stability zone of the three-dimensional flow, the final structure thus achieved depends on the initial conditions of numerical calculations. There is a great number of convergent structures of flow for the same choice of conditions (Ra^*, ϕ, A, B).

First of all, the injection of longitudinal rolls in the calculations under the shape of modes (0,m,1) allows the development of the latter as well as their combination with the unicellular flow which keeps going.

If coupled modes are injected of the type (0,m,1) + (n,0,1) they only amplify when the tilting angle is not too important or at least lower than a maximum value of 30°. Beyond this critical value, the flow changes to give only longitudinal helices. This phenomenon thus confirms the linear theory of stability treated by J.P. Caltagirone and S. Bories (1983).

In this zone of low values of the angles, the number of structures which are calculated for the same couple (Ra^*, ϕ) is very important. An example of structure achieved for the modes (0,4,1) + (3,0,1) is represented in figure 1 by the isothermal surface T = 0.5 and the isotherms drawn in the median plane z = 0.5.

When the initial conditions are injected under the shape of a white noise $a_{ijk}(o) = 10^{-5}$ piled on the distribution of pure conduction (1-z) a perturbation appears (figure 2a) at the abscissa y = 0, while the unicellular flow is already set. This transversal perturbation amplifies and is propagated to reach the opposite side y = B (fig. 2b, 2c) ; these figures show the creating of failures in the convective structure and a certain inclination to an orthogonal hanging of the rolls to the adiabatic vertical sides of the medium. The final flow on figure 2d reveals the definitive creating of rolls of longitudinal dominance and of two transversal rolls in a limited zone of the porous medium.

4. CONCLUSIONS

The numerical modelization of the natural convection flow in a tilted porous layer helps to point out the main convective structures, studied experimentally. Moreover it shows the great variety of shapes of the flows achieved by following initial conditions of calculation.

It also corfirms the stable failure observed during experiments in a porous
medium as well as in a fluid medium.

5. REFERENCES

- Bories S. and Combarnous M. - Natural convection in a sloping porous layer,
  J. Fluid Mech., 57, p. 63-75, 1973.

- Caltagirone J.P. and Bories S. - Solutions numériques bidimensionnelles et tri-
  dimensionnelles de l'écoulement de convection naturelle dans une couche poreuse
  inclinée, C.R.Acad.Sci., 190, p. 197-200, 1980.

- Caltagirone J.P. and Bories S. - Stabilité de l'écoulement de convection naturelle dans une couche poreuse inclinée, 6e Congrès Français de Mécanique, 5-9 sep-
  tembre 1983, Lyon.

- Caltagirone J.P. - Convection in a porous medium, Convective Transport and Instability Phenomena, G. Braun Ed., Karlsruhe, 1982.

- Moron J.P. - Etude des transitions entre différentes structures thermoconvecti-
  ves dans un milieu poreux incliné, Rapport interne du Laboratoire d'Aérothermi-
  que, 1981.

- Walch J.P. and Dulieu B. - Convection naturelle dans une boîte rectangulaire lé-
  gèrement inclinée contenant un milieu poreux, Int. J. Heat Mass Transfer, 22,
  p. 1607-1612, 1979.

- Weber J.E. - Thermal convection in a filled porous layer, Int. J. Heat Mass
  Transfer, 18, p. 474-475, 1975.

(b)

(a)

Figure 1 (a,b), Isotherm surface T = 0.5 and isotherms in the planes z = 0.5 obtained for Ra* = 100, B = 4, φ = 30° and for initial conditions corresponding to the modes (0, **4**, 1) + (**3**, 0, 1).

Figure 2(a,b,c,d) : Isotherms in the planes z = 0.5 obtained for Ra* = 100, A = 6, B = 4, ϕ= 20°an for initials conditions corresponding to a white noise a$_{ijk}$ = 10^{-5}.

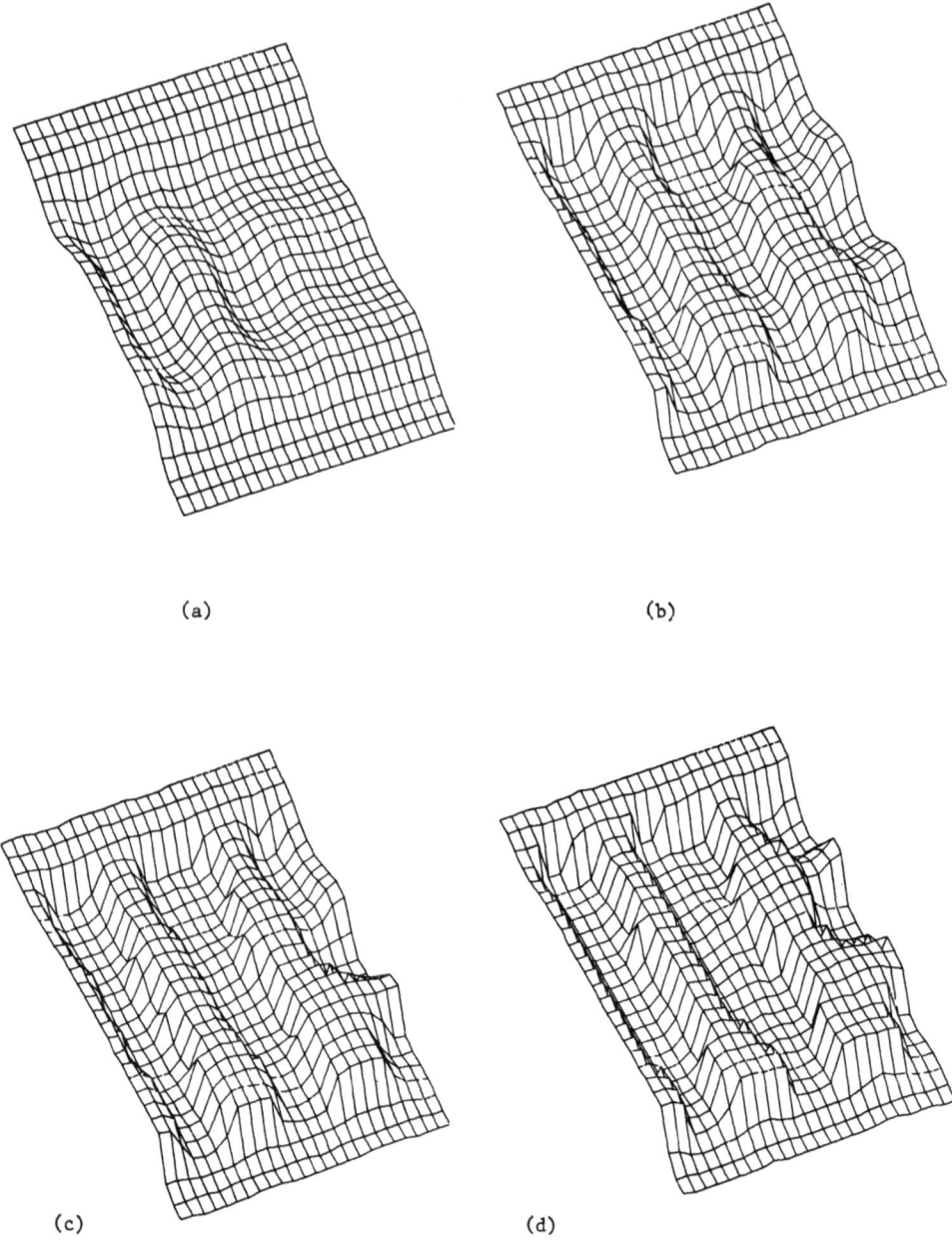

(a)

(b)

(c)

(d)

Figure 2 (a,b,c,d) : Isotherm surface T = 0.5 obtained for $Ra^* = 100$, A = 6
B = 4, $\phi = 20°$ and for initial conditions corresponding to a white noise $a_{ijk} = 10^{-5}$

MODELISATION AND SIMULATION OF CONVECTION IN EXTENDED GEOMETRY

Paul MANNEVILLE

DPh-G/PSRM Orme des Merisiers
CEN-Saclay 91191 Gif sur Yvette-Cedex (France)

1. INTRODUCTION

In convective cells with horizontal dimensions L_ℓ comparable with
their depth d (confined geometry) it is now well known that the
transition to turbulence takes place through the destabilisation of a
small number of modes with well defined spatial structures and that
"weak turbulence" means short term deterministic but long term
stochastic evolution due to the instability of trajectories in phase
space and to the sensitivity to initial conditions [1]. The situation
is much less clear in wide cells ($L_\ell \gg d$, extended geometry) [2].
Indeed while the physical mechanism allows for the destabilisation of
periodic structures (with a typical wave-length of the order of 2d) it
does not put restrictions on the actual position and orientation of
these structures.   This   task   is left to the lateral boundaries, the
effect of which will be weak and slow to manifest since the cells are
"wide".   This   leaves   room   for   position   ("phase") disorder with
"textures" and   localized defects such as dislocations or   grain
boundaries [3].   Naively one can understand weak turbulence in wide
systems as a permanent unsteady state of these structures which do not
succeed in finding a time independent arrangement of defects. Such a
situation apparently occurs close to the instability threshold in
Rayleigh-Bénard (RB) experiments using liquid   helium   in   wide
cylindrical cells [4].   In fact this runs up against the widespread
belief that, close to the threshold, there should be a parameter range

of steady convection and the evolution towards the steady state should
be relaxational. The first property is suggested by a thorough
stability analysis of infinite perfect rolls [5] while the second
derives from the variational structure of the amplitude equation at
lowest order and should remain valid for sufficiently slightly
textured structures [6].

In fact the variational structure of the amplitude equations is
lost even at lowest order as soon as one takes into account the
generation of vertical vorticity for "stress free" velocity boundary
conditions (BC) [7]. Of course free BC over-estimate the effects of
the corresponding induced "drift flow" on the roll structure but one
may expect to find their trace even for realistic "no slip" (or
"rigid") BC in spontaneously disordered structures with a sufficiently
large density of defects.

In order to gain insight into the nature of weak turbulence in
extended geometry, before building a theory, it may be interesting to
learn as much as possible from experiments and more especially from
numerical experiments when laboratory experiments do not allow easily
for a visualization of the structures. Unfortunately, 3-dimensional
(3-D) simulations are often performed in periodic boxes the width of
which is of the order of the critical wave-length λ_c or a few λ_c.
This situation is closer to confined geometry and actually leads to
results consistent with the picture appropriate for this case [8]. To
our knowledge 3-D simulations in realistic conditions of interest for
wide cells are still lacking. The reason seems to be that they
resolve the vertical dependence of the fluctuations too accurately and
that handling this somewhat irrelevant information impedes giving
sufficient attention to the horizontal dependence which is of greater
interest. This urges for a modelisation which eliminates the vertical
dependence and includes only the relevant horizontal variations [9,10]
($2 and appendix A). Simulations are then possible at sizes

sufficiently large ([11,12] $3 and appendix B) to get valuable infor-
mation on the nature of weak turbulence at large aspect ratios ($4).

2. A 2-D MODEL OF 3-D CONVECTION FOR STRESS FREE BC

Usually 3-D simulations of convection involve a decomposition of
the fluctuations on a complete basis of time-dependent independent 3-D
modes the amplitude of which remains time-dependent [8]. Then the
numerical solution consists in an initial value problem for a suitably
truncated set of such amplitudes. All this works much like a "black
box". In order to gain some insight into the relevant couplings we
use a similar idea of a projection but we do by hand and with some
(perhaps badly uncontrolled) approximations what is performed more
exactly but numerically by the computer. Here we restrict ourselves
to a projection on a basis of z-only dependent functions, the
amplitude of these vertical modes remaining functions of the
horizontal coordinates (x,y) and time (t) [9]. In order to be sure to
keep the most important features of convection close to the threshold
we combine this idea with that of an expansion in powers of the
convection intensity which is reminiscent of the standard amplitude
equation formalism [6]. This allows us to truncate the expansion at a
very low order, keeping only two relevant fields: a rapidly varying
variable W featuring the main convective field (vertical velocity or
temperature at mid-height in the cell) and a slowly varying
hydrodynamic field featuring the drift flow [7] given by its stream
function ψ.

The starting point is of course the complete, i.e. nonlinear, set
of Boussinesq equations which reads [5]:

$$\delta_t \Delta w + NLW = Pr(\Delta^2 w + \Delta_R \theta) \qquad (1a)$$

with

$$NLW = \Delta_R(\vec{V}.\vec{\nabla}w) - \delta_3[\delta_x(\vec{V}.\vec{\nabla}u) + \delta_y(\vec{V}.\vec{\nabla}v)] \qquad (1a')$$

and

$$\partial_t \theta + \vec{V}.\vec{\nabla}\theta = \Delta\theta + \text{Ra}.w \tag{1b}$$

In these equations $\vec{V}$ denotes the velocity fluctuation with components u,v,w and θ the temperature fluctuation. Δ is the 3-D Laplacian and Δ_R the horizontal (2-D) Laplacian. The Rayleigh number Ra is the usual control parameter and the Prandtl number Pr weights the respective role of the temperature and the velocity in the nonlinear dynamics.

Horizontal components u,v of the velocity can be split into an irrotational part deriving from a potential φ and a rotational part linked to the vertical vorticity $\mathcal{S}_z$ and given in terms of a stream function ψ, i.e. $u = \partial_x\varphi - \partial_y\psi$ and $v = \partial_y\varphi + \partial_x\varphi$. The potential derives from the continuity equation:

$$\Delta_R \varphi = - \partial_z w \tag{2a}$$

and one has $\mathcal{S}_z = \Delta_h\psi$ so that ψ is governed by the vertical vorticity equation:

$$(\partial_t - \text{Pr}\Delta)\Delta_h\psi = \partial_y(V.\nabla u) - \partial_x(V.\nabla v) \tag{2b}$$

Boundary conditions (BC) remain to be added to these equations. Here we shall assume good conducting plates at the top and the bottom of the cell. Accordingly we have $\theta = 0$ at the ordinate of the horizontal walls. For the velocity field, realistic conditions require $\vec{V}=0$ but this leads to rather cumbersome calculations. An approximate treatment using polynomials instead of the exact solutions is given in appendix A. In this section devoted to a general presentation of the modelisation procedure we shall use idealistic stress free BC which are easier to handle since exact solutions can be expressed in terms of trigonometric lines which form a closed set with respect to the full nonlinear evolution equations. Since a detailed derivation has been published elsewhere [10], here we shall only sketch the main steps and give the result.

With horizontal walls at z=0 and z=1 we can assume expansions of the form:

$$(w,\theta) = \sum_n (W_n, \Theta_n) \sin(n\pi z)$$

$$(\varphi,\psi) = \sum_n (\phi_n, \psi_n) \cos(n\pi z)$$

Then we insert these expansions in the full equations and separate the different harmonics generated by the nonlinear terms. We get an infinite set of coupled partial differential equations for the quantities W_n, Θ_n, ϕ_n and ψ_n. Remembering that we stay close to the threshold we assume that the z-dependence of the vertical velocity is basically that of the first harmonic for w, i.e. $\sin(\pi z)$ with slaved higher order harmonics. Thus we consider first the linearized problem for the first harmonics. We solve it and replace the solution in the nonlinear terms reaching the formally quadratic level. Combining first harmonics obviously generates second harmonics for which approximate expressions can be obtained without too stringent approximations. As shown by Siggia and Zippelius one should not forget that this combination also leads to z-independent terms in the vorticity equation and that this precisely bring in the "drift flow" at the origin of qualitatively new phenomena. Further insertion of the solutions at first and second order leads to the formally cubic level. Collecting all terms resonant with the fundamental mode $\sin(\pi z)$ and dropping all unnecessary indices gives the sought model:

$$\tau_o(\partial_t + \vec{V}_R \cdot \vec{\nabla}_R) W =$$

$$[\varepsilon - \frac{\xi_o^2}{4 q_c^2}(\Delta_R + q_c^2)^2]W - g[(\vec{\nabla}_R W)^2 + q_c^2 W^2]W \qquad (3a)$$

Parameters entering (3a) have the well known values: $q_c = \pi/\sqrt{2}$, $\varepsilon = (Ra-Ra_c)/Ra_c$ with $Ra_c = 27 \pi^4/4$ and

$$\tau_o = \frac{2}{3\pi^2}(1+Pr^{-1}), \qquad \xi_o^2 = \frac{8}{3\pi^2} \quad \text{and} \quad g = \frac{1}{6\pi^4}.$$

On the right hand side of (3a) the linear term generates periodic structures with a spatial period of the order of $\lambda_c = 2\pi/q_c$ and the nonlinear term directly leads to the Landau type of term $|A|^2 A$ in the amplitude equation since for nearly straight rolls, i.e. $W \propto$ $a.\sin(q_c x)$, we have: $(\vec{\nabla}_R W)^2 + q_c^2 W^2 = q_c^2 a^2 \propto |A|^2$. On the left hand side of (3a) $\vec{V_R} \cdot \vec{\nabla_R} W$ accounts for the advection of the structure by the drift flow given by the vorticity equation:

$$(\partial_t - Pr \Delta_R) \Delta_R \psi = \frac{1}{q_c^2} (\partial_x W \partial_y \Delta_R W - \partial_y W \partial_x \Delta_R W) \tag{3b}$$

with

$$\vec{V_R} = (U,V), \quad U = -\partial_y \psi \quad \text{and} \quad V = \partial_x \psi.$$

It is readily checked that the r.h.s. of (3b) vanishes exactly for strictly straight rolls and is of the order of $\varepsilon^{7/4}$ for slightly curved rolls. This leads to a drift flow of the order of ε so that in (3a) the advection term is of the same order as the Landau-like term.

Partial differential equations (3a,b) have to be supplemented by BC at the borderline of the horizontal domain of interest. Assuming that all velocities vanish at that place leads to

$$W = \vec{n} \cdot \vec{\nabla_R} W = 0 = \psi = \vec{n} \cdot \vec{\nabla_R} \psi \tag{3c}$$

where $\vec{n}$ is the unit vector normal to the boundary. At this point the model is ready for analytical or numerical study.

3. NUMERICAL SIMULATIONS ON THE STRESS FREE MODEL

One of the goals of the derivation of simplified models is to use them in numerical simulations on domains much larger than what can be studied with the complete 3-D equations [11]. Here we present some preliminary results obtained in a rectangular geometry with aspect ratios $\Gamma_x =11.5$ and $\Gamma_y =15.9$, i.e. in a domain able to contain 16 rolls parallel to the short side and 11 rolls parallel to the long one. The conditions of the numerical experiment and the algorithm used are given in appendix B. A rather systematic exploration of the parameter space (ε,Pr) has been performed using either random low level initial conditions or the result of a preceding simulation as starting point. A more thorough account of these experiments is given elsewhere [12].

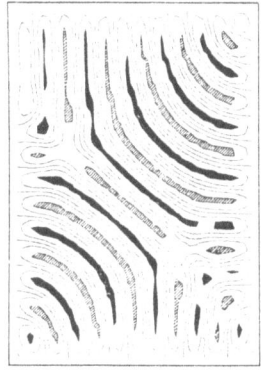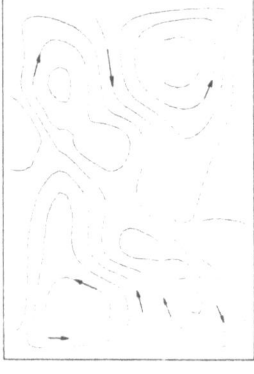

fig.1: Final steady state of a simulation at Pr=2 used as initial condition (t=0) for the simulation at Pr=1.6.

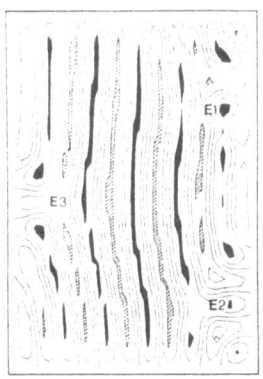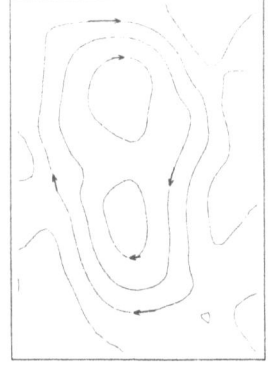

fig.2: Intermediate state at t=500.

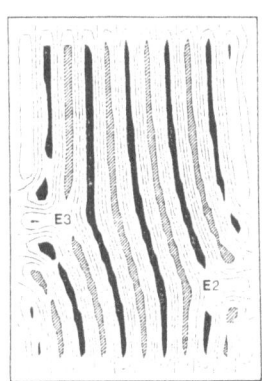

fig.3: Intermediate state at t=700.

A series of simulations at ε=0.5 and decreasing Pr is particularly interesting. Sarting with low level random initial conditions at Pr=5 we get a texture with two disclinations in diagonally opposite corners which remains steady down to Pr=2. Such textures are typical of large Pr. A turbulent transient takes place when Pr is further decreased to 1.6. Starting with the texture displayed in fig.1 (t=0) the system evolves chaotically towards the steady state of fig.4 (t=2100) passing through structures displayed in fig.2 (t=500) and in fig.3 (t=700) which is topologically analogous to that in fig.1 . In these figures the left drawing features the W-field. Regions in black correspond to hot uprising flow and hatched regions to cold sinking fluid. The flow lines of the drift flow are given in the right drawing. Arrows indicate the direction

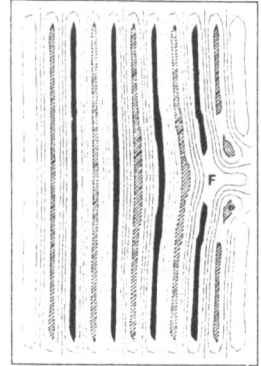 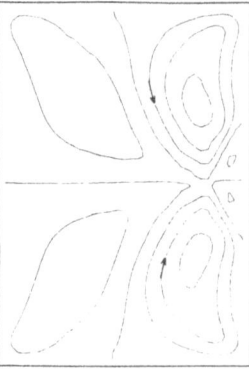

fig.4: Final steady state at t=2100.

of the flow. For a detailed analysis of this transient see ref.10 . A good idea of the evolution can be obtained from the consideration of fig.5 below which displays the variations of two integral quantities calcula-ted from the numerical solu-tion: the convective heat flux H=<w θ > which reduces itself here to <w^2> and the kinetic energy stored in the drift flow K=<$\vec{V}_R^2$/2> $\sim$ <($\vec{\nabla}_R\psi$)^2 > where <...> denotes the horizontal average $\int_S$(...)dxdy. Every accident on these curves can be put in relation with a local event in the cell (see below). Agitated periods with high K and low H alternate with calm periods with high H and comparatively low K. Remark that the final steady state corresponds to the highest flux H and the lowest kinetic energy K ever reached. Events denoted as D1,D2...D5 are dislocation nucleations each one corresponding to a marked increase of K.

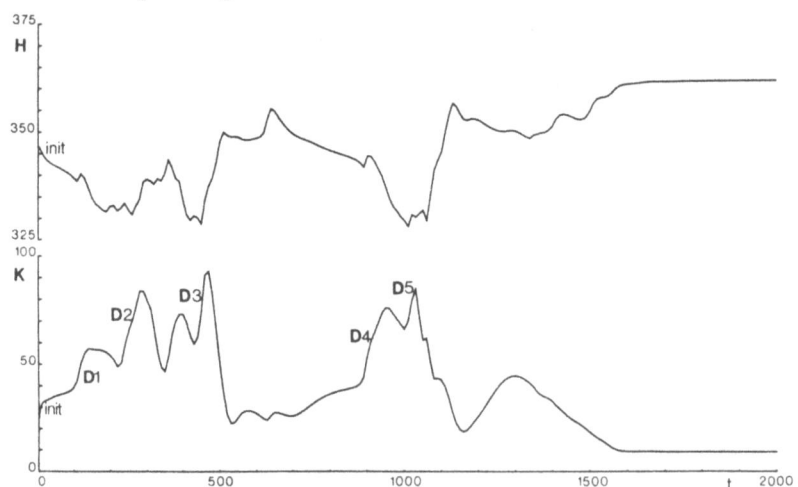

fig.5: Evolution of the convective heat flux H (top) and of the kinetic energy K contained in the large scale drift flow (bottom).

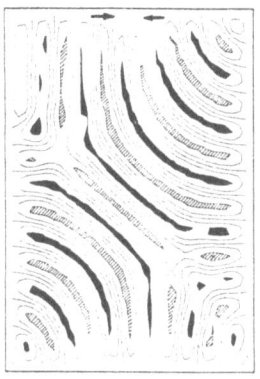

fig.6: The convergence of the drift flow induces a compression of the rolls.

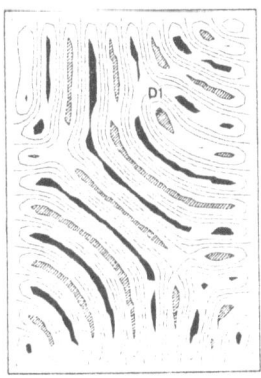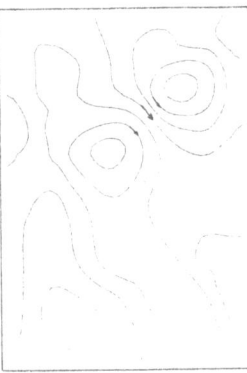

fig.7: The drift flow tends to push the dislocation further inside the structure.

The process is the following: -first a convergence of the drift flow induces a compression of the rolls as illustrated in fig.6 (note that previous simplified models without drift flow, i.e. at Pr=∞ , cannot show such a behaviour) -then when the wave-length has so decreased that the rolls are no longer stable close to the boundary a dislocation nucleates, which eliminates a pair of rolls -finally the flow concentrates itself and assumes a dipolar shape so as to reinforce the tendency for the dislocation to move inwards, to "climb", see fig.7.

Therefore it seems that the general evolution is controlled by the question of the compatibility between the flow configuration and the curvature of the texture. Indeed from the analysis of Siggia and Zippelius, one can say that the drift flow is in equilibrium with the curvature of the rolls when it is in a direction such that it tends to decrease the curvature. Now let a texture be given, for example at the moment when a parameter is modified. If the drift flow is compatible with the topological constraints associated with the structure, then the evolution is merely a relaxation towards a steady state. But this will not be the general case, i.e. the drift flow will not be in equilibrium with the curvature everywhere. Under the influence of the advection term in (3a) a slow evolution will take place inducing roll deformations which will end in defect nucleations. The subsequent and comparatively fast defect motion will involve strong localized curvature modifications which will "reset" the search for a better compatibility. This "frustration" of dynamical nature seems to come and add itself to the geometrical frustration at the heart of earlier approaches of textures [3]. The resulting basically non-relaxational behaviour could be of importance for the under-standing of weak turbulence at large aspect ratios.

Finally a consequence of the aspect of the flux variations with discontinuities or slope discontinuities at every event is that the power spectrum (taken by Fourier transformation on the interval $70<t<1350$) follows roughly a power law $|\hat{H}(\omega)|^2 \propto \omega^{-n}$ with $n \sim 3$ (recall that random discontinuities would give $n=2$ and slope discontinuities $n=4$). This value is not very significant by itself since, due to the relatively short duration of the transient, no averaging could be performed but the general trend may give a hint of how low frequency power spectra can introduce themselves in weak turbulence.

4. <u>CONCLUSION</u>

Models of convection such as the one of §2 give an interesting starting point for a theoretical analysis devoted to the elucidation of concepts which seem important for an understanding of weak turbulence at large aspect ratios. In particular one can study in nearly realistic conditions the problems of pattern selection and defect motion using the standard amplitude equation formalism. As a matter of fact one can easily recover from (3a,b) the modified amplitude equations derived by Siggia and Zippelius for nearly straight rolls in 3-D convection with stress free BC. However it should be kept in mind that, due to the approximations made in order to have models sufficiently simple to remain tractable, one should not expect more than semi-quantitative results at best.

Another and perhaps more promising use of these models is for numerical simulations in domains of lateral extent incomparably larger than those accessible to 3-D simulations. This study is at its very beginning. The above example of a turbulent transient observed at low Prandtl number (Pr=1.6) shows what one can obtain from such simulations. First it seems that they are able to reproduce the often quoted laboratory observation of alternating agitated and calm periods. A careful examination of the evolution allows to interpret this phenomenon in terms of "dynamical frustration" linked to the compatibility between the large scale drift flow which tends to advect the rolls and the overall curvature of the current structure which generates the vertical vorticity. Calm periods then correspond to the slow distortion of the structure by the flow and agitated periods to the nucleation and motion of dislocations responding to distortions getting too large. Whether this feed-back mechanism is responsible for the occurrence of weak turbulence at large aspect ratios is not yet known but this is a very tempting conjecture in view of our preliminary results.

APPENDIX A: THE CASE OF NO-SLIP BOUNDARIES

Here we develop an extension of the stress free model ([10] and $2) to the realistic "no slip" case. The derivation rests on a Galerkin expansion which generalizes the separation of harmonics used previously.

The basis functions are z-polynomials [13] chosen so as to fulfil boundary conditions and to reflect parity properties of the fluctuations. Thus the origin for z is taken at mid-height in the cell (horizontal plates at $z=\pm 1/2$). Next, for the vertical velocity w no slip BC read: $w(\pm 1/2)=\partial_{3}w(\pm 1/2)=0$ so that the polynomials must be taken of the form $(1/4-z^{2})^{2}$ $P_{n}(z)$ where $P_{n}(z)=C_{0}+C_{1}z+...+C_{n}z$. In the same way BC for θ: $\theta(\pm 1/2)=0$ are automatically fulfilled by $(1/4-z^{2})Q_{n}(z)$. From the continuity equation (2a) it is obvious that the basis functions for the potential φ are the z-derivatives of those for w. They average to zero over the thickness. On the other hand the rotational part given by ψ is not submitted to this constraint. Thus we must take $(1/4-z^{2})R_{n}(z)$ and it is clear that the lowest order term is just the Poiseuille flow thought to represent most of the drift [7].

The first step is the complete solution of the linearized problem and more especially the threshold determination. Assuming $\partial_{t}=\partial_{y}=0$ and $\partial_{x}=iq$ we solve the resulting differential system in z according to the Galerkin method prescriptions, i.e. insert the assumed expansions, multiply the first equation by $(1/4-z^{2})^{2}z^{n}$, the second by $(1/4-z^{2})z^{n}$ and integrate over z from $-1/2$ to $1/2$. Note that useful integrals are of the form $I_{n} = \int_{-1/2}^{1/2} (1/4-z^{2})^{n} dz$ and $J_{n} = \int_{-1/2}^{1/2} (1/4-z^{2})^{n} z^{2} dz$ and that they can be obtained from the recurrence relations:

$$I_{o} = 1, \quad I_{n} = \frac{n}{2(2n+1)} I_{n-1} \quad \text{and} \quad J_{n} = \frac{1}{2(n+1)} I_{n+1}.$$

Due to parity properties we get two infinite sets of linear algebraic equations one for even polynomials and the other for odd polynomials, the coefficients of which are functions of q and Ra. For the sake of simplicity we restrict ourselves to the lowest order of the even system which gives an excellent approximation of the threshold value (this is due to the fact that the trial functions truncated at lowest order are very close to the exact solution in the sense of the variational formulation underlying the linearized problem). When truncated in that way, the marginal stability problem reads:

$$Pr \left[(q^{4}+24q^{2}+504)W_{o} - (9/2)q^{2}\Theta_{o} \right] = 0$$

$$(3/14)Ra.W_{o} - (q^{2}+10)\Theta_{o} = 0$$

The compatiblity condition gives the approximate form of the marginal stability curve:

$$\mathrm{Ra} = f(q) = \frac{28}{27} \frac{(q^4 + 24 q^2 + 504)(q^2 + 10)}{q^2}$$

which in turn gives the approximate critical conditions: $q = q_c = 3.1165$ ($\simeq$ exact) and $\mathrm{Ra} = \mathrm{Ra}_c = 1750$ (only 2.5 per cent larger than the exact value 1708). The curvature of the marginal stability curve at threshold is linked to the coherence length ξ_o given by $\xi_o^2 = f''(q_c)/2\,\mathrm{Ra}_c$. Here we find $\xi_o^2 = 0.1497$ in good agreement with the exact value 0.148.

Now, returning to the complete time dependent linearized problem which reads:

$$(\Delta_R - 12)\,\partial_t W_o = \mathrm{Pr}[(\Delta_R^2 - 24 \Delta_R + 504)W_o + (9/2)\Delta_R\Theta_o] \qquad \text{(A1a)}$$

$$\partial_t \Theta_o = (\Delta_R - 10)\Theta_o + (3/14)\mathrm{Ra}.W_o \qquad \text{(A1b)}$$

we eliminate Θ_o, replace Δ_R by $\mathcal{L} - q_c^2$ and Ra by $\mathrm{Ra}_c(1+\varepsilon)$, assume $\partial_t \sim \mathcal{O}(\varepsilon)$ and $\mathcal{L} \sim \mathcal{O}(\varepsilon^{1/2})$, expand in powers of $\mathcal{L}$ and, making use of the relations between coefficients at threshold, get the same linear part as in (3a) at lowest non-trivial order but with $1/\tau_o = 38.2927\ \mathrm{Pr}/(1+1.9425\ \mathrm{Pr})$ again in excellent agreement with the exact value $38.4429\ \mathrm{Pr}/(1+1.9544\ \mathrm{Pr})$ [6].

Other components of the linearized solution can be obtained from W_o. At lowest order we assume $\Delta_R = -q_c^2$, $\partial_t = 0$ and $\mathrm{Ra} = \mathrm{Ra}_c$ in (A1b) which gives $\Theta_o = T_o W_o$ with $T_o = \frac{3}{14}\frac{\mathrm{Ra}_c}{10+q_c^2} = 19.023$. In the same way from the continuity equation (2a) we obtain $\phi_o = \frac{1}{q_c^2}W_o$ and thus $\vec{V}_{Ro} = \frac{1}{q_c^2}\vec{\nabla}_R W_o$, the corresponding vertical dependence being: $\partial_z^2(1/4-z^2)^2 = -4z(1/4-z^2)$.

Inserting the linear solution in the nonlinear terms leads to the formally quadratic level which is solved by projection of (1a) onto $z(1/4-z^2)^2$ and (1b) onto $z(1/4-z^2)$. Assuming that this order is slaved to the previous one (i.e. $\partial_t \equiv 0$) we get

$$[J_4 \Delta_R^2 - (40J_3 - 4J_2)\Delta_R + 120J_2]W_1 + J_3 \Delta_R\Theta_1 = \frac{1}{\mathrm{Pr}}\Delta_R[c_1\mathcal{S} + c_2\mathcal{R}] \qquad \text{(A2a)}$$

$$(J_2 \Delta_R - 6J_1)\Theta_1 + \mathrm{Ra}_c J_3 W_1 = c_3(3\mathcal{S} + \mathcal{R}) \qquad \text{(A2b)}$$

where we have defined:

$$\mathcal{S} = (\vec{\nabla}_h W_o)^2 + q_c^2 W_o \qquad \text{and} \qquad \mathcal{R} = (\vec{\nabla}_h W_o)^2 - q_c^2 W_o$$

and where the J's are the numerical constants defined earlier while the c's can be expressed in terms of the J's. As already stated for nearly straight rolls $\mathcal{S}$ behaves as the square of the modulus of the complex amplitude and is expected to be slowly varying. On the other hand $\mathcal{R}$ is roughly proportional to $\sin(2q_c x)$ and thus rapidly varying, hence the choice of notations. This decomposition will be useful for the simplification of the nonlinear term at next order. In any case these quantities can be understood as independent sources for W_1 and Θ_1. As for stress free BC [10] an approximate solution to eqs.A2 can be obtained by assuming that all terms involving Δ_h on the l.h.s. of (A2) are negligible with respect to those arising from pure z-derivations. This comes from the fact that forcing terms on the r.h.s involve wave-vectors the length of which lies in the range $(0,2q_c)$ so that, upon averaging over the thickness, Δ_h is much smaller than z-derivatives. This leads to:

$$W_1 = \Delta_h(\sigma \mathcal{S} + \rho \mathcal{R})$$

with $\sigma = d_1(1+d_1'/Pr)$ and $\rho = d_2(1+d_2'/Pr)$ where constants d_i, d_i' are combinations of the c's and the J's introduced above. From now on and for the sake of simplicity we shall only give the generic form of the expressions and forget the precise values of the coefficients entering the formulas. They can be obtained by staightforward but tedious computations and would be of interest for a comparison with exact values known for some special cases only.

Θ_1 can be deduced from (A2b) by assuming simply $\Delta_h=0$. Generically one gets:

$$\Theta_1 = k \Delta_h(\sigma \mathcal{S} + \rho \mathcal{R}) + k'(3\mathcal{S} + \mathcal{R}).$$

The irrotational component of the horizontal flow is easily obtained from the continuity equation (2a) since Δ_h is present both in the expression of W_1 and on the l.h.s. of this equation so that it can be "simplified" to give:

$$\phi_1 = -(\sigma \mathcal{S} + \rho \mathcal{R}) \quad \text{and} \quad V_{h1}^{irr} = \vec{\nabla}_h \phi_1 .$$

The vertical dependence of this contribution is $\partial_z(z(1/4-z^2)^2)$.

After projection onto the lowest order function $(1/4-z^2)$, i.e. the Poiseuille-like flow, the vorticity equation reads:

$$[\partial_t + Pr(10- \Delta_h)] \Delta_h \psi_1 =$$
$$\frac{2}{21\, q_c^2}[\partial_{xy}((\partial_x W_0)^2 - (\partial_y W_0)^2) - (\partial_{xx}^2 - \partial_{yy}^2) \partial_x W_0\, \partial_y W_0] \qquad (A2c)$$

from which we can deduce the drift flow $\vec{V}_{h1}^{rot} = \vec{V}_D$. Note that the quantity between brackets at the r.h.s. of (A2c) is equivalent to that at the r.h.s. of (3b) and that the simple averaging proposed by Siggia and Zippelius [7] amounts to the replacement of $(10- \Delta_h)$ by $(12- \Delta_h)$ and of 2/21 by 4/35, i.e. a minor quantitative modification only.

At this level the solution is complete and we can turn to the formally cubic terms arising from NLW and $\vec{V}.\vec{\nabla}\theta$ in eqs.1a,b. A projection onto the lowest order basis functions for w and ϕ gives:

$$NLW = \Delta_h[\vec{\nabla}_h W_0 (k_1 \Delta_h + k_2)\vec{\nabla}_h \phi_1 + k_3 W_0\, \Delta_h \phi_1] + \vec{\nabla}_h W_0 (k_1' \Delta_h + k_2') \vec{\nabla}_h \phi_1$$
$$+ k_4(\Delta_h + k_5)W_0\, \Delta_h \phi_1 + k_6 \Delta_h(\vec{V}_D . \vec{\nabla}_h W_0) + k_7(\partial_x W_0\, \partial_y \Delta_h \psi_1 - \partial_y W_0\, \partial_x \Delta_h \psi_1)$$

$$\vec{V}.\vec{\nabla}\theta = m_1 W_0 (3\mathcal{S} + \mathcal{R}) + m_2 W_0\, \Delta_h \phi_1 + m_3 \vec{\nabla}_h W_0 \vec{\nabla}_h (\sigma'\mathcal{S} + \rho'\mathcal{R})$$
$$+ m_4 \vec{\nabla}_h W_0 \vec{\nabla}_h\, \Delta_h(\sigma''\mathcal{S} + \rho''\mathcal{R}) + m_5 \vec{V}_D . \vec{\nabla}_h W_0$$

where the k's and m's are Pr-independent constants and the primed σ's and ρ's assumes the same Pr dependence as σ and ρ. The contribution of the drift flow $\vec{V}_D$ has been made explicit. These nonlinear terms remain to be added on the r.h.s. of eqs.A1a,b. The final form of the model equation governing the main convective variable W_0 is then obtained upon elimination of $\textcircled{\scriptsize o}_0$ as before. The cubic interaction term is now quite involved. In order to get some simplification we assume that $\Delta_h \sim -q_c^2$ when acting on W_0, ~ 0 when acting on $\mathcal{S}$ or $\vec{V}_D$ and $\sim -4q_c^2$ when acting on $\mathcal{R}$. All this would be exact for strictly parallel rolls at q_c but is very approximate here. This leads to:

$$\tau_0\, \partial_t W_0 + \tau_0'\, \vec{V}_D . \vec{\nabla}_h W = [\epsilon - \frac{\xi_0^2}{4q_c^2}(\Delta_h + q_c^2)^2]W_0 - g(Pr)[(\vec{\nabla}_h W_0)^2 + q_c^2 W_0^2]W_0 \quad (A3)$$

Generically τ_o' turns out to be slightly different from τ_o and $g(Pr)$ assumes the form $\alpha + \beta/Pr + \gamma/Pr$, which appears different from the stress free case presented in §2.

Remark: From the expression of the linear operator in (A3) it is easily checked that the most unstable wave-vector $q(m.u.)$ is independent of ε and equal to q_c. This result no longer holds for actual convection yielding $q(m.u.)-q_c \propto q_c \varepsilon$ [14]. This comes from the fact that for $q-q_c \sim \varepsilon^{1/2}$ and $\partial_t \sim \varepsilon$ all linear terms are of the order of ε and that for $q-q_c \sim \varepsilon$ higher order terms must be included. After adding the first nontrivial correction, the modified linear operator reads:

$$(1-Q(Pr)\mathcal{L})\tau_o \partial_t W_o = [\varepsilon(1-\mathcal{L}/q_c^2) - \frac{\xi_o^2}{4q_c^2}\mathcal{L}^2]W$$

with $Q(Pr)=(1+1.0483Pr)/10.332(1+1.9425Pr)$ and $\mathcal{L} = \Delta_h+q_c^2$ as defined earlier. This leads to $q(m.u.)-q_c=c(Pr)$ in good semiquantitative agreement with exact numerical results. In the same way, the modified linear operator of the stress free model reads:

$$(1-\frac{4}{3\pi^2}\mathcal{L})\tau_o \partial_t W_o = [\varepsilon(1-\mathcal{L}/q_c^2) - \frac{\xi_o^2}{4q_c^2}\mathcal{L}^2]W$$

which reproduces the exact behaviour $q(m.u.)$ close to the threshold: $q(m.u.)=q_c(1+\varepsilon/4)$. Note that using these modifications improves the approximation to the marginal stability curve at small q (divergence as q^{-2}) but makes it worse at large q (parabolic branch in q^2 instead of q^4).

Let us conclude this appendix in noting that the generic form of the quadratic nonlinearities is so complicated that drastic simplifications have to be performed at the final step in order to keep a tractable model. Thus using exact eigen-modes of the linearized Boussinesq equations, though leading to better values at lowest order, has no particular advantages as far as the reliability of the model is concerned for textures with a large density of defects.

APPENDIX B: THE NUMERICAL SCHEME

To obtain results presented in §3 partial differential equations 3a,b have been integrated using space and time finite differences.   In order to avoid stringent numerical stability conditions [15] the linear operators (which involve the highest space derivatives) have been treated implicitly while the nonlinear terms were kept explicit. Denoting the evolution problem as $\dot{X} = F_\ell(X) + F_{n\ell}(X)$ we have solved $(X^{k+1} - X^k)/\delta t = F_\ell(X^{k+1}) + F_{n\ell}(X^k)$ which is $\mathcal{O}(\delta t)$.   A second order scheme would be [15]:

$$(X^{k+1} - X^k)/\delta t = (F_\ell(X^{k+1}) + F_\ell(X^k))/2 + (3F_{n\ell}(X^k) - F_{n\ell}(X^{k-1}))/2.$$

In any case, after space discretization the evolution problem is reduced to a linear algebraic system of equations of the form:

$$aX_{ij}^{k+1} + b_1(X_{i\,j+1}^{k+1} + X_{i\,j-1}^{k+1} + X_{i+1\,j}^{k+1} + X_{i-1\,j}^{k+1}) + b_2(X_{i+1\,j+1}^{k+1} + X_{i-1\,j+1}^{k+1} + X_{i+1\,j-1}^{k+1} + X_{i-1\,j+1}^{k+1})$$

$$+ c(X_{i+2\,j} + X_{i-2\,j} + X_{i\,j+2} + X_{i\,j-2}) = Y_{ij}(X^k) \tag{B1}$$

The simulation is performed on a rectangle $(L_x = (N+1)h) \times (L_y = (M+1)h)$ with interior points at (i,j), $i=1,N$, $j=1,M$.   Boundary conditions (3c) are taken into account by adding fictitious rows and columns (for $i=-1,0,N+1,N+2$ and $j=-1,0,M+1,M+2$)   and imposing the conditions $X_{0j} = X_{N+1\,j} = X_{i\,0} = X_{i\,M+1} = 0$ and $X_{-1\,j} = X_{1\,j}$, $X_{N+2\,j} = X_{N\,j}$, $X_{i\,-1} = X_{i\,1}$, $X_{i\,M+2} = X_{i\,M}$.

The NM-system is in fact a bloc pentadiagonal system $P\,X = Y$ with P an M×M-bloc matrix of the form

$$P = \begin{bmatrix}
A' & B & C & 0 & \cdots & & & \\
B & A & B & C & 0 & \cdots & & \\
C & B & A & B & C & 0 & \cdots & \\
 & & \cdots & & & \cdots & & \\
 & \cdots & 0 & C & B & A & B & C \\
 & \cdots & & 0 & C & B & A & B \\
 & \cdots & & & 0 & C & B & A'
\end{bmatrix}$$

where each element is a NxN matrix.

From (B1) we have: $C = c\,U$ where U is the NxN-unit matrix while A and B matrices are penta- and tri-diagonal matrices of the form:

$$A = \begin{bmatrix}
a+c & b & c & 0 & \cdots & & \\
b & a & b & c & 0 & \cdots & \\
c & b & a & b & c & 0 & \cdots \\
 & & \cdots & & \cdots & & \\
 & & \cdots & 0 & c & b & a & b \\
 & & \cdots & & 0 & c & b & a+c
\end{bmatrix}
\qquad
B = \begin{bmatrix}
b & b & 0 & \cdots & & \\
b & b & b & 0 & \cdots & \\
0 & b & b & b & 0 & \cdots \\
 & & \cdots & & & \\
 & & \cdots & 0 & b & b & b \\
 & & \cdots & & 0 & b & b
\end{bmatrix}$$

The term c which appears in the first and last diagonal element in A comes from the boundary conditions. For the same reason the first and last diagonal elements of the first and last diagonal matrices A' in P read a+2c.

The resolution of the pentadiagonal block system is easily performed using standard L-R decomposition which for notational convenience is sketched here for ordinary matrices and not for block-matrices [16]. Let $PX = L(RX) = Y$ i.e. $LZ = Y$ and $RX = Z$, where L is a lower tridiagonal matrix and R an upper tridiagonal matrix with 1 on the principal diagonal, then the Z's are given by the forward recurrence:

$$L_{11}Z_1 = Y_1; \quad L_{22}Z_2 + L_{21}Z_1 = Y_2; \quad L_{ii}Z_i + L_{ii-1}Z_{i-1} + L_{ii-2}Z_{i-2} = Y_i \text{ or}$$

$$Z_1 = L_{11}^{-1}Y_1; \quad Z_2 = L_{22}^{-1}(Y_2 - L_{21}Z_1); \quad Z_i = L_{ii}^{-1}(Y_i - L_{ii-1}Z_{i-1} - L_{ii-2}Z_{i-2}).$$

and the X's by the backward recurrence:

$$X_M = Z_M; \quad X_{M-1} = Z_{M-1} - R_{M-1\,M}X_M; \quad X_i = Z_i - R_{ii+1}X_{i+1} - R_{ii+2}X_{i+2}.$$

Now the matrix elements of L and R are easily obtained from the recurrences:

$$P_{ii-2} = L_{ii-2}; \quad P_{ii-1} = L_{ii-2}R_{i-2\,i-1} + L_{ii-1}; \quad P_{ii} = L_{ii-2}R_{i-2\,i} + L_{ii-1}R_{i-1\,i} + L_{ii},$$

$$P_{ii+1} = L_{ii-1}R_{i-1\,i+1} + L_{ii}R_{ii+1}, \quad P_{ii+2} = L_{ii}R_{ii+2},$$

with for i=1: $L_{-11} = L_{01} = 0$, i=2: $L_{02} = 0$, i=M-1: $R_{M-1\,M+1} = 0$ and i=M: $R_{M\,M+1} = R_{M\,M+2} = 0$. Dealing with block elimination all these expressions are understood as matricial equations. The main advantage of this algorithm is that it solves the NM-system directly and not iteratively and that it is easy to implement. Its drawback is that it requires a lot of memory to save the matrices L and R which are calculated once for all at the beginning of the simulation. However for moderate aspect ratios as those considered in $3 it has proven to be quite efficient due to a full vectorization of matrix products.

Calculations have been performed in part thanks to a CPU time allocation on the CRAY-1S of the GCCVR.

REFERENCES

[1] D.Ruelle and F.Takens: Comm.math.Phys. 20(1971)167,
    see also: a) J.P.Eckmann: Rev.Mod.Phys. 53(1981)643,
    b) Workshop Common Trends in Particule and Condensed Matter
    Physics Les Houches (March 1983) Physics Reports, to appear.

[2] P.Manneville: "the transition to turbulence in "wide systems"",
    in ref.1b above.

[3] M.C.Cross: Phys.Rev. A25(1982)1065.
    E.D.Siggia and A.Zippelius: Phys.Rev.A24(1981)1036.
    Y.Pomeau, S.Zaleski and P.Manneville: Phys.Rev.A27(1983)2710.
    P.Manneville and Y.Pomeau: Phil.Mag.A48(1983)607.

[4] G.Ahlers and R.P.Behringer: Phys.Rev.Lett. 40(1978)712 and
    J.Maurer and A.Libchaber: J.Physique Lettres41(1980)L-515.

[5] a) C.Normand, Y.Pomeau, M.G.Velarde, Rev.Mod.Phys. 49(1977)581,
    b) F.H.Busse: Rep.Prog.Phys.41(1973)1929.

[6] a1) A.C.Newell and J.Whitehead: J.Fluid Mech. 38(1979)279,
    L.A.Segel: J.Fluid Mech. 38(1969)203,
    a2) J.Wesfreid,Y.Pomeau,M.Dubois,C.Normand and P.Berge:
    J.Physique 39(1978)725.
    b) M.C.Cross and A.C.Newell: "Convection patterns in large aspect
    ratio systems" Physica D (1983)

[7] E.Siggia and A.Zippelius: Phys.Rev.Lett. 47(1981)835 and
    Phys.Fluids 26(1983)2905.

[8] see for example: J.B.McLaughlin and S.A.Orszag: J.Fluid Mech.
    122(1982)123.

[9] J.Swift and P.Hohenberg: Phys.Rev. A15(1977)319 (Appendix A) and
    C.Normand: Z.Angew.Math.Phys. 32(1981)81.

[10] P.Manneville: J.Physique 44(1983)759.

[11] a) H.S.Greenside,W.M.Coughran,Jr and N.L.Schryer:
    Phys.Rev.Lett. 49(1982)726, H.S.Greenside and W.M.Coughran,Jr:
    "Non-linear Pattern Formation near the Threshold of
    Rayleigh-Benard Convection" Phys.Rev.A(1933).
    b) P.Manneville: J.Physique 44(1983)563.

[12] P.Manneville: J.Physique Lettres 44(1983)L-903.

[13] See for example: P.Glansdorff and I.Prigogine: Structure,
    Stabilite et Fluctuations Chap.11, Sect.10 (Masson Ed.,
    Paris,1971).

[14] P.Manneville: Phys.Lett. 95A(1983)463.

[15] R.D.Richtmyer and K.W.Morton: Difference Methods for Initial
    Value Problems (Wiley,New-York,1967).

[16] F.S.Acton: Numerical Methods that Work (Harper and Row,
    New-York,1970).

PATTERN EVOLUTION FROM CONVECTIVE AND ELECTROHYDRODYNAMIC INSTABILITIES

J.P. Gollub

Haverford College (and the University of Pennsylvania)

Haverford, PA 19041 U.S.A.

In this lecture, I presented the results of two experimental studies directed toward the following question: Of the many possible structures of a system of hydrodynamic rolls, which ones are preferred, and how are they chosen? Since this work has been published elsewhere, only a brief summary appears here.

Evolution of Convective Structures in Fourier Space and in Color [1-3]. In order to study convective structures in a Rayleigh-Benard system (rectangular and with large aspect ratio), we measure the velocity field in two dimensions by means of computer-automated laser Doppler scanning. The velocity measurements at thousands of locations are then encoded on a graphics device, with rolls of opposite vorticity displayed in different colors. However, for the purpose of this note, we use a grey scale centered at zero velocity, which causes rolls circulating in opposite senses to appear as light and dark regions, respectively.

For a given Rayleigh number not too far above the critical value, there are many stable or metastable flows, depending on initial conditions. Some of these are shown in Fig. 1. We find that rolls usually align perpendicular to the lateral cell boundaries. This causes the patterns to be curved, and to contain defects. Stable defects can even occur in the center of the cell (Fig. 1(c)).

Not too far above the onset of convection, a change in R induces patterns with many defects which evolve toward simple patterns by expelling or eliminating defects and reducing the curvature of the rolls. This is a noisy process, in which the local velocity evolves erratically. The bulk of the evolution generally occurs in a time comparable to the horizontal thermal diffusion time. However, some evolution continues for a considerably longer time. Sometimes a steady state is not reached in 400 hours (9,000 vertical thermal diffusion times, or ten horizontal thermal diffusion times). The observation of long transients involving pattern evolution in a domain where the infinite layer theory predicts stable rolls is a significant challenge to theoretical understanding. Similar long time-scales have been noted in numerical simulations of a two-dimensional model equation by Greenside, Coughran, and Schryer [4].

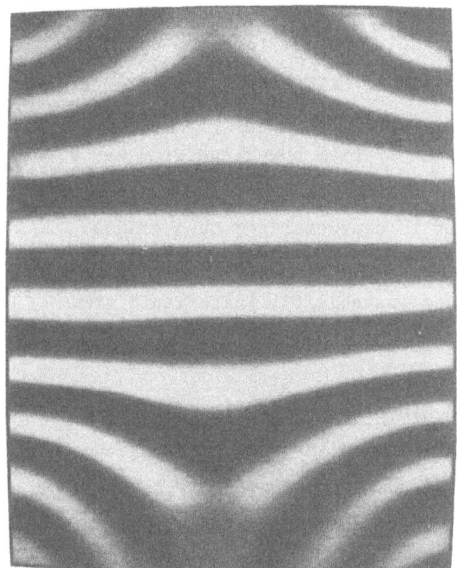

Fig. 1. Doppler maps showing several stable patterns long after the Rayleigh number was established. Each map is based on about 3000 measurements of the local velocity component parallel to the long cell axis, in a horizontal plane above the center of the fluid layer. Positive and negative values have been encoded as light and dark regions using a graphics device, and correspond to rolls with opposite sense of circulation. (a) $4R_c$; (b) $4R_c$; (c) $2R_c$.

Spatial power spectra have been computed from the Doppler data in order to follow the process of pattern evolution in Fourier space. We find that the dominant wavenumber grows during the evolution of the flow as defects are eliminated, with most of the change taking place during the first horizontal thermal diffusion time. Even in the steady state, the peak in the spatial power spectrum has a finite width that reflects the textured nature of the pattern. The dependence of the dominant wavenumber on the Rayleigh number is characterized by a gradual diminution (as Pocheau and Croquette [5] have also found) followed by a rapid decline at the threshold of the skewed varicose instability, where rapid time-dependence begins.

Commensurate and Incommensurate Structures in Electrohydrodynamic Flows [6].
Commensurate and incommensurate phases are often found in systems having two competing length scales. We have discovered several new ordered structures in a system of rolls in the presence of a spatially periodic perturbation.

The system consists of a layer of nematic liquid crystal between two transparent electrodes. An electric voltage produces electrohydrodynamic rolls whose width $l_o/2$ (about 100 microns) is comparable to the thickness of the layer. These "Williams domains" are well known. The rolls are aligned in a particular horizontal direction by a treatment of the electrode surface.

Now we add a spatially periodic voltage whose period l_1 is comparable to (but different from) l_o. This is accomplished by means of a photolithographically produced interdigitated electrode. The wavevector of this periodic perturbation is parallel to the wavevector of the rolls. We then examine the resulting structures as a function of the ratio l_1/l_o and the amplitude of the perturbation, using optical microscopy in conjunction with a digital imaging system. We also calculate two-dimensional Fourier spectra of the digitized intensity patterns. These spectra resemble the x-ray diffraction patterns of a crystal, and give us the reciprocal lattice of the structure.

When the roll size $l_o/2$ is within about 5% of $l_1/3$, we find a one-dimensional phase-locked commensurate state in which there are three rolls per period of the perturbation (Fig. 2(a)). On the other hand, when the ratio l_1/l_o is far from this locking condition, we find two dimensional patterns (Fig. 2(b)) in which the rolls are rotated, and modulated by a regular lattice of kinks. The structure in Fourier space is then a two-dimensional reciprocal lattice spanned by two basis vectors.

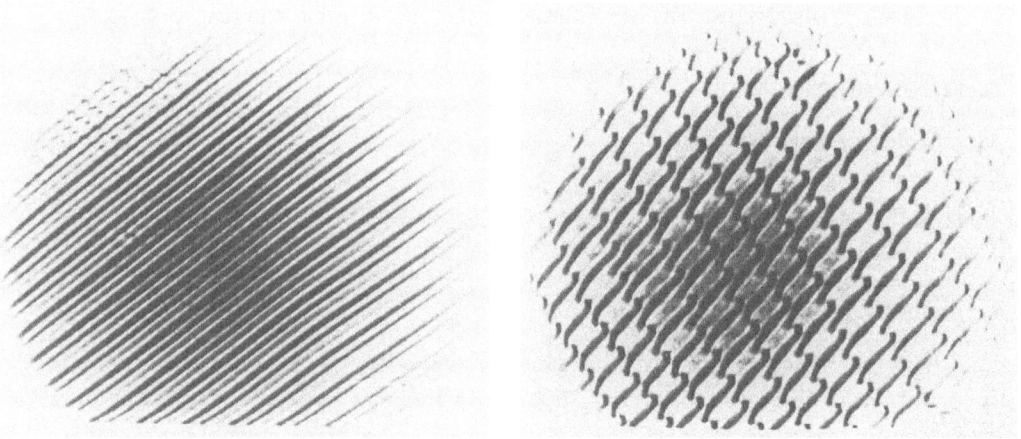

Fig. 2. Patterns of electrohydrodynamic domains resulting from competing
periodicities. (a) One-dimensional commensurate phase-locked pattern in which
there are exactly three rolls per period of the perturbation; (b) Two-dimensional
incommensurate pattern whose reciprocal lattice is spanned by two basis vectors.

These structures, which are perhaps the first commensurate and incommensurate
phases to be found in a system far from equilibrium), may be explained by a theory
due to Lubensky in which a Lyapunov function of an order parameter evolves toward
a local minimum. A more extensive discussion may be found in Ref. 6.

REFERENCES

1. J.P. Gollub, A.R. McCarriar, and J.F. Steinman, J. Fluid Mech. 125 (1982)
259-281.

2. J.P. Gollub and A.R. McCarriar, Phys. Rev. A 26 (1982) 3470-3476.

3. J.P. Gollub and M.S. Heutmaker, in Proceedings of the IUTAM Symposium on
Turbulence and Chaotic Phenomena in Fluids, Kyoto, 1983, ed. by T. Tatsumi, to
appear.

4. H.S. Greenside, W.M. Coughran Jr., and N.L. Schryer, Phys. Rev. Lett. 49
(1982) 726-729 and preprint.

5. A. Pocheau and V. Croquette, J. de Physique (in press).

6. M. Lowe, J.P. Gollub, and T.C. Lubensky, Phys. Rev. Lett. 51 (1983) 786-789.

AMPLITUDE EQUATIONS FOR NON LINEAR CONVECTION IN HIGH VERTICAL CHANNELS

Christiane NORMAND

Service de Physique Théorique
C.E.N. Saclay, CEA
91191 Gif-sur-Yvette.

1. INTRODUCTION

Elementary theories of convective systems deal with the assumption of an infinite extent in at least one direction. The models the most generally encountered are then : fluid layers of infinite horizontal extent or vertical channels of infinite height. In both cases the temperature gradient will be parallel to the gravity. These two models are the limit of what we call respectively horizontal or vertical systems, to characterize finite systems which have a dimension (horizontal or vertical) much larger than the other ones. The fundamental difference between the two systems comes out when investigating the non linear regime. In horizontal systems the non linear effects can be studied independently of the size effects and many important results valid in the weakly non linear regime have been derived in the case of an infinite horizontal layer. For instance the dependence of the amplitude of the instability on the temperature difference has been obtained first for an infinite layer[1].

The vertical systems possess the property that the non linear terms disappear from the governing equations when assumption is made of an infinite height. The linearization of the equations which results of this assumption is sometimes known as the Ostroumov limit[2] and expresses the fact that the relevant physical quantities, the convective temperature θ and the vertical convective velocity v_z, are both independent of the vertical coordinate. A way to restore the non linearities into the equations consists in taking into account the finite size effect due to the presence of horizontal boundaries limiting the fluid at the top and bottom of the vertical cell. Several aspects of this problem have been described by means of approximate methods like Galerkin expansion, in cylindrical vessels[3] or in Hele Shaw cells[4]. In this paper we carry out an entirely analytical procedure bearing some analogy with the techniques employed by Segel[5] and Newell and Whitehead[6] and also Chapman and Proctor[7]. The disparity between horizontal and vertical scales is exploited to develop an expansion scheme of the equations in powers of the (small) inverse aspect ratio of the cell, λ^{-1}($\lambda$is the ratio of the height to the characteristic horizontal length). The first non trivial result we get is a fully non linear differential equation for the vertical amplitude of the instability. The derivation has been undertaken for two systems : the plane vertical layer (2) and the vertical cylinder (3).

2. THE PLANE VERTICAL LAYER

We consider a vertical gap of width 2R confined between the planes $z = \pm \frac{h}{2}$ and infinite along the horizontal direction y ; (x,y,z) are Cartesian coordinates. The fluid has velocity $\vec{v}$ and pressure p. Its kinematic viscosity is ν and thermal diffusivity κ. The temperatures at the top and bottom boundaries are held constant and equal to T_1, T_2 with $T_1 < T_2$. The temperature of the fluid is $T = T_0 - \beta z + \theta$ where T_0 is the temperature at mid-height and $\beta = (T_2 - T_1)/h$ is the temperature gradient. If α is the coefficient of expansion, then the non-dimensional equations of motion and heat conduction are :

$$Pr^{-1}\left[\frac{\partial \vec{v}}{\partial t} + (\vec{v}.\vec{\nabla})\vec{v}\right] + \vec{\nabla}p = \Delta\vec{v} + \theta\vec{e}, \qquad (2.1a)$$

$$\frac{\partial\theta}{\partial t} + (\vec{v}.\vec{\nabla})\theta = \Delta\theta + Ra\; v_z \qquad , \qquad (2.1b)$$

$$\vec{\nabla}.\vec{v} = 0 \qquad , \qquad (2.1c)$$

where $|\vec{v}|$ is scaled with κ/R, p with $\rho_0 \nu\kappa/R^2$, time t with R^2/κ, lengths with R and θ with $\beta R\, Ra^{-1}$. The unit vector $\vec{e}$ is (0,0,1) and $v_z = \vec{v}.\vec{e}$. The dimensionless parameters are the Prandtl number, Pr, and the Rayleigh number :

$$Pr = \frac{\nu}{\kappa} \qquad , \qquad Ra = \frac{\alpha g\beta R^4}{\nu\kappa} \qquad (2.2a,b)$$

It will be noted that the horizontal boundaries are now at $z = \pm\lambda$ and the vertical boundaries at $x = \pm 1$. The boundary conditions on $\vec{v}$ and θ at $z = \pm\lambda$ are rigid and conducting :

$$\vec{v} = \theta = 0 \qquad .$$

In order to simplify the following calculations we shall only consider plane motions for which, $v_y = 0$ and all the remaining quantities are independent of the y coordinate. These kind of disturbances have been examined previously by Ostrach[8] and Yih[9]. The incompressibility condition is automatically satisfied by introducing the stream-function ϕ :

$$v_x = \frac{\partial\phi}{\partial z} \qquad v_z = -\frac{\partial\phi}{\partial x} \qquad , \qquad (2.3a,b)$$

and after elimination of the pressure in (2.1a) one gets :

$$Pr^{-1}\left(\frac{\partial\Delta v_z}{\partial t} + N_{\vec{v}}\right) = \Delta^2 v_z + \frac{\partial^2\theta}{\partial x^2} \qquad (2.4a)$$

with

$$N_{\vec{v}} = -\partial_x(\vec{v}.\vec{\nabla})\Delta\phi \qquad . \qquad (2.4b)$$

To recover the classical result of Ostroumov in the limit, $h \rightarrow \infty$, it is appropriate to make the change of variable $z = \lambda Z$, and to expand all the physical quantities in two contributions :

$$f = f^{(o)} + \tilde{f} \qquad (2.5)$$

where a vector notation $f = \{\phi, \theta, Ra\}$ has been introduced. The superscript 0 in (2.5) refers to quantities relative to the infinite problem, whereas the tilde denotes that the corresponding quantity can be expanded in power of λ^{-1}. After substituting

(2.5) in (2.1b) and (2.4a) we collect the terms according to their order in λ^{-1}. The lowest order gives :

$$[\partial_x^4 - Ra^{(o)}]v_z^{(o)} = 0 \tag{2.6a}$$

$$\theta^{(o)} = -\partial_x^2 v_z^{(o)} \tag{2.6b}$$

This is precisely the equation for an infinite plane vertical layer. The solutions for the odd modes are :

a. *Conducting lateral boundary*

$$v_z^{(o)} = A \sin qx \quad , \quad \theta^{(o)} = A q^2 \sin qx \tag{2.7a,b}$$

with for the lowest mode, $q = \pi$, and A is a constant.

b. *Insulating lateral boundary*

$$v_z^{(o)} = A\left(\frac{\sin qx}{\sin q} - \frac{\sh qx}{\sh q}\right) \quad , \quad \theta^{(o)} = q^2 A\left(\frac{\sin qx}{\sin q} + \frac{\sh qx}{\sh q}\right) \tag{2.8a,b}$$

with thq = -tgq, and q = 2.365 for the lowest mode.

For the even modes the flux closure condition implies the vanishing of the transverse heat flux

$$\frac{\partial\theta}{\partial x} = 0 \quad \text{at} \quad x = \pm 1$$

Therefore the solutions of (2.6) are the same for either type of thermal boundary condition :

$$v_z^{(o)} = A\left(\frac{\cos qx}{\cos q} - \frac{\ch qx}{\ch q}\right) \quad , \quad \theta^{(o)} = Aq^2\left(\frac{\cos qx}{\cos q} + \frac{\ch qx}{\ch q}\right) \tag{2.9a,b}$$

with tgq = thq and q = 3.927 for the lowest mode.

The presence of boundaries at the top and bottom of the vertical layer makes a significant difference to the above statement. However, provided $\lambda \to \infty$, the form :

$$v_z^{(o)} = Aw(x) \quad , \quad \theta^{(o)} = A\Theta(x) \quad , \quad \phi^{(o)} = A\Phi(x) \quad ,$$

may be adapted to the finite problem by allowing A to be a slowly varying function of Z as well as time. Then to the second order of perturbations one gets :

$$[\partial_x^4 - Ra^{(o)}]\tilde{v}_z = (1+Pr^{-1})A_t w_{xx} + Pr^{-1} N_{\vec{v}} - N_\theta + \frac{3A_{ZZ}}{\lambda^2}\Theta \tag{2.11}$$

$$+ \tilde{Ra} Aw$$

where :

$$N_\theta = \lambda^{-1} AA_Z(\Phi\Theta_x - \Theta\Phi_x)$$

$$N_{\vec{v}} = \lambda^{-1} N_{\vec{v}}^{(1)} + \lambda^{-3} N_{\vec{v}}^{(3)}$$

with :

$$N_{\vec{v}}^{(1)} = -\partial_x(\phi_Z\partial_x - \phi_x\partial_Z)\partial_x^2\phi$$

$$N_{\vec{v}}^{(3)} = -\partial_x(\phi_Z\partial_x - \phi_x\partial_Z)\partial_Z^2\phi$$

In the foregoing, except for the components of the velocity v_x, v_z, the subscript denotes derivation with respect to the corresponding variable $\left(\text{i.e. } A_z = \frac{\partial A}{\partial z}\right)$. Equation (2.11) is solved provided that its r.h.s. is orthogonal to the kernel of the homogeneous adjoint operator. Neglecting all the terms of order higher than λ^{-2} in (2.11) the compatibility condition demands :

$$A_t I_1 (1+Pr^{-1}) = 3\lambda^{-2} A_{zz} I_1 + \tilde{Ra}\, AI_2 + \lambda^{-1} AA_z (I_3 + Pr^{-1} I_4) \qquad (2.12)$$

where :

$$I_1 = \int_{-1}^{+1} |\Theta(x)|^2\, dx \qquad I_2 = \int_{-1}^{+1} \Theta(x) w(x)\, dx$$

$$I_3 = -\int_{-1}^{+1} \Theta(x)[\Phi(x)\Theta_x - \Theta(x)\Phi_x]\, , \quad I_4 = -\int_{-1}^{+1} \Theta(x)[\Phi\partial_x^4 \Phi - \Phi_{xx}^2]$$

The integrals, I_1 and I_2 are positive. The quantities I_3 and I_4 vanish when the odd modes are considered and take a finite value for the even modes :

$$I_3 = 2I_4 = \frac{32}{5}\, (q\ thq)^3$$

In this latter case, $I_3 \neq 0$, $I_4 \neq 0$, all the terms in eq. (2.12) can be made of the same order of magnitude in λ, by an appropriate normalization. Let :

$$Ra = \frac{3 I_1 \delta}{I_2 \lambda^2} \qquad , \qquad t = \frac{\lambda^2}{3}(1+Pr^{-1})\tau$$

and

$$A \to \lambda^{-1}\left(\frac{3 I_1}{I_3 + I_4\, Pr^{-1}}\right) A$$

Then instead of (2.12) we get :

$$A_{zz} + \delta A + AA_z = A_\tau \, , \qquad -1 < z < +1 \qquad , \qquad (2.13)$$

or

$$A_{\xi\xi} + A + AA_\xi = A_\tau \, , \qquad -\delta^{1/2} < \xi < \delta^{1/2} \qquad (2.14)$$

where $\xi = \delta^{1/2} z$, A is scaled with $\delta^{1/2}$ and τ with δ^{-1}. The boundary conditions associated to this equation have been derived elsewhere[10] and are : $A = 0$ at $\xi = \pm\delta^{1/2}$. We now set $A_\tau = 0$. The solutions to (2.14) can be reduced to quadratures :

$$\frac{A^2}{2} + A_\xi - Log(A_\xi + 1) = C \qquad (2.15)$$

where C is a constant. A numerical solution of (2.15) using a space discretization gives the form of A (fig. 1). We see that $2\delta^{1/2} > \pi$ for a solution to be possible. When $2\delta^{1/2}$ is close to π, $A(\xi)$ is sinusoïdal of small amplitude and as C increases the maximum of A moves to the left when $A > 0$. The non-symmetrical form of A suggests that the configuration of the convective currents results of a mixing between the odd and even eigenmodes of the linear problem (figs. 2 and 3). If one tries to fit the numerical solution by a superposition of the two lowest eigenmodes :

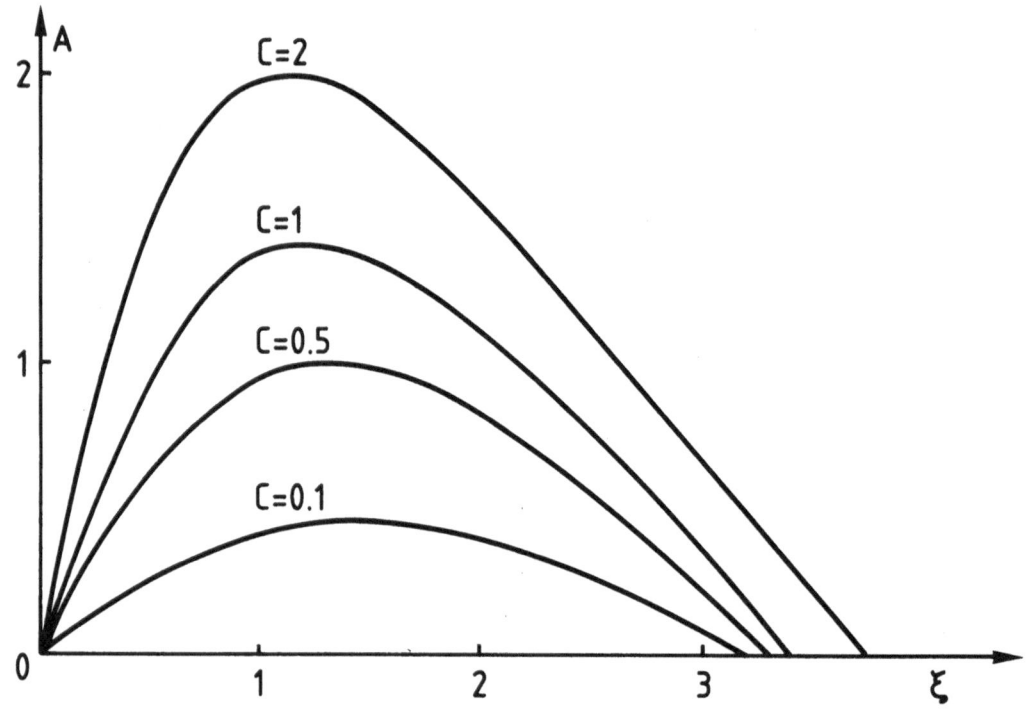

Figure 1

Numerical solutions of equation, $A_{\xi\xi} + A + AA_{\xi} = 0$ with $A = 0$ at $\xi = \pm\delta^{1/2}$. Solutions are only possible for $\delta^{1/2} > \pi/2$.

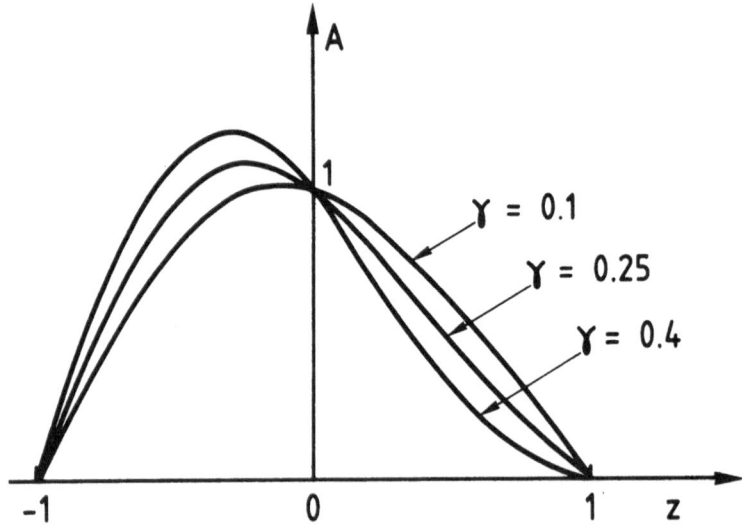

Figure 2

$A(Z)$ is approximated by the function $A = |A|\left(\cos\frac{\pi}{2}Z - \gamma\sin\pi Z\right)$ where the normalization $|A| = 1$ is taken. For increasing values of γ the maximum amplitude is shifted toward the left. For larger values of the coupling constant, γ, a second cell can even appear on the right.

$$A = A_1 \cos \frac{\pi}{2} Z + A_2 \sin \pi Z \qquad -1 < Z < +1 \qquad (2.16)$$

The Galerkin approximation gives the result that :

$$A_2 = -\frac{4}{\pi}\left(\delta - \frac{\pi^2}{4}\right) \quad , \qquad A_1^2 = \frac{4}{\pi}(\delta - \pi^2)A_2$$

It has been checked for $\delta^{1/2}$ = 1.7 and 1.85 that the maximum value of A given by (2.16) is consistent with the corresponding numerical values deduced from fig. 1.

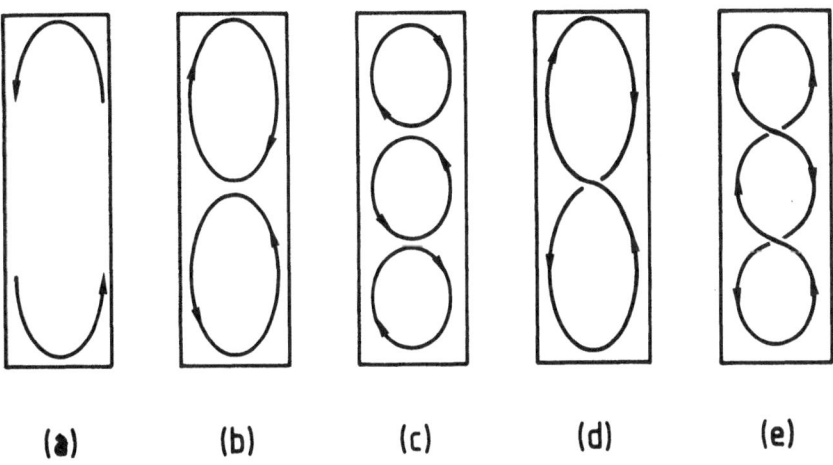

(a)   (b)   (c)   (d)   (e)

Figure 3

(a), (b) and (c) Schematic representation in the plane (x,z) of the first three vertical modes having a constant phase. (d) and (e) vertical modes with a non constant phase. Such patterns in the form of a "eight" have been observed by Micheau et al in high evaporative cooling cells.

We shall come back now to the case where $I_3 = I_4 = 0$ in (2.12) and then the non linear term , AA_Z, disappears from the differential equation satisfied by A. The method to introduce the non linearities in that case consists in expanding $\tilde{f} = \{\tilde{\phi}, \tilde{\theta}, Ra\}$ in powers of λ^{-1}

$$\tilde{f} = f^{(1)} \lambda^{-1} + f^{(2)} \lambda^{-2} + \dots \qquad (2.17)$$

After substituting in (2.11) and equating terms of the same order in λ^{-1}, we get to the first order :

$$[\partial_x^4 - Ra^{(o)}]v_z^{(1)} = Pr^{-1} N_{\vec{v}}^{(1)} - N_\theta^{(1)} + Ra^{(1)}v_z^{(o)} \qquad (2.18)$$

The solvability condition for this equation is :

$$Ra^{(1)} \sim I_3 + Pr^{-1}I_4 = 0 \qquad (2.19)$$

and the solutions for $v_z^{(1)}$ and $\theta^{(1)}$ take the form

$$v_z^{(1)} = AA_Z w^{(1)}(x) \quad , \qquad \theta^{(1)} = AA_Z \Theta^{(1)}(x)$$

where the explicit expressions for $w^{(1)}(x)$ and $\Theta^{(1)}(x)$ are given in[10] for conducting and insulating vertical boundaries. We shall just mention that in the conducting case, the solutions up to this order are independent of the Prandtl number. Then, the equation to be solved at the next order is :

$$[\partial_x^4 - Ra^{(o)}]v_z^{(2)} = -A_\tau \Theta^{(o)} + Pr^{-1} N_{\vec{v}}^{(2)} - N_\Theta^{(2)} + 3A_{zz}\Theta^{(o)} + Ra^{(2)}v_z^{(o)} \qquad (2.20)$$

This is the solvability condition of this equation which gives the equation satisfied by A :

$$A_\tau = A_{zz} + \delta A + A^2 A_{zz} + \mu\, AA_z^2 \quad , \quad -1 < Z < +1 \qquad (2.21)$$

where all the parameters have been eliminated, except μ, which depends in general upon the Prandtl number. We set $A_\tau = 0$ and $Z = \delta^{-1/2}\xi$ and examined the solutions of :

$$A_{\xi\xi} + A + A^2 A_{\xi\xi} + \mu\, AA_\xi^2 = 0 \qquad -\delta^{1/2} < \xi < \delta^{1/2} \qquad (2.22)$$

which can be reduced to :

$$\text{Log}(1+A^2) + \frac{1}{\mu}\, \text{Log}(1+\mu\, A_\xi^2) = Q \qquad (2.23)$$

where Q is a constant. For conducting lateral boundary in the limit of infinite Prandtl number μ takes the value $\mu = 0.6$ when $\gamma = \pi$. It must be noticed that for $\mu = 1$, solutions of (2.22) are possible if : $\delta^{1/2} = C^{1/2}\mathbb{E}\left(\frac{C-1}{C}\right)$ where $C = \exp Q$ and $\mathbb{E}$ is the complete elliptic integral of the second kind. Numerical solutions for $\mu = 1$ are shown on fig. 4, for the even modes since eq. (2.22) does not allow the mixing between modes of different parity.

We have shown in this section that for a plane vertical layer of fluid the vertical amplitude A(Z) satisfies a non linear differential equation where the non linear terms contain derivatives toward Z and are either quadratic or cubic in A according to the symmetry of the horizontal modes. These results can be transposed to the case of a vertical circular cylinder.

3. THE VERTICAL CIRCULAR CYLINDER

We introduce the following representation for the velocity :

$$\vec{v} = \vec{\nabla}\times\psi\vec{e} + \vec{\nabla}\times\vec{\nabla}\times\phi\vec{e} \qquad . \qquad (3.1)$$

After operating with $\vec{e}.(\vec{\nabla}\times$ and $\vec{e}.(\vec{\nabla}\times\vec{\nabla}\times$ on the equation of motion (2.1a) we obtain the equations for ϕ, ψ and θ :

$$(Pr^{-1}\partial_t - \Delta)\Delta_2\psi = Pr^{-1} N_{\vec{v}}^{(a)} \qquad (3.1a)$$

$$(Pr^{-1}\partial_t - \Delta)\Delta\Delta_2\phi + \Delta_2\theta = -Pr^{-1} N_{\vec{v}}^{(b)} \qquad (3.1b)$$

$$(\partial_t - \Delta)\theta = -Ra\Delta_2\phi - (\vec{v}.\vec{\nabla})\theta \qquad (3.1c)$$

where :

$$\Delta = \Delta_2 + \partial_z^2 \quad , \quad \Delta_2 = \partial_r^2 + r^{-1}\partial_r - r^{-2}\partial_\varphi^2$$

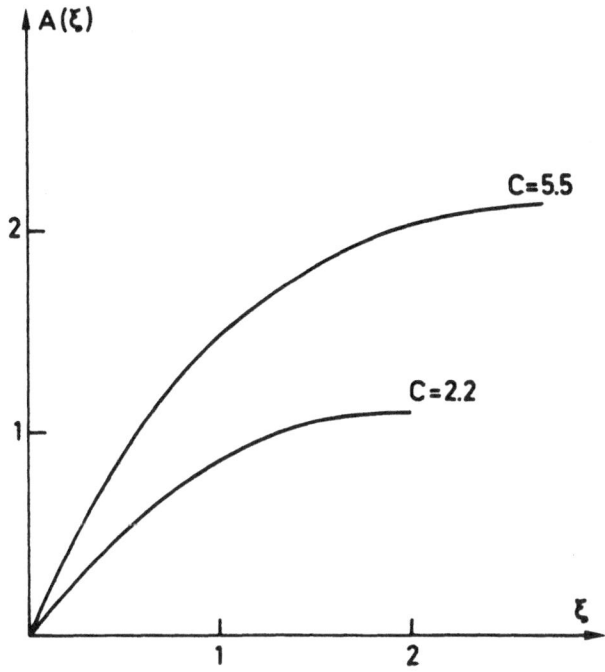

Figure 4

Numerical solutions of $A_{\xi\xi}+A+A^2A_{\xi\xi}+\mu AA_{Z}^2 = 0$ *with* $A = 0$ *at* $\xi = \pm\partial^{1/2}$ *and* $\mu = 1$.

As in the previous section we develop $f = \{\phi,\psi,\theta,Ra\}$ in two contributions :

$$f = f^{(o)}+\tilde{f}$$

and set $z = \lambda Z$. In the limit $\lambda \to \infty$, equations (3.1) reduce to the Ostroumov problem :

$$\Delta_2\psi^{(o)} = 0 \tag{3.2a}$$

$$[\Delta_2-Ra^{(o)}]v_z^{(o)} = 0 \tag{3.2b}$$

$$\theta^{(o)} = -\Delta_2 v_z^{(o)} \tag{3.2c}$$

The rigid boundaries conditions at $r = 1$ imply that $\psi^{(o)}$ is identically null. The solutions for the vertical component of the velocity and the temperature divide in two classes[11] following the thermal boundary condition and the angular mode :

a) Conducting lateral boundary and $n \neq 0$

$$v_z^{(o)} = A\ e^{in\varphi}\ J_n(kr) \quad, \quad \theta^{(o)} = Ak^2\ e^{in\varphi}\ J_n(kr) \tag{3.3}$$

with $J_n(k) = 0$.

b) Insulating lateral boundary or $n = 0$

$$v_z^{(o)} = A\ e^{in\varphi}\left[\frac{J_n(kr)}{J_n(k)} - \frac{I_n(kr)}{I_n(k)}\right] \quad, \quad \theta^{(o)} = Ak^2\ e^{in\varphi}\left[\frac{J_n(kr)}{J_n(k)} + \frac{I_n(kr)}{I_n(k)}\right] \tag{3.4}$$

with :

$$I_n(k)J_n'(k)+J_n(k)I_n'(k) = 0 \quad ,$$

J_n and I_n are the nth Bessel and modified Bessel functions of the first kind. When λ is large but finite we allow A to be a slowly varying function of Z as well as time and expressions (3.3), (3.4) are the starting point of an expansion scheme in power of λ^{-1}. The derivation of the amplitude equation closely follows the method described in the previous section and will not be repeated here. For more details see[10]. The results divide in two categories :

i) axisymmetrical mode, n = 0

The solvability condition for the non homogeneous equation satisfied by $\tilde{v}_z$ gives the amplitude equation :

$$A_{ZZ}+\delta A+AA_Z = A_\tau \tag{3.5}$$

$$A = 0 \quad \text{at} \quad Z = \pm 1$$

This result is identical to what we have obtained in previous section for the even modes.

ii) Non-axisymmetrical modes

For diametrically antisymmetrical modes, $n \geq 1$, one allows A to be a complex function, $A = |A|e^{i\alpha}$, where both $|A|$ and the phase α are slowly varying function of Z and t. Then the convective velocity, for instance, is of the form :

$$v_z = (A e^{in\varphi} + A^* e^{-in\varphi})w(r) \tag{3.6}$$

The solvability condition at order λ^{-1} is trivially satisfied and at order λ^{-2} it demands :

$$A_{ZZ}+\delta A+\mu_o A|A|^2_{ZZ}+\mu_1 A_Z|A|^2_Z+\mu_2 A|A_Z|^2+\mu_3 A^*(AA_Z)_Z = A_\tau \tag{3.7}$$

where μ_i are constant coefficients which depend on the Prandtl number. Explicit expressions for the μ_i's are yet unknown which make difficult the study of eq. (3.7). However some preliminary remarks can be done concerning the occurrence of oscillatory modes with a time-dependent phase like :

$$A = W(Z)expi[\alpha(Z)+\sigma\tau]$$

The imaginary part of (3.7) reads :

$$\ddot{\alpha} W+2\dot{\alpha} \ddot{W}+ 2\mu_1 \dot{\alpha}\dot{W}^2W +\mu_3 W(4\dot{\alpha} \dot{W}\dot{W}+ \ddot{\alpha} W^2) = \sigma W \tag{3.8}$$

where the dot means derivation with respect to Z.

Multiplying by W and integrating by parts over Z from -1 to $+1$ gives :

$$\sigma = 2\mu_1\left[\int_{-1}^{+1} \dot{\alpha} W^3\dot{W} dz\right] / \int_{-1}^{+1} W^2 dz \tag{3.9}$$

We get as a consequence that $\sigma=0$ if either $\mu_1 = 0$ or if the phase is constant along the vertical axis $(\dot{\alpha}=0)$. Until now all the theoretical studies of convection in vertical cylinders have neglected the spatial variation of the phase. Fig. (5) shows the difference between a diametrically antisymmetrical mode n = 1, with a constant phase (a) and (b) with an arbitrary helical variation of the phase. Experimentally there is some evidence for helicoïdal convective patterns bearing some analogy with fig. (5) and associated with an oscillatory behaviour[12]. During this colloquium J.C. Micheau has drawn my attention to the fact that a similar feature also occurs in the field of photochemical instability[13]. In order to estimate the capacity of equation (3.7) to describe the reality complementary calculations must be done.

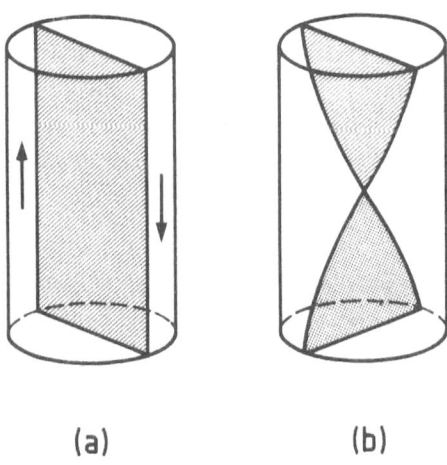

(a)   (b)

Figure 5
Schematic representation of a diametrically antisymmetrical mode in a cylinder. a) With constant phase along the axis. We have drawn the vertical nodal plane which divides the cylinder in zones of respectively downward and upward flow. b) Example of an arbitrary helical variation of the phase : the nodal plane is twisted.

4. CONCLUSION

The problem of weakly non linear convection in vertical (planar or cylindrical) layers of fluid is not easily tractable by means of analytical methods. For instance the perturbation expansion in powers of the amplitude of the instability (or ε-expansion) does not work. Then, using the disparity between the small horizontal scale, R, and the large vertical scale, h, we have developed an expansion scheme in powers of $\lambda^{-1} = 2R/h$. The governing equations are solved at each order in λ^{-1} and the first solvability condition which is not trivially satisfied yields the evolution equation for the slowly varying function $A(Z,\tau)$. Three typical equations are obtained owing to the symmetry of the horizontal mode of flow.

a. Even modes in a plane vertical layer or axisymmetrical modes in a cylinder :

$$A_{ZZ} + \delta A + AA_Z = A_\tau \tag{4.1}$$

b. Odd modes in a plane vertical layer :

$$A_{ZZ} + \delta A + A^2 A_{ZZ} + \mu AA_Z^2 = A_\tau \tag{4.2}$$

c. Antisymmetrical modes in a cylinder :

$$A_{ZZ} + \delta A + \mu_0 A |A|_{ZZ}^2 + \mu_1 A_Z |A|_Z^2 + \mu_2 A |A_Z|^2 + \mu_3 A^* (AA_Z)_Z = A_\tau \tag{4.3}$$

where $A = |A| e^{i\alpha}$. When the phase α is constant equation (4.3) reduces to (4.2).

Numerical solutions of (4.1) reveal a mixing between the first even and odd modes of the linear problem (fig. 1). This kind of configuration seems to have been observed by Olson and Rosenberger[14] in cylindrical cells. On the contrary solutions of (4.2) show no mixing between modes of different parity. Preliminary results on eq. (4.3) indicate that solutions with a rotating phase cannot be excluded and this point is confirmed by experiments.

Two types of non linearities (cubic or quadratic) are present in eqs. (4.1,2,3) and are directly connected to the symmetry of the horizontal modes. A kind of related phenomenon has been pointed out by Palm[15] and concerns the formation of rolls or hexagones in horizontal layers of fluid according to the vertical symmetry of the layer.

To conclude we emphasize the fact that no variational formulation exists for eqs. (4.1,2,3) which makes difficult the stability analysis of the stationary solutions.

REFERENCES

[1] Gor'kov, L.P. "Stationary convection in a plane liquid layer near the critical heat transfer point", Soviet Physics JETP, 6 (1958) 311-315.

[2] Ostroumov, G.A. "Natural convective heat transfer in closed vertical tubes", Izv. Estatstv. Nauch. Inot. Perm. Univ. 12 (1947) 113.

[3] Charlson, G.S. and Sani, R.L. "Finite amplitude axisymmetric thermoconvective flows in a bounded cylindrical layer of fluid", J. Fluid Mech. 71 (1975) 209-229.

[4] Lyubimov, D.V. Putin, G.F. Chernatynskii, V.I. "On convective motions in a Hele-Shaw cell", Sov. Phys. Dokl 22 (1977) 360-362.

[5] Segel, L.A. "Distant side-walls cause slow amplitude modulation of cellular convection", J. Fluid Mech. 38 (1969) 203-224.

[6] Newell, A.C. and Whitehead, J.A. "Finite bandwith, finite amplitude convection", J. Fluid Mech., 38 (1969) 279-303.

[7] Chapman, C.J. and Proctor, M.R.E. "Nonlinear Rayleigh-Bénard convection between poorly conducting boundaries", J. Fluid Mech. 101 (1980) 759-782.

[8] Ostrach, S. "Unstable convection in vertical channels with heating from below" NACA TN 3458 (1955).

[9] Yih, C.S., Quart. Appl. Math. 17 (1959) 25.

[10] Normand, C. "Nonlinear convection in high vertical channels". Preprint SPhT/83-24 (1983).

[11] Gershuni, G.Z. and Zukhovitskii, E.M. "Convective stability of incompressible fluids", Keter publishing house Jerusalem (1976) p. 58.

[12] Azouni, M.A. "Helical impurity distribution produced by a hydrodynamic instability", J. Cryst. Growth 42 (1977) 405.

[13] Micheau J.C., private communication.

[14] Olson, J.M. and Rosenberger F. "Convective instabilities in a closed vertical cylinder heated from below. Part.1. Mono-component gases", J. Fluid Mech. 92 (1979) 609-629.

[15] Palm, E. J. Fluid Mech. 8 (1960) 183.

STABILITY OF CELLULAR SYSTEMS IN TAYLOR-COUETTE INSTABILITY

by P. TABELING

Laboratoire de Génie Electrique de Paris
Plateau du Moulon
91190 GIF-SUR-YVETTE (FRANCE)
Laboratoire associé au C.N.R.S. aux Universités
Paris VI et Paris XI et à l'E.S.E.

1. INTRODUCTION

There are many analogies between the Taylor-Couette instability and the Rayleigh-Benard instability. For instance, the first instability of the laminar regime is, in the two cases, in form of stationary cells characterized by approximately the same wavelength and the same value of the control parameter (1).

However, the analogies between two problems become rarer when one considers the stability of the stationary periodic structures. In the Rayleigh-Benard problem, the stability diagram of the convective rolls is rather complicated. Many modes of instability can be excited : zig-zag, Eckhaus, cross-rolls, varicose instability, oscillatory instability, etc... (2). The abundance of the modes which can destabilize the structures in form of rolls is essentially due to the important number of invariance properties initially present in the Rayleigh-Benard problem : translationnal invariance, rotationnal invariance and, in the limit of zero Prandtl numbers, galilean invariance.

The situation is very different concerning the Taylor-Couette instability. In contrast with the convective rolls, the Taylor cells form an oriented periodic system (the orientation results from a balance between the rotation effects and the effects related to the shear of the mean flow). Therefore, the Taylor instability does not possess any rotationnal invariance property which favours the modes of torsion of the cells.

It follows that the modes which destabilize the Taylor cells are less abundant than in the Rayleigh-Benard instability. Such modes are essentially of two types : Eckhaus (related with translational invariance) and wavy vortices. The wavy vortices are commonly observed in the experiments performed on Taylor-Couette instability. They are not directly related to an invariance property but more probably result from an interaction between a stationary mode (corresponding to the cellular system of Taylor) and an unstable oscillatory mode. If we restrict ourselves to the temporal aspect of the problem, we can describe this interaction by a dynamical system with two degrees of freedom. Davey (3) has shown how such a system can be deduced from the governing equations of flow. However, Davey's analysis does not take very interesting spatio-temporal phenomena into account, such as those shown recently by KING (4). At the

present time, the theorical schemes which describe the onset and the properties of
the wavy vortices are still very incomplete.

In this short paper, we present some simple results about the stability of stationary Taylor cells against non axisymmetric disturbances. This study is restricted to the
vicinity of the instability point. The main result is the establishment of an amplitude equation which governs the dynamics of slowly varying modes near treshold, by
using a method similar to that of Newell and Whitehead (5). We shall then be able to
demonstrate that the cellular structure is stable against a representative class of
non axisymmetric disturbances (zig-zag).

2. ESTABLISHMENT OF THE AMPLITUDE EQUATION AND DISCUSSION

The problem which we consider herein is the viscous flow of an incompressible fluid
between two concentric cylinders of radii R_1 and R_2. The inner cylinder rotates at
angular velocity Ω while the outer one is at rest. We restrict ourselves to the case
where the gap $d = R_2 - R_1$ is small compared to the inner radius. At low values of
angular velocity Ω, the flow is laminar, and the velocity is azimuthal. The system
of equations, which governs the perturbed state, reads (at the lowest order in term
of ratio d/R_1) :

$$\frac{\partial u_x}{\partial t} + T^{1/2} V(x) \frac{\partial u_x}{\partial y} + (\mathbf{u}.\nabla) u_x - \frac{1}{2} u_y^2 = -\frac{\partial p}{\partial x} + \Delta u_x + T^{1/2} V(x) u_y, \quad (1)$$

$$\frac{\partial u_y}{\partial t} + T^{1/2} V(x) \frac{\partial u_y}{\partial y} + (\mathbf{u}.\nabla) u_y = \Delta u_y + T^{1/2} u_x, \quad (2)$$

$$\frac{\partial u_z}{\partial t} + T^{1/2} V(x) \frac{\partial u_z}{\partial y} + (\mathbf{u}.\nabla) u_z = -\frac{\partial p}{\partial z} + \Delta u_z, \quad (3)$$

and

$$\frac{\partial u_x}{\partial x} + \frac{\partial u_y}{\partial y} + \frac{\partial u_z}{\partial z} = 0, \quad (4)$$

in which we have $x = (R - R_1)/d$ (where R is the distance from the common axis of the
cylinders), and $z = Z/d$ (where Z is the axial coordinate). The velocity scale in x
and z directions is ν/d (where ν is the kinematic viscosity) while it is (ν/d)
$(R_1/2d)^{1/2}$ in y direction ; lastly, $V(x) = 1 - x$ is the reduced laminar velocity
profile and $T = 2\Omega^2 R_1 d^3/\nu^2$ is the Taylor number. In the neighbourhood of the instability point, perturbation field $\underline{U}_1$, associated with the system of Taylor stationary cells, can be put in the form :

$$\underline{U}_1 = Re (A \underline{u}_1^{(1)}(x) e^{iq_o z})$$

where q_o is the optimal wave-number, $\underline{u}_1^{(1)}(x)$ is an eigenfunction of the linearized
problem and A is a slowly varying function in time and space. It is possible to show,
by scaling arguments, that amplitude function A must be sought under the form

$$A = A (\varepsilon^{1/2} y, \varepsilon^{1/2} z, \varepsilon^{1/2} t, \varepsilon t, ...) \quad (5)$$

where $\varepsilon = (T^{1/2} - T_o^{1/2})/T_o^{1/2}$ is the discrepancy to threshold in which T_o is the
critical value of the Taylor number. The length scales introduced in the problem

are $0(\varepsilon^{-1/2})$; the characteristic time scale $\tau_a = 0(\varepsilon^{-1/2})$ represents a much smaller characteristic time that that related to the critical slowing down $\tau_r = 0(\varepsilon^{-1})$. The physical meaning of time τ_a is related to advective effects : the slow non axisymmetric modes are advected by the mean flow at a velocity equal to that of the inner cylinder (in term of order of magnitude) ; this value is obtained by calculating ratio l_y/τ_a, where l_y is the characteristic length scale in y direction ; τ_a is therefore a typical time of propagation.

The spatial and temporal correlation lengths are very different from those involved in the Rayleigh-Benard problem. Such differences are essentially due to the anisotropic character of the Taylor problem and to the presence of advection terms. It is possible to represent the differences between the spatial characteristic lengths by considering, in the Fourier space, the wave-vector associated with an arbitrary slow mode, at a given (small) value of ε (see Figure 1). In the case of the Rayleigh-Benard instability, the wave-vector is confined in a region which is stretched along the circle defined by $q_x^2 + q_y^2 = q_o^2$; in the case of the Taylor instability, the wave-vector is confined in a region which has comparable dimensions in the two directions of the plane.

Using expressions (4) and (5), it is possible to develop multiple'scale technique, by introducing the usual expansions of the operators into series in increasing powers of $\varepsilon^{1/2}$. The method which we have used is similar to that of Newell and Whitehead (5). Expanding the perturbation velocity field into series of powers of $\varepsilon^{1/2}$, and each term of such an expansion into Fourier series leads to an infinite hierarchy of equations which can be solved iteratively. At second order in $\varepsilon^{1/2}$, a solvability condition arises under the form :

$$\frac{\partial A}{\partial t} + T_0^{1/2} v_0 \frac{\partial A}{\partial y} = 0, \tag{6}$$

where v_0 is a constant which can be expressed in terms of various scalar products which involve the eigenfunctions of the linearized problem and those of the adjoint problem. Numerically we have obtained $v_0 = 0.526$.

The amplitude equation corresponds to a solvability condition at order $\varepsilon^{3/2}$. We obtain :

$$\frac{\partial A}{\partial t} = c_0 \varepsilon A + c_1 \frac{\partial^2 A}{\partial z^2} + i c_2 \frac{\partial^2 A}{\partial y'\, \partial z} + c_3 \frac{\partial^2 A}{\partial y'^2} - c_4 A |A|^2, \tag{7}$$

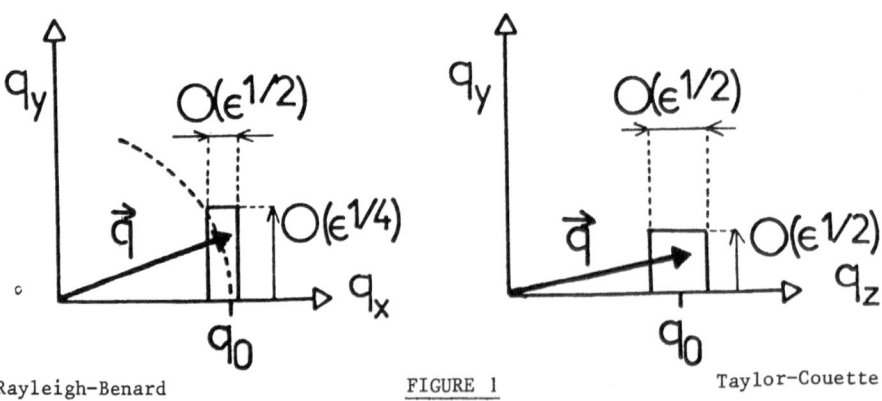

Rayleigh-Benard   FIGURE 1   Taylor-Couette

where $y' = y - T_0^{1/2} v_0 t$ is an angular variable measured in a new frame of reference which rotates at velocity $T_0^{1/2} v_0$, and C_0, C_1, C_2, C_3 and C_4 are real positive coefficients. Numerically we have obtained $C_0 = 26.16$, $C_1 = 0.984$, $C_2 = 0.395$, $C_3 = 2.61$ and $C_4 = 40.2$ (with normalization condition $U_{1y}^{(1)}(\frac{1}{2}) = 1$).

Equation (7) definies a non-potential problem, in contrast with the case of the Rayleigh-Benard instability. The absence of variationnal principle is due to term $\partial^2_{y'z} A$ which favours the persistence of azimuthal oscillations. Since at the lowest order, the expression for helicity H is

$$H = \frac{1}{v} \int_v \underline{u} \cdot \underline{rot}\,(\underline{u})\, dV \propto \Im\left(\int A^* \frac{\partial^2 A}{\partial y\, \partial z}\, dy\, dz\right),$$

on can also relate the absence of variationnal principle with the existence of modes with non-zero helicity.

The equilibrium solutions of Equation (7) are in the form :

$$A = A_0\, e^{i(\mu y' + \zeta z + \omega t)}$$

where ζ represents the compression (or dilatation) of the cellular system, and ω is related to μ and ζ by the dispersion relation $\omega = C_2 \zeta \mu$. Spatially, the equilibrium solutions are in form of spiraling vortices (they are just the modes calculated by Krueger et al (6)).

Two types of modes can destabilize the system of stationary cells : the longitudinal modes ($\partial_{y'} = 0$) and the transversal modes ($\partial_z = 0$). The former are associated with Eckhaus instability, while the latter correspond to zig-zag disturbances. It is easy to prove, by using Equation (7), that the cellular system is stable against zig-zag disturbances ; this result can also be obtained by using expansions into phase gradients (7). The resulting equation for the phase variable is purely diffusive and the transversal diffusivity is positive ; this ensures stability of the cellular system against zig-zag instability.

3. CONCLUSION

We have obtained some results about the stability of cellular systems in the Taylor problem, and we have shown some important differences between this problem and that of Rayleigh-Benard. The stability of the stationary cells against transversal phase disturbances is a crucial property of the Taylor-Couette instability. It is directly related to the absence of rotationnal invariance. It should be possible to prove it by using more general arguments.

REFERENCES

1 CHANDRASEKHAR S., "Hydrodynamic and Hydromagnetic Stability, (Oxford University Press, Oxford, 1961).

2 BUSSE, F.H., CLEVER, R.M., J. Fluid. Mech., 91, (1979), 319.

3 DAVEY, A., DI PRIMA, R.C., STUART, J.T., J. Fluid. Mech., $\underline{31}$ (1968) 17.

4 KING, Ph. D, University of Texas, Austin, 1983.

5 NEWELL, A.C., WHITEHEAD, J.A., J. Fluid. Mech., $\underline{36}$, (1969), 239.

6 KRUEGER, E.R., GROSS, A., DI PRIMA, R.C., J. Fluid. Mech., $\underline{24}$ (1966), 521.

7 TABELING, P., J. Physique - Lett., $\underline{44}$, (1983), L-665.

SPATIAL DISSIPATIVE STRUCTURES AND HYDRODYNAMIC INSTABILITIES

M. Gimenez[*], J.C. Micheau

[*] Service de Chimie-Physique II, Université Libre de Bruxelles

1050 BRUXELLES, BELGIUM

Intéractions Moléculaires et Réactivité Chimique et Photochimique,

ERA au CNRS n° 264, Université Paul Sabatier, 31062 TOULOUSE, FRANCE

It is now well known that temporal, spatio-temporal and stationary spatial dissipative structures can emerge in a homogeneous
chemical reacting medium. They can be seen as the coupling of a complex
chemical kinetic network with a diffusion process. Reactions with bromate oscillators used by Zhabotinsky (1), Showalter (2), Orban (3),
giving rise to spatial and temporal oscillations (4), are good examples
of this theory. However, in contrast to oscillatory chemical reactions,
there are few experimental examples of stationary spatial chemical
structures. Some authors reported the apparition of such structures
during the irradiation of an homogeneous medium (5), (6), (7). They
appear when a 1-7 mm layer of solution is irradiated in a 70 mm Pétri
dish. Two types of systems have been studied : photochromogenic
systems corresponding to an irreversible colour change during irradiation, and photochromic ones leading to a reversible colour change in
the medium.

First Möckel (5) in 1977 observed the appearance of striations during
the irradiation of the photochromogenic systems KI/CCl_4/starch in
water and CBr_4/diphenylamine in organic solvents. Later Avnir (6)
described other systems with different colour revelators. The origin
of these structures remained unclear. We used either photochromogenic

and photochromic systems which were all chosen because of their different reaction mechanisms. Some examples of the systems described in
the literature are shown in table I.

Compounds	Colour change	Solvent	References
CCl_4/KI/starch	Uncoloured:violet	Water	(5)
Various aromatic amines+halogen source	Uncoloured:coloured	Water	(6a)
Chromogenic developper 3	Pink:violet	Water	(7)
Mercury dithizonate	Orange:blue	Toluol	
Diphenylamine+various halogenated solvents	Uncoloured:coloured	Organic solvents: hydrocarbons/ alcohols	(6b)
Various aniline derivatives	Uncoloured:coloured	Halogenated solvents	

Table I : Various photochromogenic and photochromic systems
giving rise to spatial stationary dissipative structures

During the irradiation of the chromogenic developper 3 and mercury dithizonate, a thin layer of coloured product appears first and
then breaks down into inhomogeneous zones (Fig. 1). These appear
whether the irradiation is from above or from below, and they look

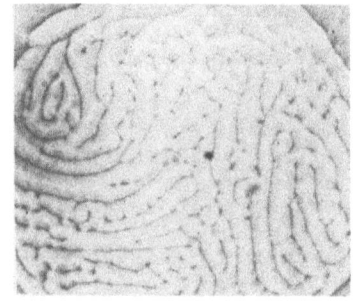
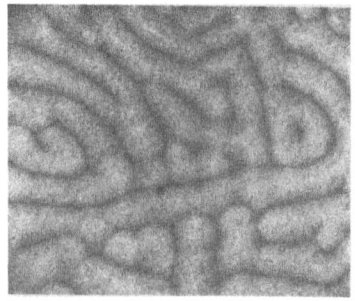

a   b

Fig. 1 : Spatial dissipative structures : a) chromogenic
developer 3 in water. b) mercury dithiozonate in
toluol. Size :⊢———⊣1 cm.

like "vermiculated rolls". The chemical mechanisms for colour production are very simple in both cases, so that the origin of the instability has to be sought among physical causes, rather than in a coupling
of a complex kinetic network with a simple physical phenomenon such
as diffusion.

We have shown that evaporation of the solvent in which the structures emerge, and therefore the room temperature and relative humidi-
ty are the main factors influencing their characteristic wavelength
and shape. Without evaporation no patterns appear. Gravimetric measurements of the evaporation rate allow an estimation of the temperature
gradient across the fluid layer. For water and toluol we find respectively $1°cm^{-1}$ and $1.25°cm^{-1}$. Using these values we have calculated the
critical depth for the surface tension (h_c^{st}) and buoyancy (h_c^b) driven
instabilities (8) (Table II).

	Water	Toluol
h_c^{st} (mm)	0.8	0.7
h_c^b (mm)	4.7	2.7

Table II : Theoretical estimations of critical depths
for the onset of hydrodynamic convection in
water and toluol.

These results show that hydrodynamic instabilities can take place
in the system. Our experiments are always conducted with fluid layers
greater than the critical one. For thicknesses $< h_c^b$ the surface tension
forces play a greater role than buoyancy effects, whereas both mechanisms reinforce one another for larger depths.

The following experimental facts strongly suggest that these
patterns are associated with convective motions in the layer.
1 - Using the Schlieren technique (Fig. 2) we have shown that patterns
were present in the system *before* irradiation, and that *after*

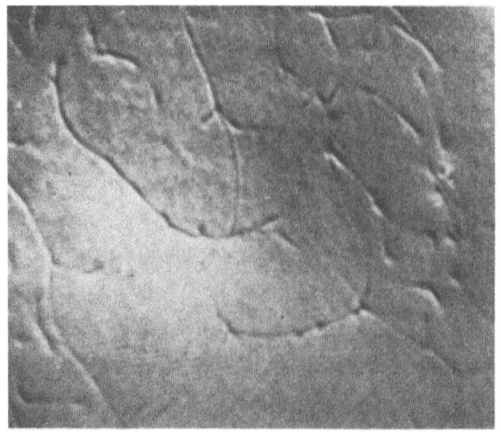

Fig. 2a : Surface pattern in
Möckel's reaction before
irradiation Size :⊢⊣5 mm.

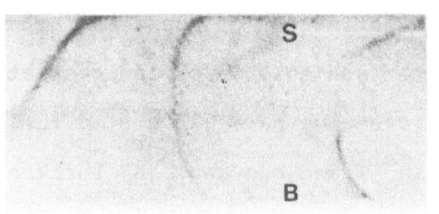

Fig. 2b : Bulk patterns in an
aqueous solution of chromogenic
developer 3 before irradiation
S : surface ; B : Bottom.
Size :⊢⊣1 mm.

illumination the coloured structures appeared where the "pre-patterns"
had already been localized.

2 - There exists a linear relationship between the average wavelength
of the structures and the depth of the layer (Fig. 3.).

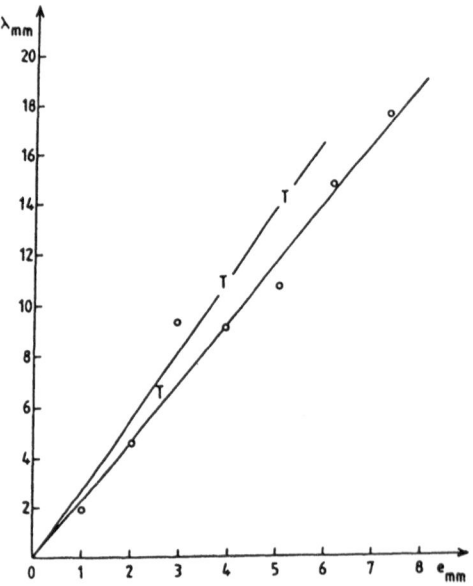

Fig. 3 : Linear relationship
between the pattern wavelength
and the thickness of the fluid
layer (e).
O : chromogenic developer 3 in
water
T : mercury dithizonate in
toluol

3 - When one deposits a layer of ink (green) on the bottom of a Petri
dish filled with an aqueous solution of chromogenic developer under
irradiation, one observes red sinking vermiculated rolls with green
ones rising in between them. (Fig. 4).

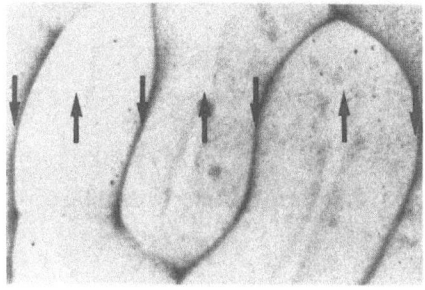

Fig. 4 : Surface patterns showing
convective movements in an aqueous
solution of chromogenic developer
during irradiation ↓: down,
red ;   ↑: up, green.
Size :⊢——⊣1 mm.

Any open surface reaction (such as photochemistry, reaction at gaz-
liquid interfaces (9)-(10), vaporisation and adsorption) yielding or
consuming a coloured product will lead to the formation of spatial
structures. A striking example is provided by the system Ru(bipyridi-
ne)$_3^{2+}$/methyl-viologen^{2+}/triethanolamine (pH=8). With visible light
the deep blue cation radical of methylviologen is formed revealing
the prepatterns. After completion of the photochemical reaction the
light source is turned off and the solution is homogenized by stirring.
Because the cation radical is sensitive to oxygen, yellow reverse
patterns on a deep blue background are observed thereafter, resulting
from the slow surface reaction of the cation radical with oxygen.

   More generally a chemical reaction exhibiting a complex non-linear kinetic network, such as : bromate or glycolitic oscillators (11),
can give rise (in addition to spatio-temporal center waves, the origin of which is purely chemical) to quasi-stationary mosaic striped
structures (2),(3). The shape of such mosaic structures has a striking
similarity to that of the prepatterns. Here again their origin lies in
the coupling with hydrodynamic motions. We conclude that not all the

structures we have studied, including those arising during Möckel's reaction, are induced by light. The photochemical reaction merely serves to reveal the prepatterns induced by evaporative cooling. Having characterized similar prepatterns in the solutions used in most of the photochemical reactions described by Avnir, we feel that the same phenomenon is responsible for the onset of the structures in those systems. This conclusion can be extended to the study of oscillatory behaviours observed by various authors (Table III). The most beautiful are those reported by Laplante and Pottier during the photolysis of dimethyl anthracene in chloroform (17). Here again the use of an optical method could provide a convenient means of connecting the hydrodynamic movements in the solution with the chemical oscillations.

Compounds	Solvent	Monitoring	Oscil.period	Ref.
Rhodamine B	DCE, DMF, THF	U.V.	6 mn in DCE	12
1,5-naphthyridine	cyclohexane	Fluorescence	3-4 mn	13
Zn TPP	H_2O/SDS micellar	"	8 mn	14
Acetone	CH_3CN	"	1-2 mn	15
9,10-dimethylanthracene	$CHCl_3$	"	20-25 s	16-17
N-methylanthranilo hydroxamic acid	CH_3OH	"	3-10 s	18
Biacetyl-oxygen	CH_3CN	"	24 s	19

Table III : Photochemical oscillations of various fluorescent
compounds during photolysis. (DCE : dichloroethane,
SDS : sodium dodecylsulfate)

However these experiments are conducted in fluorescence cells where the height of the system exceeds its radius and the corresponding convective behaviour is rather different.

The simplicity and the flexibility of photochemical imaging could provide an alternative technique for studying the evolution, in real time, of convective patterns, particularly in thin layers where standard optical methods are less efficient.

The coupling of chemical and photochemical reactivity with hydrodynamic instabilities has not often been studied, but we consider that it will lead to new sources of dissipative phenomena.

Acknowledgements

Part of this work has been supported by the NATO research grant 0244/83. M.G. acknowledges a scientific and technical grant from the European Communities Commitee. We are pleased to thank Drs G. Dewel and P. Borckmans (Brussels) for many helpful critical discussions.

References

1 - A.M. Zhabotinsky and A.N. Zaïkin, J. Theor. Biol. 40, 45-61 (1973)

2 - K. Showalter, J. Chem. Phys., 73, 3735-42 (1980)

3 - M. Orban, J. Am. Chem. Soc. 102, 4311-14 (1980)

4 - A. Pacault and C. Vidal, J. Chim. Phys. 79, 691-707 (1982)

5 - P. Möckel Naturwiss, 64, 224 (1977)

6 - a) M. Kagan, A. Levi & D. Avnir Naturwiss, 69, 548-49 (1982)

    b) D. Avnir, M. Kagan and A. Levi Naturwiss, 70, 144-45 (1983)

7 - M. Gimenez and J.C. Micheau Naturwiss, 70, 90 (1983)

8 - J.C. Micheau, M. Gimenez, P. Borckmans and G. Dewel Nature 305, 43-5 (1983)

9 - D. Avnir and M. Kagan Naturwiss. 70, 361-3 (1983)

10 a) P. Möckel Naturwiss. 66, 575-6 (1979)

    b) S.C. Muller and Th. Plesser to be published in "Lectures Notes in Biomathematics" (1983)

11- A. Boiteux and B. Hess Ber. Bunsenges Phys. Chem. 84, 392-8 (1980)

12- R.W. Bigelow J. Phys. Chem. 81, 88-9 (1977)

13- I. Yamazaki, M. Fujita and H. Baba Photochem. Photobiol. 23, 69-70 (1976)

14- S. Toushiya, H. Kanai and M. Seno J. Am. Chem. Soc. 103, 7370-1 (1981)

15- T.L. Nemzek and J.E. Guillet J. Am. Chem. Soc. 98,(4) 1032-4 (1976)

16- R.J. Bose, J. Ross and M.S. Wrighton J. Am. Chem. Soc. 99, (18) 6119-20 (1977)

17- J.P. Laplante and R.H. Pottier J. Phys. Chem. 86, 4759-66 (1982)

18- E. Lipczynska-Kochany and H. Iwamura Chem. Letters 1825-8 (1983)

19- I.R. Epstein, M. Morgan, C. Steel and O. Valdes-Aguilera (in press)

CHEMICAL STRUCTURES FAR FROM EQUILIBRIUM

P.Borckmans,G.Dewel and D.Walgraef

Chercheurs Qualifiés au F.N.R.S.
Service de Chimie-Physique II,
Université Libre de Bruxelles,
Campus Plaine, C.P.231,
B-1050 BRUXELLES, BELGIUM.

1. INTRODUCTION

Perhaps the most fascinating and intriguing aspect of natural phenomena is that complex systems far from equilibrium may undergo symmetry
breaking instabilities leading to pattern formation or coherent temporal
behaviour over macroscopic space and time scales.[1] In contrast to those
exhibited in an hydrodynamical context,the spatial structures which
appear in some chemically active media have long been considered as
curiosities, spurious effects or even as bad experimentation, and aroused
at first very little interest. It should however be clear from the scope
of the current literature that an enormous amount of theoretical, com-
putational and experimental work is now being reported where chemically
reacting systems exhibit multiple steady states, periodic solutions,
wave phenomena and pattern formation. In addition, many of these reaction
systems have great scientific and technological importance [2]. Their
possible relation to numerous important biological phenomena is also often
emphasized.

The concept of chemical instability is used specifically when the
chemical reactions provide the driving force behind the instability.

The analysis of oscillatory, even chaotic behaviours of autocatalytic
chemical networks occuring in *continuously stirred flow tank reactors*
is now well advanced [3]. The literature presents however an evergrowing
number of such systems in *unstirred batch reactors* which exhibit inhomogeneous concentration structures [4,5] .

From the experimental point of view analyzing spatial chemical structures presents certain specific difficulties. Indeed one has to work at
a distance from equilibrium in closed unstirred reactors without any
feeding of the reagents. The system therefore tends to drift in the space
of its parameters towards the absolutely stable point representing the
final equilibrium state as given by the law of mass action and the structures thus tend to appear as transients. However in some cases - and this
includes our problem - this final equilibrium state is not approached

for a long time, and the interesting phenomena happen on shorter time
scales. It is then possible (and appropriate), from the theoretical
point of view, to make approximations which produce a simpler set of
kinetic equations, usually a subset from the original set.(The simplest
example of such approximations appropriate for instance if some of the
substances are initially distributed fairly homogeneously and in concentrations large compared to the amount by which they change over the
short time scales otherwise of interest, is that these particular concentrations may be taken as approximately constant in the equations for
the rates of change of the others). These approximate chemical kinetic
equations, even though they describe accurately the course of the reactions on the short time scale, need not then have an absolutely stable
point, and might for instance have a stable limit-cycle solution.
Furthermore the detailed mechanism of the reaction networks exhibiting
spatial structures is complex, involving numerous (often dozens) inter-
mediates and usually not completely resolved.

All this led to the creation of prototype schemes (Brusselator, Ore-
gonator, ...), model kinetic equations $\dot{c} = F(c)$, as a compromise between
a minimum of chemical realism and mathematical tractableness. (Here c
is a vector whose components represent the concentrations of the various
substances which participate in the reaction, while F(c),the rate at
which reactions occur, is a vector function, eventually non linear, of
these concentrations.)

Obviously some account of transport processes is required for an
understanding of the development of spatial structuration, and the
simplest conceivable addition to the purely chemical kinetic equations
consists of diffusional currents. Therefore *reaction-diffusion* systems
form an important class of models capable of accounting for major features
of organization. The phenomenological rate equations for the local concentrations of the intermediate species then take   the form

$$\dot{c} = F(c,\lambda) + D\nabla^2 c \tag{1}$$

where D is a positive definite matrix of diffusitivities and λ stands
for a set of parameters describing the external constraints (e.g. concentrations of buffer products). We have thus limited ourselves to the
case where the hydrodynamical fluxes other than diffusion are sufficiently
small not to affect chemistry appreciably. For instance we will consider
neither couplings with the heat equation (as occurs for instance in combustion [6]), nor with the Navier-Stokes equations (as occurs for instance in the problem of the structure of flame fronts [7]). It is also
certainly desirable to avoid the onset of convection which might confu-
se the study of chemical stuctures [8] (see also the contribution of

J. Micheau and M.Gimenez in these proceedings).

We will now focus on two kinds of structuration phenomena occuring in chemical systems and involving two different types of symmetry breaking processes.

2. AUTOWAVES

Experimentally it seems that autowaves may be observed in nearly every thin layer of chemical system that exhibits bulk temporal oscillations [4] or that is excitable. The most conspicuous example is that of the Belousov-Zhabotinsky reaction. The reacting solution placed in a covered Petri dish in the presence of ferroin as indicator alternates in colour periodically between red and blue reflecting the oscillations (limit cycle) of chemical composition. In the course of time, *leading centers* appear which have higher frequency than that of the bulk oscillation. As a result circular waves are sent out in succession from these centers, developing into *target patterns* [9] (concentric alterning red and blue rings). The spacing of the rings and the period of oscillation are quite uniform over a given target but are not usually the same in different targets, nor is the propagation speed. Where adjacent targets meet, the colliding waves annihilate one another so that cusped structures are formed. Also no reflections occur at the boundaries of the medium. This behaviour, as indicated by N. Wiener [10] is a consequence of the parabolic nature of the reaction-diffusion equations. In the presence of disturbances (impurities, concentration inhomogeneities, hydrodynamical perturbations, ...) rotating *Archimedian spiral waves* may appear, isolated or by pairs with opposite winding number [11]. Multi-armed spirals may even be nucleated by special experimental procedures [12]. The wavelength is a constant for all spirals of the same chirality.

Theoretically [13], one considers a reaction-diffusion system possessing a stable homogeneous limit cycle in some range of its parameters. The local fluctuations about this state may be described by the variables $(R(r), \theta(r))$ where θ parametrizes points along the cycle, while R measures displacements normal to it (when the cycle is a circle, R and θ are the usual amplitude and phase fluctuations. We keep this terminology even for the more complicated situations [14]). The long ranged phase fluctuations relax diffusively, hence much more slowly than the amplitude fluctuations which are of the relaxational type. This is a consequence of the spontaneous breakdown of the phase symmetry at the Hopf bifurcation leading to the limit cycle. The dynamics of these phase fluctuations which govern the long time behaviour of the system is then given by a stochastic Burgers type equation

$$\partial_t \theta(r,t) = \mu \, \nabla^2\theta + \nu \, (\nabla\theta)^2 + \eta(r,t) \tag{2}$$

where the noise term depends on the shape of the cycle $(X(t),Y(t))$

$$< \eta(r,t) \, \eta(r',t') > \, = \, \delta(r-r')\delta(t-t') \, \frac{8\pi^2\Gamma}{T\int_0^T d\tau \, [\,(\partial_\tau X)^2 + (\partial_\tau Y)^2\,]}$$

$$= 2(\Gamma d/\kappa) \, \delta(r-r')\delta(t-t') \tag{3}$$

and where μ and ν are functions of the intrinsic parameters of the system (diffusion constants, rate constants, buffer concentrations, ...). We consider $\mu > 0$ which guarantees linear stability of the limit cycle while $\mu < 0$ would lead to phase turbulence [15]. Hence the asymptotic probability for the phase fluctuations may be written

$$P \cong \exp \frac{\kappa}{\Gamma} \int dr (\nabla\theta)^2 \tag{4}$$

Accordingly, we expect in a thin layer chemical system behaviours analog to those exhibited by two dimensional equilibrium systems (i.e. non existence of true long range order [16]) except that the dynamics is here not purely diffusive.

On the one hand, phase singularities may appear at isolated points where the amplitude of the oscillations vanishes. They are characterized by their topological charge $N = \frac{1}{2\pi}\oint dr\nabla\theta$ and their associated fluctuations induce concentration waves in the form of N armed rotating Archimedean spirals originating at the singularity

$$\theta(r,t) = \omega_0 t + N \arctan (y-y_i)/(x-x_i) + \max (0,\nu k^2(t-t_0)-k.|r-r_i|)$$

It may be shown [13] in the asymptotic regime that one armed spirals are the most likely to occur while N armed spirals should decay into N one armed spiral. This tendency has been observed by Agladze and Krinsky [12] who were the first to be able to generate N (>1) armed spirals in a chemically active medium. In relation to the analogy with equilibrium situations, we note that the presence of isolated spirals would indicate the complete desynchronization of the chemical oscillator.

On the other hand, target pattern may also develop and their origin is still controversial. They may indeed be induced by impurities but the question of their possible intrinsic origin remains open. We think that because of the strong coupling between phase and amplitude fluctuations on the short length scales, sufficiently localized phase fluctuations may induce, at least temporarily, local frequency shifts of the oscillations (i.e. small regions where the system oscillates faster than the bulk). These leading centers generate outgoing concentric waves [17] corresponding to isoconcentration lines defined by

$$\theta(r,t) = \omega_0 t + \max (0,\nu k^2(t-t_0)-k.|r-r_i|)$$

where t_0 is the ignition time of the center and $\underline{r}_i$ its location, ω_0 is the bulk frequency. The wavenumber k depends on the intensity α and the range L_0 of the generating fluctuation

$$kL_0 \sim \exp -(2\pi d/\alpha) \tag{5}$$

However the asymptotic regime is practically never reached in real experiments because of the transience referred to in the introduction. Nevertheless, if one focuses on the *early stages* of the evolution after the stirring which maintained the homogeneity of the bulk oscillation has been interrupted, the average number of centers as a function of the wavenumber is given by [13]

$$n(k,t) = [\exp (\pi K(t)/\Gamma \ln^2 kL_0)-1]^{-1} \tag{6}$$

with

$$K(t) \simeq \kappa[1 + (\Gamma D_0/\kappa)(L_0^2/ L_0^2+2\mu t)^2] \tag{7}$$

while the average total number of centers behaves as

$$n(t) \sim \Gamma/\pi K(t) \tag{8}$$

where D_0 ($\geqslant 1$) is the effective diffusion coefficient associated to the stirring mechanism. These results which are functions of the parameters of the bulk oscillation are a signature of the intrinsic character of the centers. They are at least in qualitative agreement with the experimental observations [9].(see Figure)

One may along the same lines show that on such times spiral waves are less likely to appear.

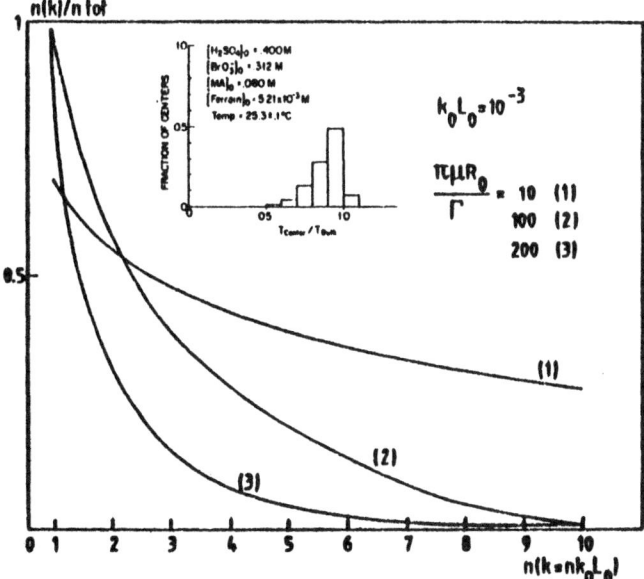

Fraction of centers as a function of their wavenumbers for different values of the parameters.The inset (see ref.9) is an experimental result for this quantity as a function of the period of the centers.

3. CELLULAR STRUCTURES

Very few experimental examples of *mosaic* (cellular) structures have been reported in chemical systems (see the contribution of J.Micheau and M.Gimenez for photochemical structuration). No quantitative systematic experimental study has yet been undertaken mainly because of the transient nature of these structures for the reasons alluded to in the introduction. It seems however that hydrodynamical processes, other than diffusion, play an important role in most of those experiments.

However from the theoretical point of view such stationary concentration structures are indeed solutions of the governing reaction-diffusion equations (Turing's instability). As a matter of fact, close to the bifurcation point one can obtain [19] a contracted description of the reaction diffusion system in terms of the order parameter only which is associated to the unstable mode (which is usually some combination of the intermediate chemical concentrations).

Its equation of motion takes the familiar form

$$\partial_t \sigma_q = \omega_q \sigma_q - \sum_{q_1}{}' v(q_1,q)\, \sigma_{q_1} \sigma_{-q_1+q}$$

$$- \sum_{q_1 q_2}{}' u(q_1,q_2,q)\, \sigma_{q_1} \sigma_{q_2} \sigma_{-q_1-q_2+q} \tag{9}$$

In (9), the summation is restricted to wavenumbers close to q_c. In the case of the Rayleigh-Benard instability, the quadratic term is due to non-Boussinesq effects (e.g. temperature dependent transport coefficients) a similar term is always present in the case of the Marangoni or Turing instabilities. Let us note that contrary to most hydrodynamical instabilities, the critical wavelength q_c is determined by intrinsic properties of the chemical system: rate constants, diffusion coefficients. When (9) presents a gradient structure (this property is sometimes satisfied in the vicinity of the bifurcation point), one may define a Lyapounov functional. The pattern selection problem is thus analogous to that occuring in hydrodynamical problems.

We will now distinguish between iso- and anisotropic systems.

a. Isotropic systems

In this case

$$\omega_q = \epsilon - (q^2 - q_c^2)^2 \quad ; \quad \epsilon = b - b_c / b_c \tag{10}$$

and if one furthermore describes local fluctuations in the usual way by a gaussian white noise and when the non linear coupling terms do not depend on the angles between the interacting wavevectors, the Lyapounov functional takes the simple form (Brazovsky's model)

$$V = \frac{1}{\Gamma} [\sum' \omega_q |\sigma_q|^2 + \frac{v}{3!} \sum \sum' \sigma_{q_1} \sigma_{q_2} \sigma_{-q_1 -q_2}$$

$$+ \frac{u}{4!} \sum \sum \sum' \sigma_{q_1} \sigma_{q_2} \sigma_{q_3} \sigma_{-q_1 -q_2 -q_3}] \tag{11}$$

This functional then plays the role of a generalized potential far from equilibrium. Such functionals have been derived in the case of the Rayleigh-Benard instability, the Turing instability and hydrodynamic instabilities in nematic liquid crystals.[20,19,21]

Each pattern can be characterized by m pairs of wavevectors $(q_i, -q_i)$. For the sake of simplicity we consider explicitly in the following only the structures which minimize the potential (11): $|q_i| = q_c$. In space variables the corresponding order parameter becomes

$$\sigma(r) = 2 \sum_{i=1}^{m} a_i \cos q_i \cdot r \tag{12}$$

Two classes must be considered for $d = 2$:

A. The structures described by m independent pairs. In that case the quadratic terms in the equations of motion do not contribute. The pattern appears supercritically through a second order like phase transition; we get indeed from the stationarity condition of equation (9)

$$a_m = \begin{cases} 0 & b < b_c \\ [\frac{2(b-b_c)}{(2m-1)u}]^{1/2} & b > b_c \end{cases} \tag{13}$$

B. The structures the wavevectors of which satisfy the triangular condition $q_1 + q_2 + q_3 = 0$

$$\sigma_3(r) = 2a_3 [\cos q_c x + \cos \frac{q_c}{2}(x + \sqrt{3}y) + \cos \frac{q_c}{2}(x - \sqrt{3}y)] \tag{14}$$

Depending on the sign of the cubic term v in (9), the maxima of concentration respectively define a triangular $(v < 0)$ or a honeycomb lattice $(v > 0)$. These patterns are the chemical analogues of the Benard l-hexagones (upward motion in the center) or g-hexagons (downward motion in the center) in non-Boussinesq fluids [22]. These structures appear subcritically through a first order like transition. The amplitude jumps indeed discontinuously to a finite value. Such behaviour has been verified experimentally by Berge and coworkers in water near its 4°C anomaly [23] and by Pantaloni and coworkers for the Marangoni instability in silicone oil [24]. The physically possible patterns correspond to the fraction of all the stationary solutions which is stable with respect to arbitrary disturbances of infinitesimal amplitude. In the case of model (11) all the solutions of class A are unstable with the exception of the case m = 1 corresponding to a stationary wave periodic in one direction (rolls). In this case the stability analysis strongly restricts the manifold of possible solutions; only two structures remain: rolls and hexagons. The same principle of selection through stability can still be applied to more complex problems where a variational formulation is not possible.

There is a range of parameters (from b_c to $b_c(1 + 8v^2/u)$ where rolls and hexagons coexist. In this regime, the potential may be used to calculate the relative stability of the coexisting patterns and determine the value of the control parameter at which one structure becomes more stable than the other. Such transitions between rolls and hexagons have been experimentally studied and hysteresis effects have been detected at this transition [23,25].

The bifurcation diagram of Brazovsky's model (u,v costant) is unfortunately not universal. Indeed the angular dependence of the coupling terms can sometimes play an important role in the pattern selection. To illustrate this point we now consider the case of the variational model introduced by Sivashinsky [26] to describe planforms of the buoyancy driven instability with nearly insulated layers. This model is defined by the following equation for the order parameter $\sigma(r,t)$

$$\partial_t \sigma = [\epsilon - (\nabla^2 + q_c^2)^2]\sigma + 3[(\frac{\partial\sigma}{\partial x})^2(\frac{\partial^2\sigma}{\partial x^2}) + (\frac{\partial\sigma}{\partial y})^2(\frac{\partial^2\sigma}{\partial y^2})]$$

$$+ (\frac{\partial\sigma}{\partial x})^2(\frac{\partial^2\sigma}{\partial y^2}) + (\frac{\partial\sigma}{\partial y})^2(\frac{\partial^2\sigma}{\partial x^2})$$

$$+ 4(\frac{\partial^2\sigma}{\partial x\partial y})(\frac{\partial\sigma}{\partial x})(\frac{\partial\sigma}{\partial y}) . \tag{15}$$

In this case the only stable structure in class A defined above is the square pattern. The amplitude of the sqares can in general be determined by the following stationarity condition:

$$\epsilon a_i - g_D a_i |a_i|^2 - g_{ND} a_i |a_j|^2 = 0 \tag{16}$$

(i and j correspond to the two orthogonal directions of the structure). From (15) we obtain $g_D = 3q_c^4$ and $g_{ND} = 2q_c^4$; the squares of amplitude $a_i = a_j = (\epsilon/5q_c^4)^{1/2}$ appear through a second order like transition (exchange of stability with the homogeneous state) whereas the rolls are now unstable. More generally the squares will be selected whenever in (16) we have the inequality $g_D > g_{ND}$ as a result of the dependence of the nonlinear terms on the angles between the interacting wavevectors (here q_i and q_j). In Sivashinsky's model hexagons are marginally stable and could appear in the supercritical region. The present situation exhibits also analogies with the experimental results found in the case of the Rayleigh-Benard instability in homeotropic nematics (H = 0) [27] and of a nematic subjected to an ellipital shear when the ellipticity (E) is equal to one [28]. In both cases one has a direct transition to a square structure but hexagons are often met with the squares in the convective geometry. In these situations the hexagons can be stabilized by non-Boussinesq effects resulting for instance from the rapid variation of elastic and viscous coefficients with temperature.

b. Anisotropic systems

In such systems there is an *intrinsic* mechanism which raises the orientational degeneracy by inducing preferred directions for wavevectors characterizing the structures. We illustrate this property on the rather academic problem of the Turing instability in the Brusselator model in a two dimensional *uniaxial* medium (the principal axis is parallel to 0_x). The kinetic equations for the concentrations of the intermediate species α and β can be written

$$\partial_t \alpha = A - (B+1)\alpha + \alpha^2\beta + D_\perp^\alpha \nabla^2\alpha + D_a^\alpha \frac{\partial^2\alpha}{\partial x^2}$$

$$\partial_t \beta = B\alpha - \alpha^2\beta + D_\perp^\beta \nabla^2\beta + D_a^\beta \frac{\partial^2\beta}{\partial x^2} \tag{17}$$

where A and B are kept constant (see introduction) and B is the control parameter while $D_\perp^i$ represents the diffusion coefficient in the direction perpendicular to the principal axis and $D_a^i = D_\parallel^i - D_\perp^i$ is a measure of the anisotropy of the corresponding diffusion current.

This model can be considered as a caricature to describe recombination processes between excitations (e.g. phonons, vacancies, interstitials) in irradiated condensed matter systems [29].

As in the isotropic case [19] this model displays a symmetry breaking instability. Indeed the homogeneous solution $\alpha_0 = A$, $\beta_0 = B/A$ becomes unstable for $B > B_c = (1+A\eta)^2 < 1+A^2$ where

$$\eta(\phi) = \left(\frac{D_\perp^\alpha + D_\parallel^\alpha \cos^2\phi}{D_\perp^\beta + D_\parallel^\beta \cos^2\phi} \right)^{1/2} \tag{18}$$

against inhomogeneous fluctuations of wavelength $q_c(\phi)$:

$$q_c^2(\phi) = A[(D_\perp^\alpha + D_\parallel^\alpha \cos^2\phi)(D_\perp^\beta + D_\parallel^\beta \cos^2\phi)]^{-1/2} \tag{19}$$

and making an angle ϕ with the principal axis. The preferred orientation ϕ_0 is obtained by minimizing $B_c(\phi)$. One finds that if

$$D_\perp^\alpha / D_\perp^\beta < (D_\perp^\alpha + D_a^\alpha) / (D_\perp^\beta + D_a^\beta)$$

then $\phi_0 = \pi/2$ and the axis of the rolls is parallel to the principal axis, whereas when

$$D_\perp^\alpha / D_\perp^\beta > (D_\perp^\alpha + D_a^\alpha) / (D_\perp^\beta + D_a^\beta)$$

then $\phi_0 = 0$ and the axis of the rolls is thus perpendicular to the principal axis. In this simple example, the anisotropy in the transport coefficients induces preferred directions for the critical wavevector (easy axis) and this selection appears already in the linear analysis. In these anisotropic systems the frequency ω_q now contains a symmetry breaking term. In the example discussed above the frequency takes the form (at the lowest order):

$$\omega_q = \frac{b - b_c}{b_c} - (q^2 - q_c^2)^2 - Aq^2 \sin^2 \phi \tag{20}$$

When $A > 0$, this expression corresponds to an easy axis parallel to the principal axis ($\phi_0 = 0$). A similar expression can be derived in the case of the Rayleigh-Benard instability in liquid metals in presence of a horizontal magnetic field which tends to favour longitudinal rolls, i.e. rolls having their axis parallel to the field [30]. In that case ϕ is the angle between the wavevector and a direction perpendicular to the magnetic field. In the case of the instability of a nematic subjected to an elliptical shear it is possible to change the sign of the symmetry breaking term because the ellipticity measures the importance of the anisotropy of the shear and fixes the orientation of the roll structure. When one goes from $E < 1$ to $E > 1$ experimental results show a jump in the orientation of the rolls at $E = 1$ [31].

Anisotropies can also play an important role in the selction of the possible structures. Let us consider a system where in absence of anisotropy the angular dependence of the coupling term tends to select a square structure (cf.eq. 15,16). On the other hand we have seen that the introduction of an anisotropic effect induces roll structures making a well defined angle. The amplitudes of the rolls and squares of such a system can be determined from the following equations (cf.eq. 16):

$$\varepsilon a_1 - a_1 [g_D|a_1|^2 + g_{ND}|a_2|^2] = 0$$

$$(\varepsilon - A)a_2 - a_2 [g_D|a_2|^2 + g_{ND}|a_1|^2] = 0 \tag{21}$$

The anisotropy A tends to align the critical wavevector parallel to the direction "1". We furthermore take $g_D > g_{ND}$ so that when $A = 0$ the square pattern is selected. A stability analysis shows that rolls induced by the anisotropy become unstable for $\varepsilon > \tilde{\varepsilon} = g_D A / (g_D - g_{ND})$ where one has a direct transition to a square structure. In this model there is thus a competition between anisotropic and nonlinear effects. First, rolls induced by the anisotropy are selected but as the bifurcation parameter is increased, the nonlinearities become more important and favor the onset of a square structure. Here also this model presents similarities with the experimental results obtained in the case of the Rayleigh-Benard instability in nematics heated from above in the presence of a horizontal magnetic field or the instability of a nematic subjected to an elliptical shear when the ellipticity of the shear is different from one ($\hat{A} \equiv E$) [32] .

It has also been shown that long range fluctuations [33] manifest themselves in the case of cellular structures because of the spontaneous breakdown of the translation and orientation symmetry. This again can induce the presence of topological defects: dislocations, disclinations, grain boundaries which may ultimately be responsible for the disorganization, or "melting" [34] of the pattern at higher constraints as is observed for instance in nematics subjected to an elliptical shear or in the Marangoni problem [35].

REFERENCES

1. G.Nicolis and I.Prigogine: "Self Organization in Non Equilibrium Systems" (Wiley, New York, 1977)
2. G.Nicolis and F.Barras;Eds.: "Chemical Instabilities - Applications in Chemistry, Engineering, Geology and Materials Science" (Reidel, Doordrecht, 1984)
3. C.Vidal and A.Pacault;Eds.: "Non Linear Phenomena in Chemical Dynamics" (Springer, Berlin, 1981)
4. C.Vidal and A.Pacault in "Evolution of Order and Chaos in Physics, Chemistry, and Biology "- Synergetics vol.17, H.Haken Ed.(Springer, Berlin, 1982) pp 74-99
5. G.R.Ivanitsky, V.I.Krinsky, A.N.Zaikin and A.M.Zhabotinsky, Soviet Scientific Review,D, Biology Reviews $\underline{2}$ (1981) 279-324
6. See for instance Section II of reference 2.
7. P.Pelcé and P.Clavin, J.Fluid Mech. $\underline{124}$ (1982) 219-237
   G.I.Sivashinsky, Ann.Rev.Fluid Mech. $\underline{15}$ (1983) 179-199
8. G.Dewel, P.Borckmans and D.Walgraef, Proc.Natl.Acad.Sci. USA $\underline{80}$ (1983) (to appear)
9. M.L.Smoes in "Dynamics of Synergetic Systems"- Synergetics vol.6, H. Haken Ed.(Springer, Berlin, 1980) pp 80-96
10. N.Wiener and A.Rosenblueth, Arch.Inst.Cardiol.Mex. $\underline{16}$ (1946) 205-206
11. A.T.Winfree: "The Geometry of Biological Time" (Springer, Berlin,1980) Science $\underline{175}$ (1972) 634; $\underline{185}$ (1973) 937
12. K.I.Agladze and V.I.Krinsky, Nature $\underline{296}$ (1982) 424-426
13. D.Walgraef, G.Dewel and P.Borckmans, J.Chem.Phys. $\underline{78}$ (1983) 3043-3051
14. J.Neu, SIAM J.Appl.Math. $\underline{36}$ (1979) 509-512
15. Y.Kuramoto, Physica $\underline{106A}$ (1981) 128-143

16. J.M.Kosterlitz and D.J.Thouless, J,Phys. C6 (1973) 1181-1203
17. P.S.Hagan, Adv.Applied Math. 2 (1981) 400-416
                    SIAM J.Appl.Math. 42 (1982) 762-786
18. J.C.Micheau, M.Gimenez, P.Borckmans and G.Dewel, Nature 305 (1983) 43
19. D.Walgraef, G.Dewel and P.Borckmans, Adv.Chem.Phys. 49 (1982) 311-355
20. R.Graham, Phys.Rev. A10 (1974) 1762-1784
    J.Swift and P.Hohenberg, Phys.Rev. A15 (1977) 315-328
21. P.Manneville, J.Physique 39 (1978) 911-925
22. F.Busse, Rep.Progr.Phys. 41 (1978) 1929-1967
23. M.Dubois, P.Bergé and E.Wesfreid, J.Physique 39 (1978) 1253-1257
24. J.Pantaloni, R.Bailleux, J.Salan and M.Velarde, J.Non Equilib.Thermodyn. 4 (1979) 201-217
25. K.C.Stengel, D.S.Oliver and J.R.Booker, J.Fluid Mech. 120 (1982)411
26. V.L.Gertsberg and G.I.Sivashinsky, Progr.Theor.Phys. 66 (1981)1219
27. P.Pieransky, E.Dubois-Violette and E.Guyon, Phys.Rev.Lett. 30 (1973) 736-739
28. J.M.Dreyfus and E,Guyon, J.Physique 42 (1981) 459
    E.Guazzelli and E.Guyon, C.R. Hebd.Séan.Acad.Sci. 292 (1981) 142
29. G.Martin, Phys.Rev.Lett. 50 (1983) 250-252
30. P.Tabeling, J.Physique 43 (1982) 1295-1303
31. E.Guazzelli and E.Guyon, J.Physique 43 (1982) 985-989
32. E.Guazzelli, (1981) Thèse de 3ème cycle-Orsay.
33. Y.Pomeau and P.manneville, J.Physique Lett. 40 (1979) L609
34. D.Walgraef, G.Dewel and P.Borckmans, Z.Phys. B48 (1982) 167
35. R.Occelli, E.Guazzelli and J.Pantaloni, J.Physique Lett. 44 (1983) L567.

STRUCTURE DEFECTS IN BENARD-MARANGONI INSTABILITY

J. Pantaloni, P. Cerisier

Laboratoire de Thermophysique
Université de Provence
13397 MARSEILLE CEDEX 13

1. INTRODUCTION

The Benard-Marangoni convective instability is characterized by a monolayer of hexagonal prismatic cells with a vertical axis (1). Practically this structure is not perfectly regular : there is always a certain amount of structural defects that we intend to study qualitatively in this paper. In a preliminary work we concluded that the presence of structural defects is an intrinsic phenomenon : structural defects always exist independently of the shape of walls, aspect ratio, and the distance to the threshold.

A thin, horizontal silicone-oil layer (from 1 to 3mm) heated from below lies on a rectangular copper plate at uniform temperature, the upper surface being in free contact with the ambient air. The room temperature is stable for the three days of the experiment. A shadowgraph method (2) is used to visualize the cell pattern on a ground glass plate. So there is no perturbation of the flow as in the technique of visualization using small particles (aluminium flakes). For a fixed value of distance to threshold, photographs of the convective structure are taken with a camera at regular time intervals at the frequency $2.5 \ 10^{-2}s^{-1}$. The observation of the cine-film, projected at the normal usual speed, accelerates the movement by a factor about one thousand. So the main conclusion is that the structure is in perpetual evolution : the defects move, they change, they appear and then disappear. There is an obvious similarity between the lattices (and their defects) of hexagonal two-dimensionnal systems and this convective instability. This similarity was recognized, to our knowledge, for the first time in 1980 (3). In the same way this 2-D lattice exhibits also some similarity with 2-D structures existing in various fields in nature, such as zoology, botany, earth science, etc... For instance, let us think of the hexagonal lattices - and to the defects - of a lizard skin, of a honeycomb, of a fossil diatomea, of a heart of marguerite, of an ear of maize and of dried up soil of some lakes in Ethiopia or North Africa etc... Other similarities can be noted. For instance, as we shall see, the "birth" of a new convective cell looks like that of a protozoan.

2. DESCRIPTION OF DEFECTS

In the following we denote pentagons, hexagons and heptagons by P, H6 and H7 respectively. From a geometrical point of view the structural defects are numerous. They differ by the number of polygons

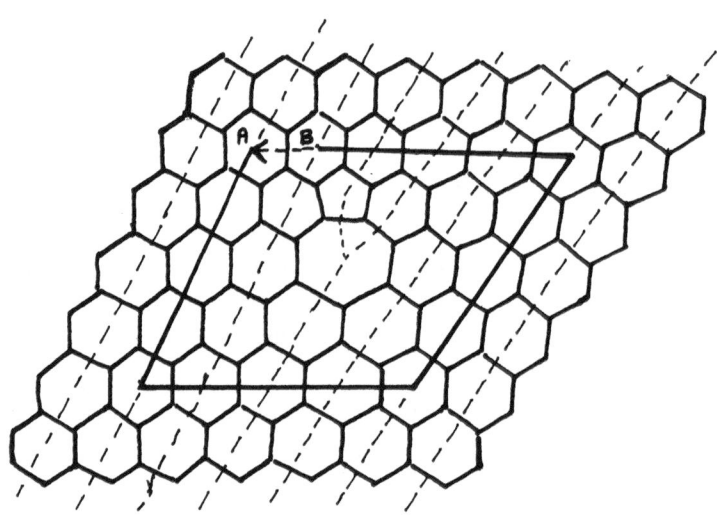

Figure 1. Basic defect (P+H7) and corresponding dislocation and Burgers vector 1.

which compose it, by their nature, their shape, and their way of aggregation. Further more, when the distance to threshold is large, the convective layer is in the preturbulence state : the defect number is high and it is getting more difficult - or even impossible - to distinguish isolated defects.

The defect structure of a 2-D system is conveniently characterized by the coordination numbers (CN) of each cell. For a perfect hexagonal lattice CN=6. The construction of Wigner-Seitz (or Dirichlet domain) provides an unambiguous definition of nearest neighbors as shown in figure 1. The lines drawn are the perpendicular bissectors of vectors to neighbour cell centre. The straight vector linking the two extremites of the path around a defect defines the Burgers vector AB.

Two kinds of defects can be considered :

1. Defects with a Burgers vector zero

The simplest one is an irregular H6. It is never alone and is not considered here. Clusters of several polygons exist, and build geometrical arrays which sometimes look like flowers (figure 2).

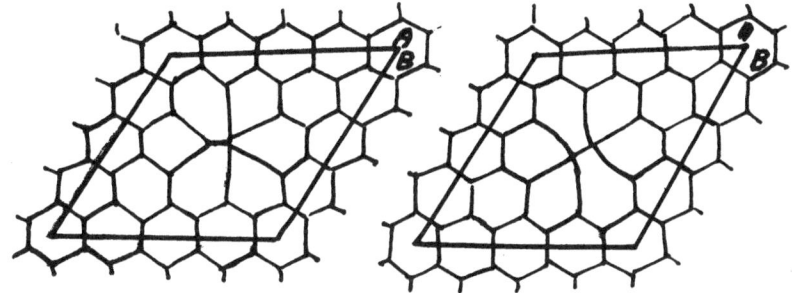

Figure 2. "Flower" defects

Generally speaking these defects have the same contour as a cluster
of several (regular or irregular) H6. Inside the polygons there are
P and irregular H6. These defects are not stable, they disappear and
the regular lattice is restored (paragraph 9).

    2.   Defects with a definite Burgers vector

       They are numerous, but the basic defect is a pair (P+H7).
It is a true dislocation, while each non six fold coordinated cell can
be viewed as a disclination. Often P and H7 are

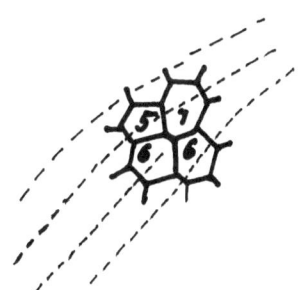

not regular and are coupled with a pair 2H6
then the four common sides of the quadrupole
are roughly perpendicular (Figure 3).
A dislocation pair is a 2(P+H7) quadrupole.
The disposition of 4a or 4b is very common.

Figure 3.   Pair (P+H7)

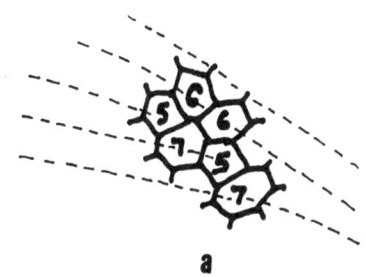

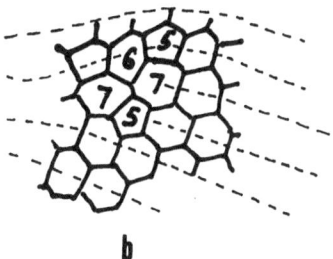

         a   b

Figure 4. Quadrupoles (P+H7)

Numerous
others clusters
are also obser-
ved. They are
for instance,
constituted by
4P, 5P,(3P+H6),
(2P+4H6),
(4P+4H6),
(figure 5).
The life time
of such clusters
is generally
short and a pair
(P+H7) is often

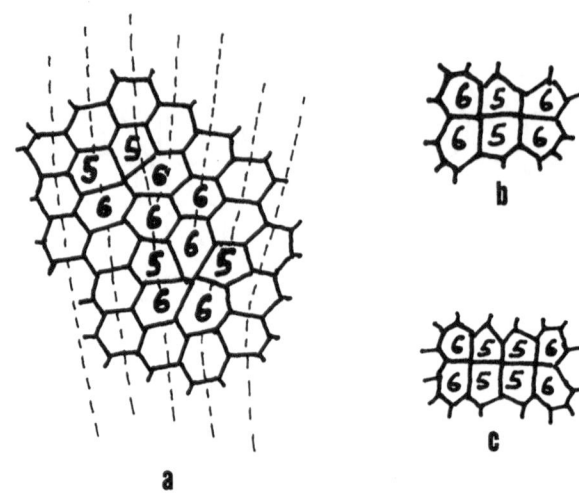

Figure 5. Multipoles

the "daughter structure" at the end of transformation. It is sometimes
difficult to ascertain the number of sides of a polygon : a very short
side and two adjacent segments can look like an angle. From a structural point of view, these clusters are dislocations (5a) and are
equivalent to one or several pairs (P+H7).

3. PROPAGATION OF A DISLOCATION

The transformation of a cluster often causes the displacement of
the dislocations. Two different processes can be considered :

1. Displacement with local variation of the cell number.

This process has its origin in the hydrodynamical nature of
the system. The cells are not independent : modifications
of the streamlines are possible, with transfer of liquid
from one cell to an adjacent cell. The dimensions of one cell are
variable (see paragraph 7) : it is often observed the coalescence of
two adjacent cells into one or the progressive disapearance of a cell
which is getting smaller and smaller. Of course the reverse is also
observed. For a fixed value of the distance to the threshold, the characteristic wave number of the structure is also fixed. So, the number
of convective cells is constant in a given vessel, but the experience
shows that the creation or disappearing of a cell can locally occur.
A synthesis of our observations is shown on figure 6. The P has a
tendency for giving a H6. So the side common to P and H7 fractures
into two segments (fig.6b). At the same time the fracture extends

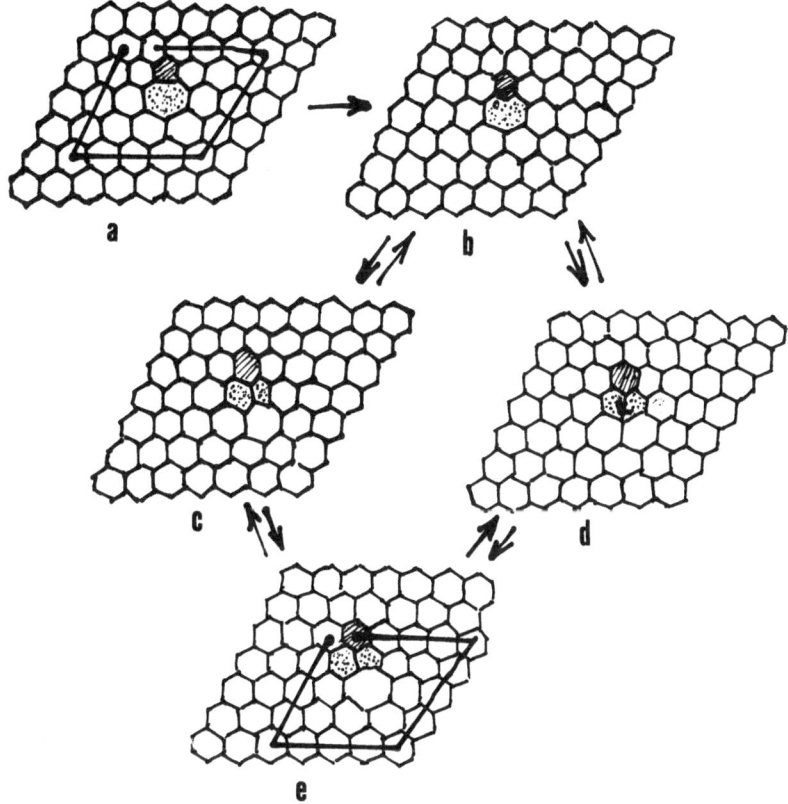

Figure 6. Displacement of a dislocation with local variation of the cells number.

through H7. There are two possibilities :

     . either the fracture moves to the opposite side with creation of a new pair (P+H7) (fig.6b, 6c)

     . or the fracture moves to the opposite angle, thus forming two P, and then, by association with two H6, a new cluster (2P+2H6) is created (6d). The latter will evolve to a new pair (P+H7)(fig.6d).

    2.   Displacement without local variation of the cells numbers.

      Some processes are very similar to the displacements of dis- locations described in the 2-D systems constituted by material par- ticles. The cell number is constant but, as the cells can be easily distorted, many processes are possible. As an illustration we show one of them on figure 7. First there is a simultaneous deformation of (P+H7) and two adjacent H6, then creation of a cluster 2(P+H6).

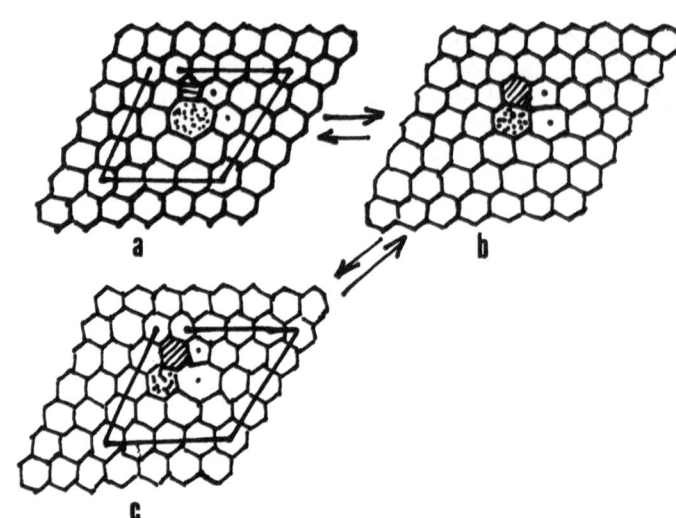

a

b

c

Then the mechanism is that described on figure 6d, 6e.

Figure 7. Displacement of a dislocation without local va- riation of the cells number.

4.   TRANSFORMATIONS OF DEFECTS

They are very numerous. Some of them have been described in the preceding paragraph. Now we introduce some examples frequently met.

1.   Creation of a new pair (P+H7)(figure 8)

First there is only one pair (P+H7)(8a). By simultaneous defor- mation of the latter and three adjacent H6, a cluster (4P+H6) is crea- ted. Then many evolutions can occur. We show (8c) one possibility with

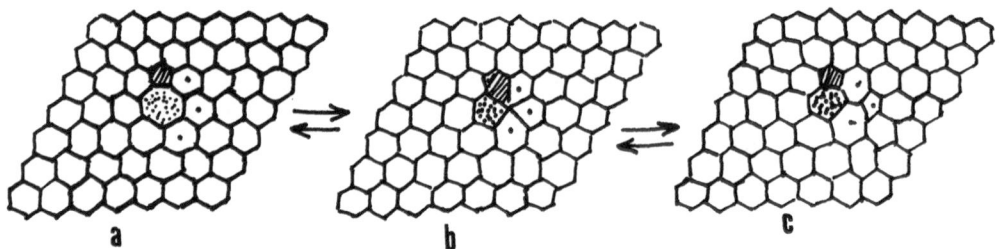

a

b

c

Figure 8. Creation of a new pair (P+H7)

creation of a second pair (P+H7). Both pairs are equivalent to a dislo- cation.

2.   Disappearance of a (P+H7) and dissociation of the second (P+H7) (figure 9).

One of the two pairs (P+H7) undergoes a deformation at the same time as three adjacent H6 and one P. (This mechanism is similar to the

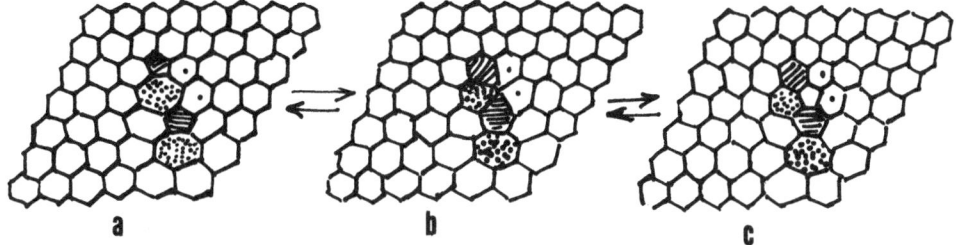

Figure 9. Disappearing of a (P+H7) and dissociation of the second pair.

preceding one but a P replaces a H6). So a 5P cluster is formed. Its centre is relatively unstable. Here two several possibilities of evolution exist ; we show one of them (9c) with creation of a P separated from the previous H7 by H6.

3. Disappearance of a cluster 6P (figure 10)

Here also the centre of this "flower" is relatively unstable. First there is a deformation of two opposite sides of the flower (10b). An unstable octogon is created. A fracture linking the new two angles produces two H6 (10c). Then the regular lattice is restored.

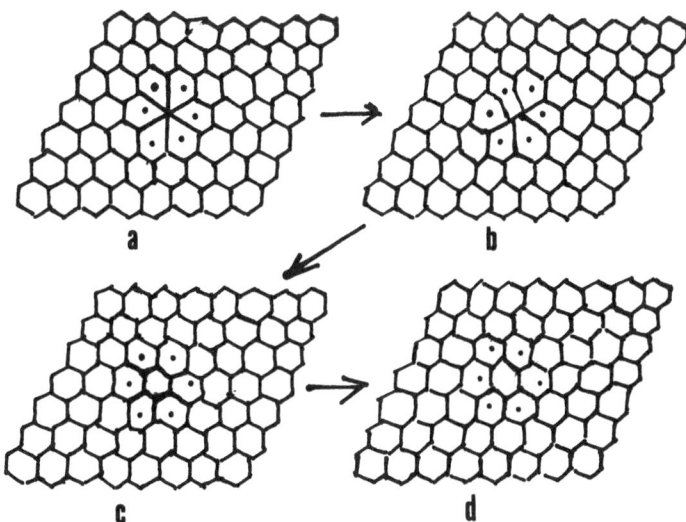

Figure 10. Disappearance of a cluster 6P.

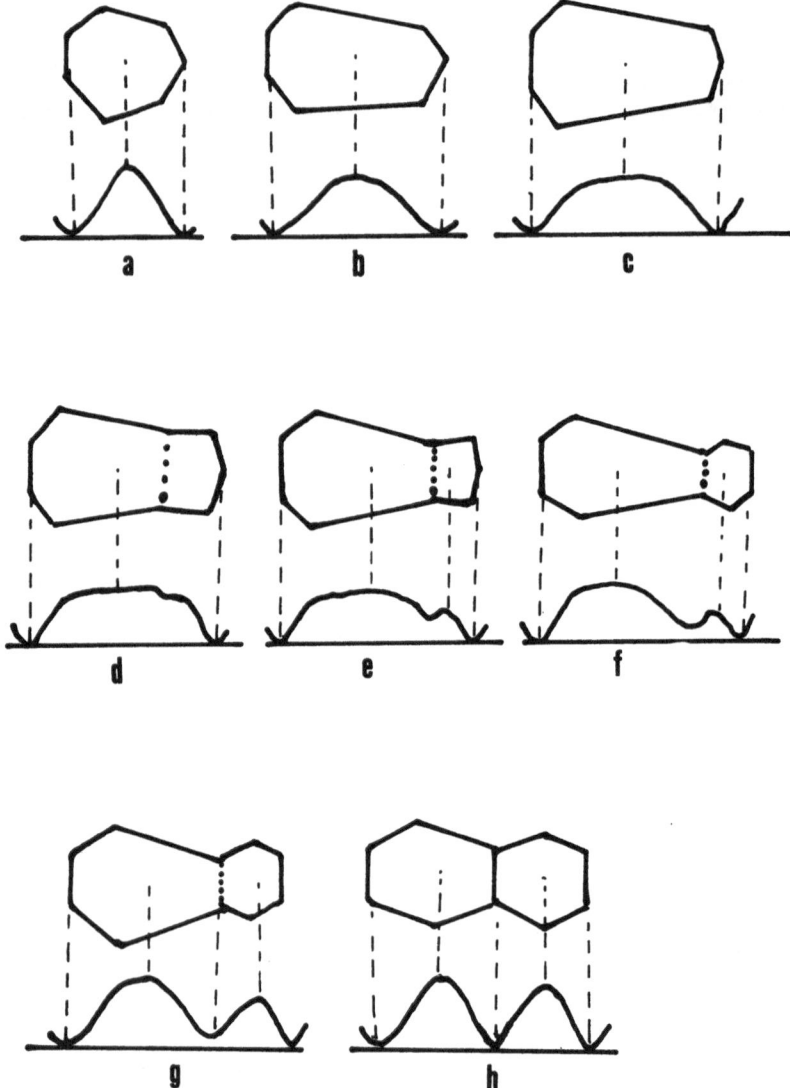

Figure 11. "Birth of a new cell. The upper line shows the shape of the cells ; the lower line shows the relief (or temperature).

5. FREE SURFACE RELIEF

The surface relief has been measured by various techniques (1).
All the defects exhibit the same deformation as one H6 of the regular
lattice (concave for small depths, and convex for large depths)(1).
The deformation is proportional to the average width of the convective
cell. The maximum deformation is that of a H6 of the regular lattice.

6. FREE SURFACE TEMPERATURE

This has been measured with an infrared camera (4). The surface
temperature field is similar to that of a H6 of the regular lattice.
The central part of the polygon, where the liquid is rising, is the
hottest zone. The dihedral angles, where the fluid is sinking, are
the coldest points. The isotherms are no longer concentric circles,
but are parallel to the cell walls (4).

7. MECHANISM OF CREATION OF A CELL

It has been studied from observation of the cell shape (shadowgraph
method, infrared camera), of the relief (interferences) and of the superficial temperature (infrared camera).

The "birth" of a new cell occurs by scission of the "mother" cell ;
this is schematically shown in figure 11·

It is clear that first, the liquid is rising along the central
axis of H7. Then H7 becomes elongated and at the same time the liquid
is rising along the central plane of H7. Somewhere on this plane a
deceleration of the flow occurs. This fact has three simultaneous
consequences in this region :

. decreasing of superficial temperature

. decreasing of relief with emergence of a "pass" more and
more distinct.

. deformation of two opposite sides of H7.

The processes continue until a "daughter cell" appears with an
individuality more and more pronounced. Then the size of the cell
increases. This phenomenon is relatively rapid (about one hour).

This process is reversible : the disappearance of a cell by the
inverse way is often observed. In the same way, the process can stop
at the stage d) or e) and come back to a) : This check occurs when a
faster birth happens in the neighbour cell.

8. CONCLUSION

By this qualitative study of Benard Marangoni convective structure it is clear that the mechanism of formation of defects can be fundamentally different from that of the defects in 2D structures of material particles (5). But the theoretical studies on material lattices can be a tool for the structure description of convective instabilities (6).

REFERENCES

(1) P. Cerisier, J. Pantaloni, Surface relief accompanying natural convection in liquid layers heated from below, this symposium.
(2) J. Pantaloni, P. Cerisier, R. Bailleux, C. Gerbaud, J. Phys. lettres 42, L-147, (1981).
(3) J. Pantaloni, P. Cerisier, Rencontres "Les Embiez", Société Française de Physique, Section Méditerranée (1980).
(4) P. Cerisier, J. Pantaloni, G. Finiels, R. Amalric, Applied Optics 21, 12, 2153 (1982).
(5) Ordering in two dimensions, Simil K. Sinha Ed. North Holland(1980)
(6) R. Occelli, E. Guazzelli, J. Pantaloni, Quantitative study of the disorganization of hexagonal convectives structures, this symposium.

NON ADIABATIC PHENOMENA

IN CELLULAR STRUCTURES

Y.POMEAU

SDR Research Center,Ridgefield,Ct,06877 Usa.

In absence of dpt of theoretical physics

CEN-Saclay,bp2,Gif/Yvette

91191,France

Abstract

Many physical phenomena,as weakly non linear convection are analysed by means of adiabatic or quasiclassical theories.Transcendendentally small terms are neglected in the corresponding expansions,although they give rise to interesting phenomena as the locking of dislocations or of grain boundaries on the fast phase

1.Introduction

The analysis of many physical phenomena uses implicitly or explicitly the concept of adiabaticity .Probably the most well known example of this is the WKB solution of the 1d wave equation

$$\Psi_{xx} + k^2(x)\Psi = 0$$

It starts as $\Psi = \Psi_0(x) e^{i\int^x k(x)dx}$ and is valid in the limit $1 k_x 1 << k^2$. If one expands this at any algebraic order in the smallness parameter one never finds any reflected wave ,that is a wave with the phase dependence $e^{-i\int^x k(x)dx}$. As shown by Dykhne[1],this is because the amplitude of this reflected wave is of order $exp(-kx_0)$,where x_0 is the location of the singularity of the complex extension of $k^2(x)$ that is the closest to the real axis .As $x_0 \approx \dfrac{k}{k_x}$ this amplitude is transcendentally small with respect to the expansion parameter.The reflection does not appear at any algebraic order because this "regular" expansion is not sensitive to the absolute phase of the fast modulation ,as viewed by the slow external parameter $k(x)$. All these phenomena are present in some form or another in the so called amplitude theory. This theory is aimed to describe weakly non linear cellular structures. It assumes that the wavelength of the structure ,say $\dfrac{2\pi}{q_0}$,is much smaller than the typical length of varia-tion of the amplitude,that is either $\epsilon^{-\frac{1}{2}}$ (parallel to the rolls)or $\epsilon^{-\frac{1}{4}}$ (perpendicular to the rolls) ϵ being the usual non linearity parameter. This amplitude equation reads[2] ,after convenient dimensionalization:

$$\chi_t = \epsilon\chi - 1\chi^2 1 + (\partial_x + i\partial_{y^2}^2)^2\chi$$

(1)

as usual t is the time,y (x) is the dimension parallel (perpendicular) to the mean roll

axis.

This equation has two formal invariances :it is invariant under phase changes

$\chi \rightarrow \chi e^{i\Phi}$ and it is formally autonomous with respect to the space variables x and

y.This double invariance reflects the single translational invariance of the original

equations of fluid mechanics.Thus it is broken,as is the formal phase invariance of the

WKB theory,by effects non analytic in the smallness parameter ϵ. And this breaking

gives rise to specific phenomena of locking of large scale structures with the 'fast

modulation.

I shall expand below on these points for the case of the grain boundary. I will first

describe the solution of the amplitude equation for this grain boundary and explain

the calculation of the force locking this g.b.on the fast modulation.

2.Grain boundary in the weakly non linear approach

Motivated by experiments[3],we have studied [4]the following type of grain boundary

:in the plane (x,y)rolls of amplitude χ_a parallel to the y-direction,fill the half plane

$x > 0$ and meet near $x \approx 0$ orthogonal rolls of amplitude χ_b perpendicular to the y-

direction.

In the steady situation that I shall consider from now on, the amplitudes χ_a and χ_b

are solutions of two coupled ordinary differential equations:

$$\epsilon\chi_a - \chi_a^3 - g\chi_b^2\chi_a + \chi_{a,xx} = 0$$

(2.a)

and

$$\epsilon \chi_b - \chi_b^3 - g \chi_a^2 \chi_b - \chi_{b,xxxx} = 0$$

$$(2.b)$$

In these equations ,one has introduced the real parameter g measuring the relative strength of the interaction between perpendicular and parallel rolls.Furthermore the amplitude have been taken to be real because these equations are purely real.We are seeking the solution of (2) that satisfies the b.c. $\chi_a \to \epsilon^{1/2}, \chi_b \to 0$ as $x \to \infty$ and $\chi_b \to \epsilon^{1/2}, \chi_a \to 0$ as $x \to -\infty$. As χ_a changes over scales of order $\epsilon^{-1/2}$ and as χ_b changes over scales of order $\epsilon^{\frac{1}{4}}$ that is much smaller than $\epsilon^{-1/2}$ as $\epsilon \to 0$ one assumes that χ_b adjusts itself adiabatically with χ_a so that one may solve (2.b)to obtain in this limit:

$$\chi_b \approx (\epsilon - g \chi_a^2)^{1/2}$$

Indeed this implies $g \chi_a^2 < \epsilon$. Thus (2.a) becomes:

$$\epsilon (1-g) \chi_a - \chi_a^3 (1+g^2) + \chi_{a,xx} = 0$$

$$(3)$$

If $g > 1$,as we shall assume it to avoid the formation of stable square pattern instead of grain boundary,the solution of (3)is

$$\chi_a = [\epsilon \frac{(g-1)}{2}]^{1/2} tg(x\eta)$$

with $\eta^2 = \epsilon \frac{(g-1)}{g^2+1}$, if $x < x^*$, x^* being defined by the turning point condition $g \chi_a^2 = \epsilon$.

And for $x > x^*, \chi_a$ is simply given by the solution of (2.a)with $\chi_b = 0$.

To estimate the force locking the grain boundary on the phase of the rolls of amplitude χ_a one has to analyse the vicinity of the turning point,that may always be chosen as $x^* = 0$ .Near this turning point one has

$$\epsilon - g\chi_a^2 \approx x\epsilon^{\frac{3}{2}} C(g)$$

$$(4)$$

,where C is some g-dependent numerical constant.

The inner equation for χ_b is obtained by putting the expansion (4) into (2.b). This gives:

$$-\chi_{b,xxxx} - \chi_b^3 + \chi_b x\epsilon^{\frac{3}{2}} C(g) = 0$$

$$(5)$$

This becomes an "universal"(=parameterless)equation by taking $[\epsilon^{\frac{3}{2}} C(g)]^{-\frac{1}{5}}$ as unit length and $[\epsilon^{\frac{3}{2}} C(g)]^{\frac{2}{5}}$ as unit amplitude. This gives :

$$-X_{b,xxxx} - X_b^3 + X_b x = 0$$

$$(6)$$

This equation has the same formal status as the Airy equation describing wavefunction near a 1d turning point. It has an unique solution with the limit behavior $\chi_b \approx x^{\frac{1}{2}}$ as $x \to \infty$ and $\chi_b \approx e^{-1 \times 1^{\frac{1}{4}}}$ as $x \to -\infty$. This last behavior is again quite similar to the one of the Airy function in the classically forbiden region.

To summarize the results of this analysis ,that is more detailed in ref[4],we notice that ,thanks to the scaling made near the turning point ,functions χ_a and χ_b vary around $x^{\cdot}$ on scales of order $\epsilon^{-\frac{3}{10}}$ although the outer region has the much larger length scale $\epsilon^{-\frac{1}{2}}$. Accordingly some complex singularities of the amplitudes are at a distance of the real axis of order $\epsilon^{-\frac{3}{10}}$. Their precise location depends on the inner equation (6) that does not seem solvable analytically due to its non linearity. Nevertheless this singularity is certainly at a finite distance of the real axis because the radius of convergence of the Laurent expansion of a solution of (6) is finite if this solution is finite for real arguments.

3 Force on the grain boundary

In that follows ,I sketch the calculation of the force acting on a grain boundary tending to lock it on the rapid phase .This calculation is done in the so called model a,to reduce as much as possible the algebraic manipulations.

The equation for this model is for steady state solution:

$$\Omega_\epsilon A = A^3$$

(7)

with $\Omega_\epsilon = \epsilon - (\Delta + 1)^2$. The corresponding amplitude equation for two sets of interacting rolls reads:

$$\epsilon \chi_a + 4\chi_{a,xx} - \frac{3}{4}\chi_a \mid \chi_a \mid^2 - \frac{3}{2}\chi_a \mid \chi_b \mid^2 = 0$$

(8.a)

$$\epsilon \chi_b - \chi_{b,xxxx} - \frac{3}{4}\chi_b \mid \chi_b \mid^2 - \frac{3}{2}\chi_b \mid \chi_a \mid^2 = 0$$

(8.b)

This is equivalent to (2) except for minor differences in numerical coefficient that may be absorbed by a convenient scaling.Thus the analysis done in the previous section applies to this particular model.

To put in evidence the transcendentally small terms one makes the following remarks.The amplitude equation as well as the equations deduced from it by continuing the expansion in ϵ are deduced from (7) by formally cancelling all slowly varying coefficients in front of e^{ix} . But actually these coefficients are only known by their series expansion and one cannot be sure that they cancel,unless this expansion gives their actual value. A simple argument shows that this is not the case. Consider a quantity like $A(x) = e^{ix}\chi(x\epsilon^\alpha)$, $\alpha > 0$, $\alpha = \frac{1}{2}$ for a dislocation, $\alpha = \frac{3}{10}$ for a grain boundary, $\chi(.)$ being the solution of the amplitude equation. Consider now the integral $\int_{-\infty}^{\infty} \chi(x\epsilon^\alpha)e^{2ix}dx$. Near $\epsilon = 0$ it is of order $e^{-x_1\epsilon^{-\alpha}}$, where x_1 is the singularity of χ the closest to the real axis.

This sort of term appears whenever one multiplies the original equation [equation (7) in the present case] by e^{ix} and integrates over the whole space in the limit $\epsilon \to 0$. This transcendentally small term has a fast phase dependence as $sin(2x_0)$ depending on the location of the slow amplitude with respect to the rapid phase. This means that they are two sets of non equivalent steady positions for the defect corresponding to the zeroes of $sin(2x_0)$ at 0 and $\frac{\pi}{2}$. These equilibrium states are not physically identical, as one may convince oneself by making drawings[6]. Stability considerations show that one of these configurations is stable the other one unstable. For a point dislocation in a 2d roll structure both correspond to the vanishing of the Peierls-Nabarro force[7].

4. Conclusion

We have already emphasized that the amplitude theory does not exhaust all the physics of weakly non linear phenomena, because at any algebraic order in the expansion parameter the rapid phase and the slow variations of the amplitude remain uncoupled. As a consequence the Peierls-Nabarro force and a similar force for g.b. are very small near the onset of convection. This could have observational consequences because in this domain of validity of the weakly non linear approach these very small forces have to be compared with other small effects as the thermal fluctuations,large scale flows,interaction with distant boundaries,etc... It is even possible that a weak phase turbulence could exist near the instability threshold where the locking forces are very small and disappear at slightly higher values of the external constraint.

REFERENCES

[1] Dykhne,JETF,38,570(1960), quoted in a footnote on page 218 of L.Landau,E.M Lifshitz **Mecanique** ed.Mir(Moscou)1960.

[2]L.A.Segel,J.Fluid Mech.38,203(1969);A.C.Newell,J.A.Whitehead,ibid., p.279.

[3] V.Croquette,A.Pocheau,to appear in J.de Phys.jan.84.

[4] P.Manneville,Y.Pomeau,to appear in Phil.Mag.A.

[5]E.D.Siggia,A.Zippelius,Phys.Rev.A24,1036(1981);Y.Pomeau,S.Zaleski, P.Manneville,ibid,A27,2710(1983).

[6] Y.Pomeau Symposium on Brain hold at Schloss Elmau(1983),to appear in the Synergetics Series (Springer-Verlag).See also fig.6 in Y.Pomeau et al.[5].

[7] J.Friedel, **Dislocations** (Addison-Wesley,Reading,Mass,1964)

SMECTICS : A MODEL FOR DYNAMICAL SYSTEMS ?

J. Prost[+], E. Dubois-Violette[++], E. Guazzelli[*] , M. Clement[**]

+ Centre de Recherche Paul Pascal - 33405 TALENCE

++ Laboratoire de Physique des Solides - Bat 510 - 91405 ORSAY CEDEX

* ERA 1000 - Université de Provence - Dept.de Physique des Systèmes
  Désordonnés - Centre de St Jérôme - 13397 MARSEILLE CEDEX 13[(*)]

**ERA 1000 - ESPCI - Laboratoire d'Hydrodynamique et Mécanique Physique
  10, rue Vauquelin - 75231 PARIS CEDEX 05

1. INTRODUCTION

Roll patterns corresponding to far from equilibrium situations
(systems above a threshold value of some control parameter) and
smectic A Liquid Crystals share in common the breaking of translational
symmetry in one direction of space. We give a phenomenological description of dissipative systems by analogy with the dynamical behavior of
smectics /1/. Such a description is valid for a "large box" with a large
number of rolls (where a homogeneous translation of the rolls will relax
infinitely slowly). In that case, the phase variable ϕ which describes
the position of the rolls is a hydrodynamical variable (relaxation
time going to infinity with some power of the wavelength of the considered mode) equivalent to the position u of the smectics layers.

The description of the phase dynamics equation by analogy with
smectics A (we shall call in the following "smectics A phase dynamics")
has been given in reference /2/. In this paper (section 2) we shall
only recall the mean features of this development (for more details, see
reference /2/). Within our smectic analogy there is a natural coupling
between the phase motion and the hydrodynamic velocity field $\underset{\sim}{v}$, called
permeation in smectics. It is not commonly introduced in model equations
of the Rayleigh-Bénard type /3/,/4/, with the exception of Zippelius and
Siggia, Cross /5 a,b,c/ (where the coupling to the vertical vorticity is
equivalent to the permeation). In the limit of large permeation, one
recovers the classical phase equation /4a/.

In section 3, experiments on the shear instability in nematics are
reported. This instability is a good example of a large box system (num-

_____
(*) The present work has been performed at the ESPCI - Laboratoire d'Hydrodynamique
et mécanique Physique.

ber of rolls ~200) where the rolls appear in a well defined direction
fixed by the external excitation. The strain field of a moving dislocation and dislocation interactions are analyzed. Moreover experiments
performed in a wedged sample (thickness d~ x and d~y) indicate dislocation motions corresponding to both glide and climb.

In section 4, our description is used to determine the strain field
around a dislocation. A comparison with experimental results is given.
The study of the static /6/,/7/ and the dynamic strain field allows one
to determine the two diffusion coefficients of the structure.

We further show (sect.5) that dislocation motion results from the Peach-
Koehler force /8/, well known in solid state physics /9/,/10/. The
effective coefficients of friction linking the velocity of the dislocation to the applied stress are found to be in general velocity dependent.
Although our description is valid for any type of instabilities we shall
in this paper, focus our attention on the case of elliptical shear instability for which experiments have been performed.

The failure of the smectic A analogy to give a proper account of
some of the experimental observations (interaction of defects and motion
of defects in a wedge shape sample) leads us to propose a "Smectics C dynamics"instead of a smectic A one. The theoretical analysis of the elliptical shear instability /11/ predicts a hydrodynamic velocity field
tilted with respect to the axis of the rolls. This defines a direction
(defined by $\pm \chi$) for one pair of rolls equivalent to the molecular axis
in a smectic C. Symmetry properties defined in reference /11/ are used
to specify the coupling between the compression $\partial\phi/\partial x$ and the orientation $\partial\phi/\partial y$ of the layers. An experimental change in the rotational
sense of the elliptical excitation reveals a change of the glide velocity
$(V_x \to -V_x)$. This change is well illustrated in the framework of the
smectic C phase dynamics given in section 6.

We also introduce the equivalent of the thermomechanical coupling
due to the presence of a wedge.

2.   DYNAMICAL EQUATIONS

a- Smectics

A dynamic of smectics close to equilibrium is described in reference
/1/. We give the dynamical equations in term of the phase $\phi = -q_o u$ which
corresponds to the dimensionless displacement of the layers in the x
direction ($q_o = 2\pi/a_o$, a_o is the layer thickness) (see fig.1). In what
follows all lengths will be scaled to the periodicity of the structure.

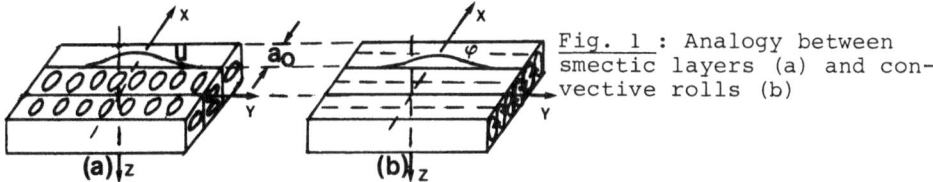

Fig. 1 : Analogy between smectic layers (a) and con-vective rolls (b)

$\tilde{x} = q_o x$ and for simplicity we shall omit the tilt superscripts

$$\rho \left(\partial V_i / \partial t \right) = -q_o \nabla_i P + q_o^2 \left(\delta F / \delta \phi \right) \delta_{ix} + \eta_{ijkl} q_o^2 \nabla_j \nabla_k V_l \qquad (1)$$

$$\left(\partial \phi / \partial t \right) + V_x = - \lambda_p q_o^2 \left(\delta F / \delta \phi \right) \qquad (2)$$

where the elastic free energy is :

$$F = 1/2 \int \left\{ B \left(\partial \phi / \partial x \right)^2 + \Lambda^2 \left(\Delta_\perp \phi \right)^2 \right\} \, dx \, dy \qquad (3)$$

and $\delta F / \delta \phi = B \left(- \partial^2 \phi / \partial x^2 + \Lambda^2 \left(\Delta_\perp \right)^2 \phi \right)$ $\qquad (4)$

$$\Lambda^2 = q_o^2 \lambda^2 \quad , \qquad \lambda^2 = K_1 / B$$

div $\underset{\sim}{V} = 0$ for incompressible fluid $\qquad (5)$

Equation (1) is the modified Navier-Stokes equation. η_{ijkl} expresses the anisotropic behavior (viscosity) of flows in smectics. The elastic force $\delta F / \delta \phi$ which appears in equation (1), originates from the permeation process as described in equation (2). λ_p is the parameter characterizing the permeation process.

For weak permeation, $\lambda_p \to 0$, the fluid motion is identical to the layer displacement (molecules do not flow across the layers) and $\partial \phi / \partial t = - V_x$.

On the contrary, in the other limit corresponding to strong permeation $\lambda_p \to \infty$, ϕ and V_x are independant (eqs. (1) and (2) are decoupled). One then recovers the classical phase equation commonly used to describe instabilities /3/,/4/,/5/ :

$$\partial \phi / \partial t = - \lambda_p q_o^2 \left(\delta F / \delta \phi \right) \qquad (6)$$

b- Roll instabilities

Equations similar to the smectic ones can now be suggested for dissipative structures. First of all we emphasize that as long as we only describe the phase dynamics (corresponding to the symmetry breaking in the x direction), we omit details on the structure over the sample thickness. The velocity field $\underset{\sim}{v}$ and the pressure P will correspond to quantities averaged over the sample thickness and then the structure of the Navier-Stokes equations corresponding to equation (1) will depend on boundary conditions. Indeed for rigid-rigid boundary conditions the z dependence of the velocity fluctuations leads to modes with wave vectors $q_z \sim \pi/d_o \sim q_o$ (d_o is the sample thickness). As long as we are concerned with the long wavelength behavior of the phase ($q \ll q_o$) this implies that the term $\eta \Delta \underset{\sim}{v}$ is in that case of the order of $q_o^2 \underset{\sim}{v}$ .One is left with a Darcy law (force$\sim \underset{\sim}{v}$). On the contrary for free-free boundary conditions, a mode at $q_z = 0$ (constant velocity through the all sample thickness) does exist ; one recovers the classical term $\eta \Delta \underset{\sim}{v}$.

The general form of equation (1) is now :

$$\rho (\partial V_i /\partial t) = - q_o \nabla_i P + q_o^2 (\delta F/\delta\phi) \delta_{ix} - \zeta_i V_i \qquad (7)$$

where $\zeta_i = \eta q_o^2$ for rigid-rigid boundary conditions $\qquad (8)$

$\qquad \zeta_i = \eta \Delta$ for free-free boundary conditions $\qquad (9)$

we note $\zeta_{//}$ for ζ_x and $\zeta_{\perp}$ for ζ_y

$$\begin{cases} \zeta_{//} = - q_o^2 ((\alpha_4/2) (\partial^2 /\partial x^2 + \partial^2 /\partial y^2) + \eta_o (\partial^2 /\partial x^2)) & (10) \\\\ \zeta_{\perp} = - q_o^2 ((\alpha_4/2) (\partial^2 /\partial x^2 + \partial^2 /\partial y^2)) \qquad \text{for smectics.} \end{cases}$$

The precise form of the elastic energy depends on the type of instability. The smectic elastic energy (equation 3) does not depend on the orientation of the smectic layers (no term $(\partial\phi / \partial y)^2$). It only takes into account compression or bend of the layers. This expression will be valid for instabilities in which the wave vector, at threshold, is degenerate (no preferred direction for the rolls). In the case of the elliptical shear instability, the direction of the rolls is fixed by the ellipticity of the excitation. The situation is that of a 2D solid and F reads :

$$F = 1/2 \int dx\, dy\, (B_{/\!/} (\partial\phi/\partial x)^2 + B_{\perp}(\partial\phi/\partial y)^2\,) \qquad (11)$$

A fourth order term $K_1 (\Delta_\perp \phi)^2$ similar to the one present in smectics must be added when one describes the distorsion field near the core of a dislocation. /7/

Let us note that for the Rayleigh-Bénard instability slightly above threshold $R > R_c$, $B_\perp = (\delta q/q_0) B_{/\!/} /4a/$ where q is the optimal wave vector taking boundary conditions into account and q_o is the optimal wave vector in an infinite medium. The term $B_\perp (\partial\phi/\partial y)^2$ is in fact a third order term $\sim B_{/\!/} (\partial\phi/\partial x) (\partial\phi/\partial y)^2$. /7/

Elimination of the pressure, with use of the incompressibility condition (eq. (5)) leads to the general dynamic phase equation

$$\rho\,(d/dt)\,(\Delta V_x) = q_o\,(\partial^2/\partial y^2)\,(\delta F/\delta\phi) - (\zeta_{/\!/}(\partial^2/\partial y^2) + \zeta_\perp\,(\partial^2/\partial x^2))\,V_x \qquad (12)$$

$$(\partial\phi/\partial t) + V_x = -\lambda_p\,q_o^2\,(\delta F/\delta\phi) \qquad (13)$$

Equations (12) and (13) describe the dynamics of the phase in absence of any dislocation. In fact, as reported in section 3, dislocations are always present in experiments. Before we describe the phase equation in presence of such singularities we want to report experimental observations.

3. EXPERIMENTS

a- The nematic instability induced by an elliptical shear
The reasons for the choice of the shear instability as an experimental model have been explained in the introduction. Let us just recall that a nematic slab is sandwiched between two rectangular plates and submitted to an elliptical shear (Ox_E, Oy_E are the axis of the ellipse)

$$\begin{cases} X_E = X_o\,\cos\,\omega t \\[2mm] Y_E = Y_o\,\sin\,\omega t \end{cases} \qquad (14)$$

X_E, Y_E being respectively the displacement of the upper and lower plates. The thickness d_o of the cell can vary between 30 and 150 µm (a new cell where the whole shear is applied to the upper plate has been constructed to perform the wedge experiments, its description is presented in §3d)

This system experiences a roll instability /11/,/12/ beyond a threshold

$$N_c = X_{oc} \, Y_{oc} \, \omega / D_o$$
(D̂₀ is the diffusivity of the nematic director)

(15)

The control parameter is $\varepsilon = (N - N_c)/N_c$. The direction of the rolls is at a fixed angle α with the axis oy_E of the elliptical shear. Since velocity and director fields are coupled in a nematic, the observation of the rolls is quite easy with polarized light.

Many edge-dislocations are present in the roll structure as shown in figure 2 and different mechanisms are responsible for their nucleation :

-mismatch between sets of rolls appearing from different parts of the sample at threshold,

-nucleation of defects around dust particles or at the sample edge /13/,

-homogeneous nucleation of a pair of dislocation above threshold /14/.

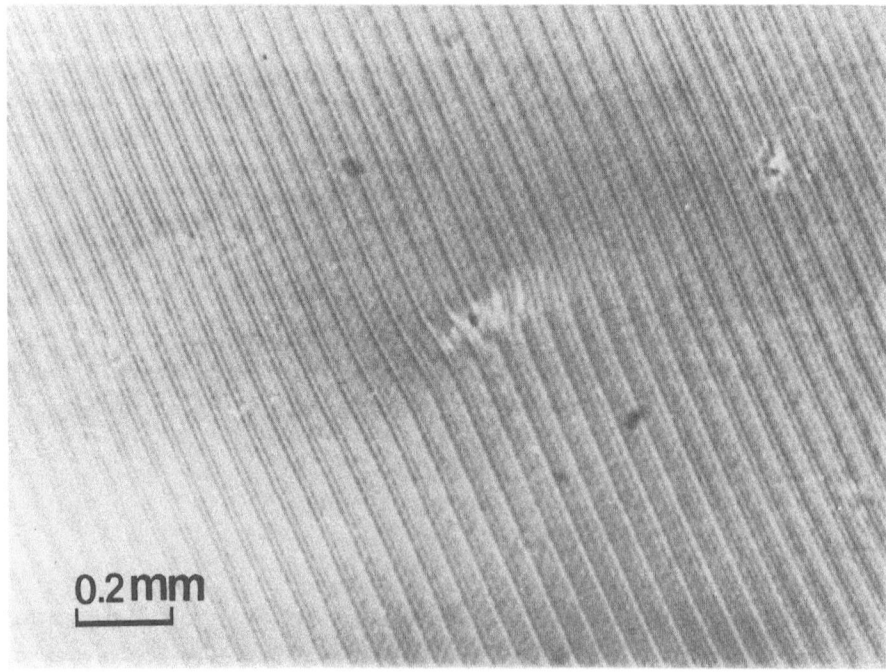

0.2 mm

Fig. 2 : An edge dislocation in a roll structure

b- <u>Strain field of an isolated moving dislocation</u>

The study of the static strain field around an edge dislocation has already been performed /6/,/7/. The distortion induced by the defect was visualized as shown in fig. 3. The fit of the strain pattern allows an estimate of $\lambda = (K_1 / B_{\parallel})^{1/2}$ and $p = (B_{\perp} / B_{\parallel})^{1/2}$

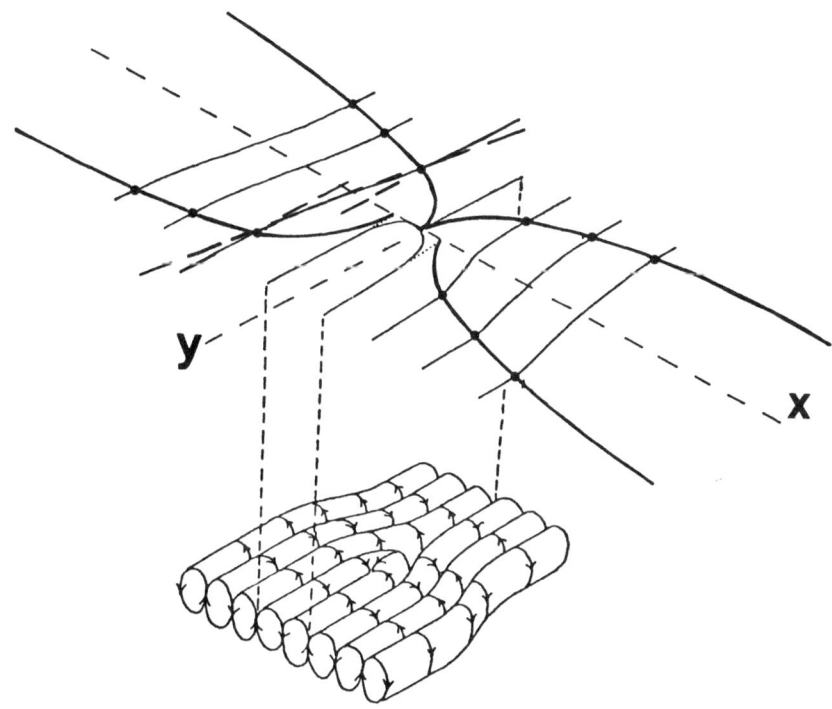

<u>Figure 3</u> : Boundaries between perturbed and unperturbed regions around a static defect /6/,/7/.
Each point is the intercept between the tangents to the roll at large distance and that to the inflection point of the same roll. Near the defect the points determine 4 branches of parabolae the aperture of which is $\sim\lambda$ (near the defect the curvature term is important in eq. (11)). At larger distances the points determine 4 lines the slope of which is $\sim 1/p$ (given by the balance between the distortion along y and the compression along x in eq. (11))

In general, dislocations are observed to move. The roll structure has been filmed with a contracted time scale (1/2 image/sec) due to the slow drift of the dislocations. In the experiments, we wait a long enough time, 10 mn period, above the threshold, such that only a small density of defects is present. In general dislocations have complicated motions due to the interaction with other defects (see § 3-c). However the movie

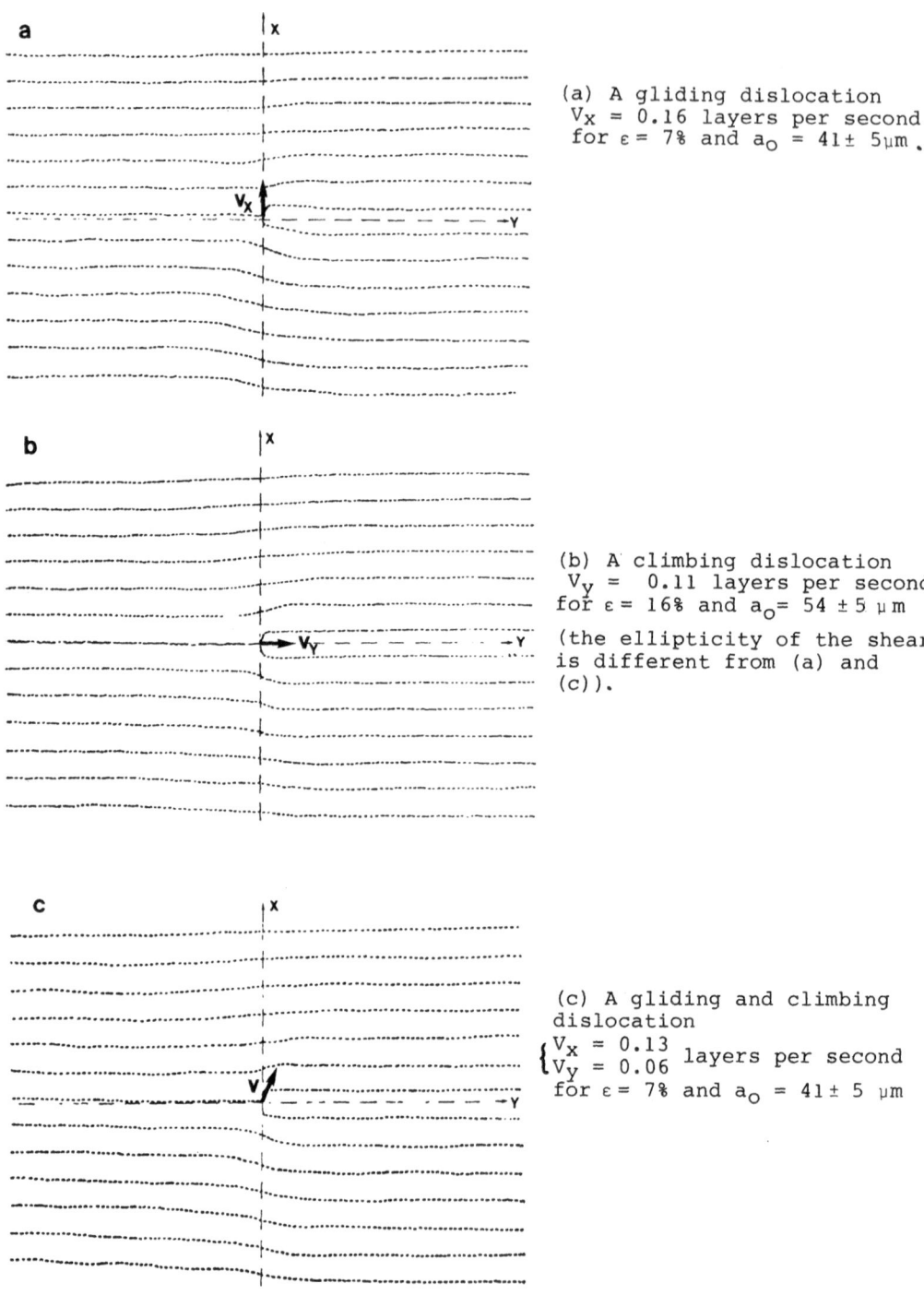

(a) A gliding dislocation
$V_x = 0.16$ layers per second
for $\varepsilon = 7\%$ and $a_0 = 41 \pm 5\,\mu m$.

(b) A climbing dislocation
$V_y = 0.11$ layers per second
for $\varepsilon = 16\%$ and $a_0 = 54 \pm 5\,\mu m$

(the ellipticity of the shear
is different from (a) and
(c)).

(c) A gliding and climbing
dislocation
$\begin{cases} V_x = 0.13 \\ V_y = 0.06 \end{cases}$ layers per second
for $\varepsilon = 7\%$ and $a_0 = 41 \pm 5\,\mu m$

Fig. 4 : Experimental roll patterns around a moving defect
The accuracy in the measurement of the velocity is about 1%.

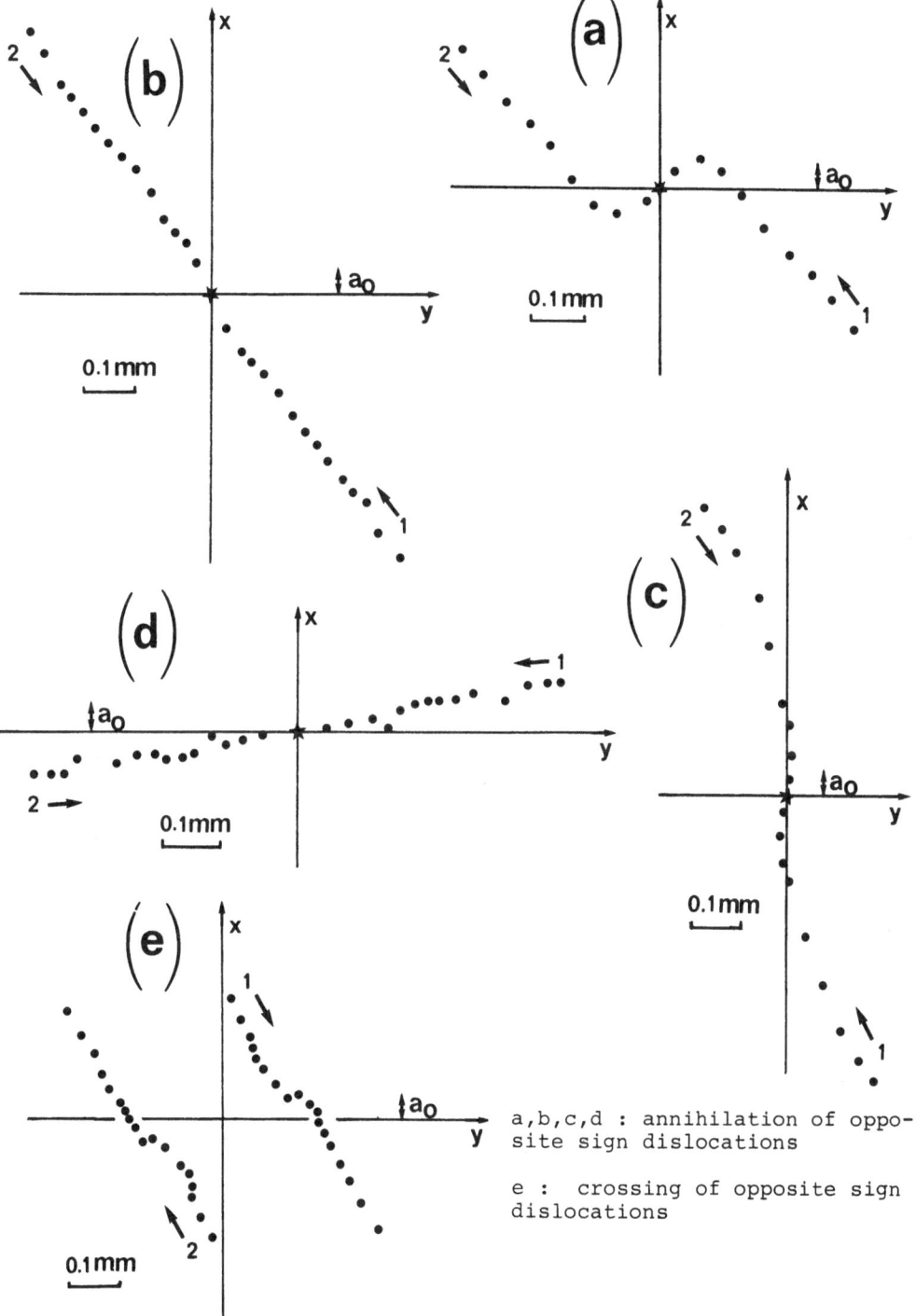

Fig. 5 : Interaction between defects of opposite sign.
The center of each defect is plotted each 8 sec (for a,b,c,e,
$\varepsilon = 7\%$, $a_o = 40 \pm 5\mu m$; for d, $\varepsilon = 16\%$, $a_o = 44 \pm 5\mu m$)

a,b,c,d : annihilation of opposite sign dislocations

e : crossing of opposite sign
dislocations

picture analysis shows the existence of a few isolated dislocations moving at a constant velocity: both glide and climb motion are observed. The corresponding deformation patterns are displayed on fig. 4 a-b-c . They correspond respectively to glide (a), climb (b) and eventually glide and climb motions (c). Comparison with theoretical results is given in the next section.

c- Interaction between defects of opposite sign

Movie picture analysis shows also the interaction between dislocations of opposite Burger-vector via their deformation field. The trajectories of the two-interacting-dislocations are displayed in the barycentric coordinates. Two dislocations of opposite sign can interact either by annihilating (fig. 5 a,b,c,d) or by passing by and accelerating (fig 5e)

d- Wedged sample

i- Experimental set-up

To create a controlled variation of thickness in the sample, a new cell has been designed .   The elliptical shear is applied to the upper circular plate[*] ((a) in fig. 6). The lower plate is attached to a set of two circular rings ((b) and (c)). These rings are cut by a plane of same inclination. It produces a variable wedge with an angle θ ranging from zero to $11.5 \ 10^{-3}$ rad. by rotating ring (b), keeping ring (c) fixed. The angle θ of the wedge is measured by the deflection of a laser beam on the upper and lower plates, the accuracy is $0.4 \ 10^{-3}$ rad. The tilt direction is set by rotating the two rings (b) and (c) in tandem. The same deflection technique is used to control the adjustment of the direction. It is set parallel (wedge I) or perpendicular (wedge II) to the diffracted image of the rolls (i.e. wedge I corresponds to rolls parallel -fig.7- and wedge II to rolls perpendicular -fig.8- to the tilt direction of the wedge).

The thickness d_o of the cell is controlled by the screw (d) . The measure of the thickness d_o is performed interferometrically.

(*)
The difference between the case where the shear is produced by applying two linear shear at right angle to the plates (eq. 14) and the one where the full motion is applied to one plate, is characterized by the presence of different convective terms. In this new geometry, the thresholds present the same shape (eq. 15) and the same structures occur (rolls,and when increasing the shear, squares).

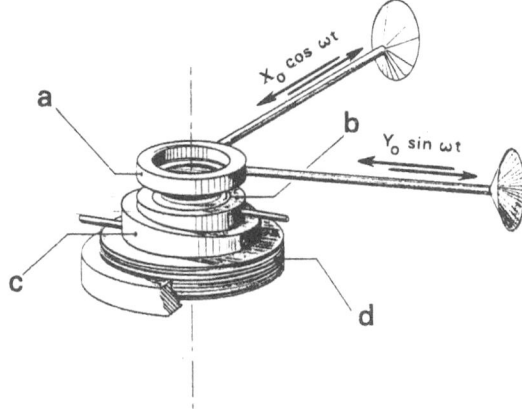

Fig. 6 : Experimental set-up to create a wedge

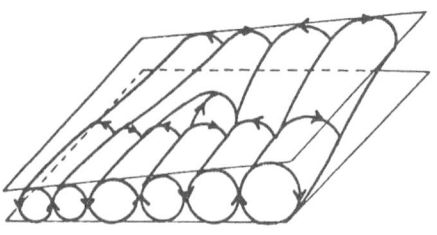

Fig. 7 : Wedge I : the rolls are kept parallel to the tilt direction of the wedge

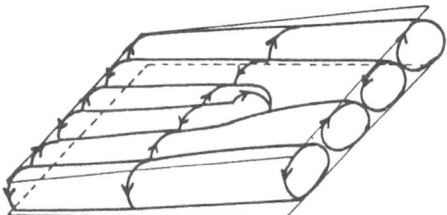

Fig. 8 : Wedge II : the rolls are kept perpendicular to the tilt direction of the wedge

ii- Wedge (I)

The schematic motion of the defects in a wedged sample of angle θ is represented in fig. 9a. The north (N) corresponds to the region of largest thickness. Defects of positive sign' (‑⊏) are nucleated on the west edge of the sample, then glide to the south (S) (smallest thickness) ; defects of negative sign (⊐‑) glide to the north (N). Rotation of the support of π , corresponding to a wedge of -θ (the south part (S) now corresponds to the region of largest thickness) gives the following result (fig. 9b) : defects of positive sign are nucleated on the west edge, then glide to the large thickness (S) ; defects of negative sign glide to the small thickness (N). The change of the rotation sense of the elliptical shear (Fig. 9c) (when the large thickness is located in the north (N)) produces the following effect : defects of positive sign nucleate on the west edge and move to the north whereas those of negative sign move to the south. Clearly the glide direction changes sign with the ellipticity.

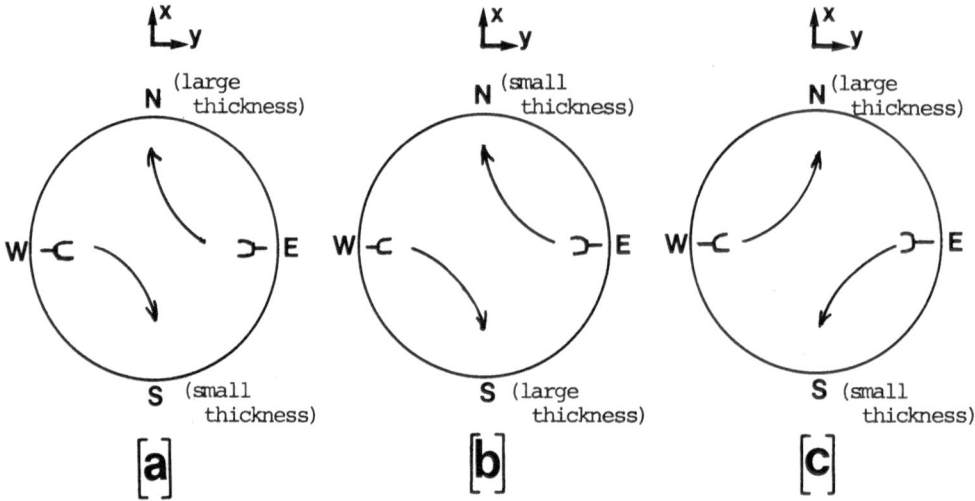

Fig. 9 : Schematic drawing of the motion of the defect in a wedged sample.

(a) : {angle of the wedge : θ
       {negative sense of rotation of the elliptical shear

(b) : {angle of the wedge :-θ
       {negative sense of rotation of the elliptical shear

(c) : {angle of the wedge : θ
       {positive sense of rotation of the elliptical shear

Typical trajectories of defects are displayed fig. 10 (observation made in the equatorial west region of the sample). Defects of positive sign are nucleated from the edge near the equator. One also observes nucleation of pairs of defects located near the equator in the central region of the cell. A defect of negative sign does not annihilate directly at the edge, a defect of positive sign is always nucleated from the edge and annihilates with it. Now let us look at the defects of positive sign nucleated by the edge : either they cannot escape from the influence of the edge and glide to annihilate at the edge (defects 0,13,15,40) or they escape from the influence of the edge, (defects 10,14,24) climb and then at about 20 rolls from the edge, glide to the small thickness side (the trajectory is bended). All these sequences occur repeatedly and the different mechanisms (annihilation , nucleation) are always located in the same region.

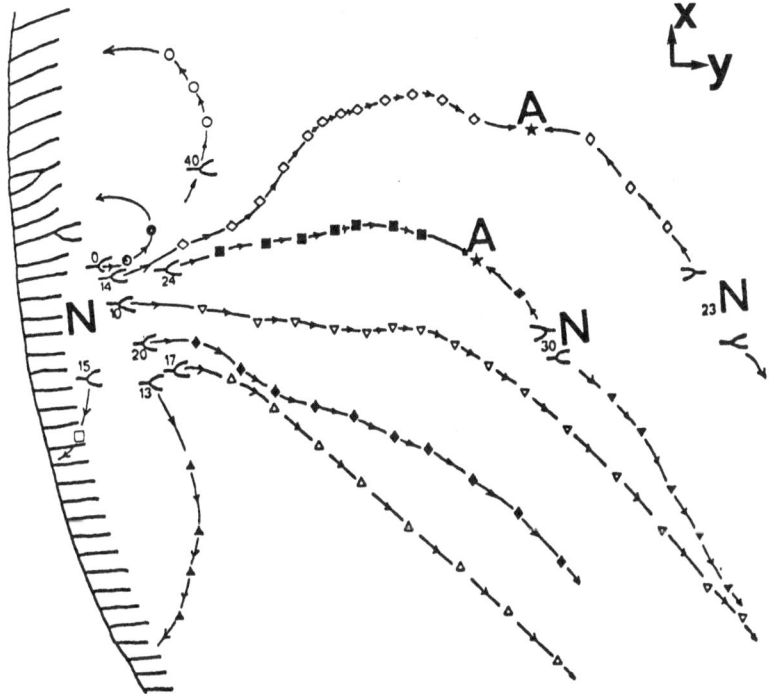

Fig. 10 : Trajectories of the dislocations in the equatorial region
(θ = 7.8 10^{-3}rad. ; ε = 8%)
The centers of the defects are plotted every 40 sec. The number indicates the time every 40 sec. (N means nucleation and A annihilation)

A systematic study has been performed as a function of θ in a square
of 20x20 rolls in the center of the cell. θ is varied up to 9.0 10^{-3}rad.
The number of defects increases with θ . We have not performed a systematic study of the velocity versus the position in the whole cell. We
chose isolated dislocations which appear to move at nearly constant
velocity in this region. The velocity multiplied by the sign ε$_d$ of the
dislocation versus θ is reported in fig. 11. For small angle, V_y
is higher than V_x but up to 3.3 10^{-3} rad, V_y is constant and V_x increases.

We have noticed that for large angles, due to a small dependence of
the roll threshold with the thickness d$_o$ of the cell, the large thickness
and the small one are not exactly at the same ε = (N-N$_c$)/N$_c$. In the
large thickness region the system is in fact subcritical whereas the
rolls are very rigid in the small thickness region (larger ε) (we
also remark a slow drift of the roll from the large thickness to the
small one).

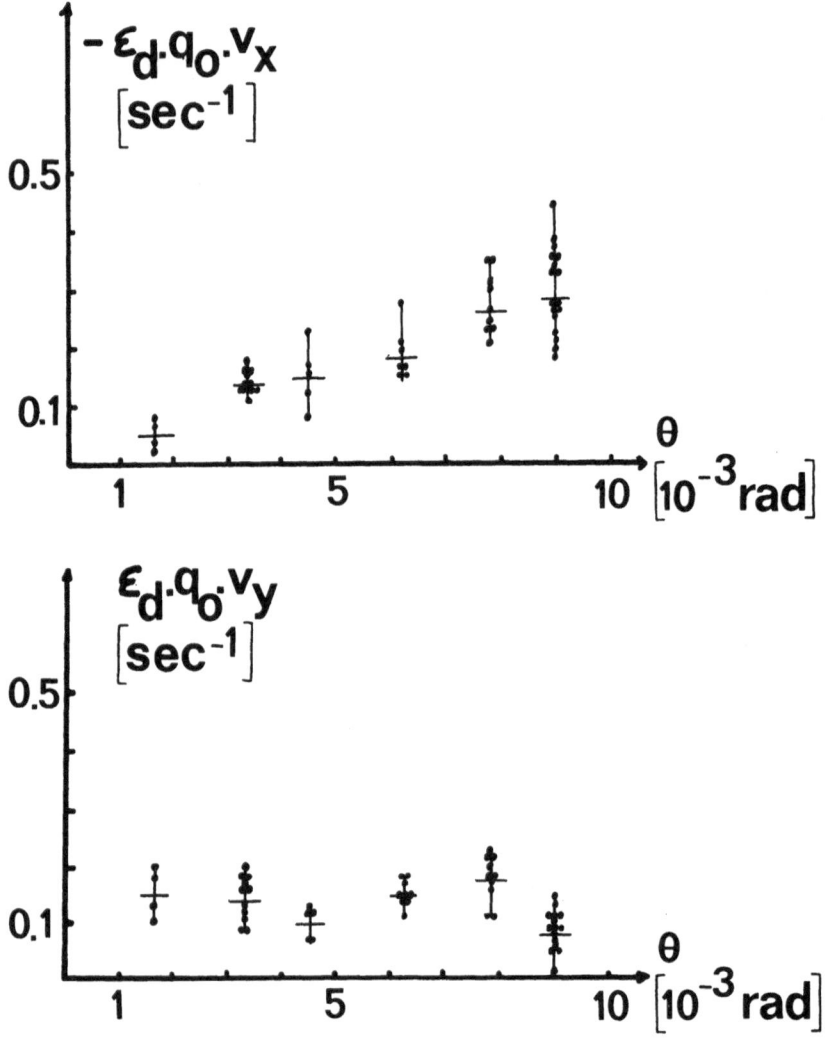

Fig. 11 : Velocity of the defects multiplied by the sign ε_d of the defect versus θ, in the center of the cell (ε =8%). Full dots (•) indicate the velocity for the different studied dislocations ; crosses (+) represent the mean values over these dislocations.

iii- Wedged II

In this configuration, dislocations of the same signs are necessary to adjust the wavelength of the rolls (fig. 12). The defects climb (a slight glide is also observed) to the small thickness region (for a wedge of - θ , the defects are of opposite sign and also climb to the small thickness region). Their number increases when one increases θ . They seem to organize in a regular pattern (nearly triangular network).

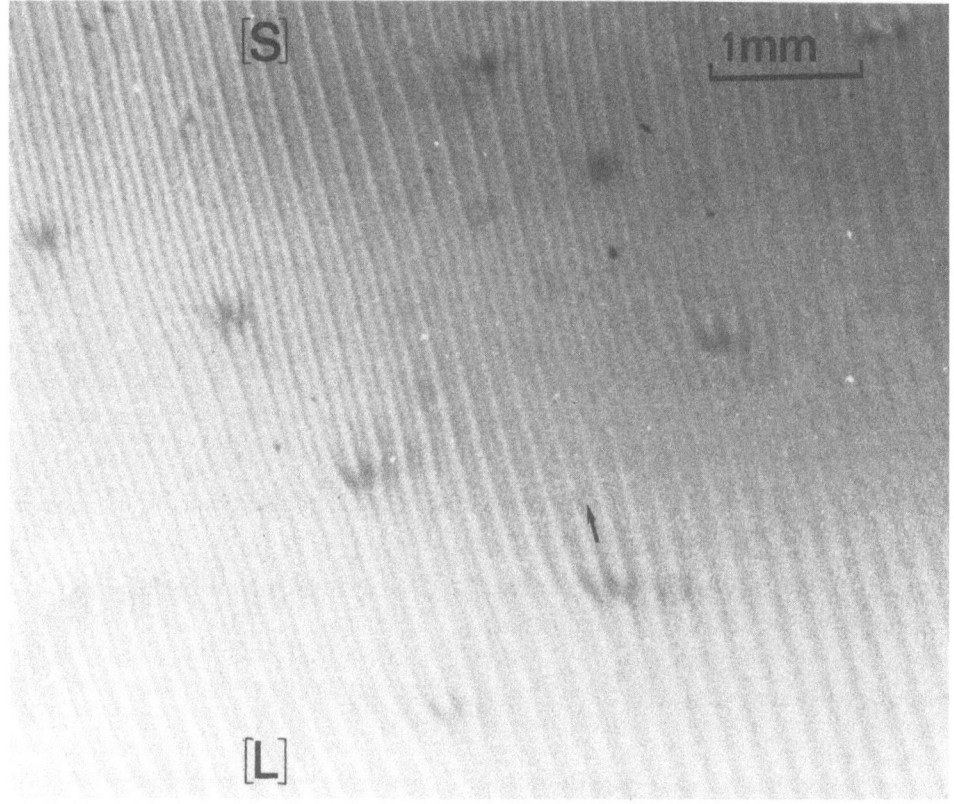

Fig. 12 : Dislocations in the wedge II ($\theta = 4.5 \ 10^{-3}$ rad. ,
$\overline{d}_O = 135 \pm 5$ μm, $\varepsilon = 8\%$)
The arrow indicates the climbing direction of the defect.
(S) : small thickness region
(L) : large thickness region

All these experiments (wedge I and II) have been performed after
a delay time of 15 mn during which most of the defects annihilate. But
we have also performed experiments after 24 hours which show that the
behavior is not relaxational : Phenomena are identical to those which
occur after 15 mn. Thus, the system does not evolve toward a steady
defect free state, but spontaneously emits dislocations (the slowest
characteristic time in the experiment is $\tau_\phi \sim L^2/D_{/\!/} \sim 6$ hours, phase
diffusion time)

4. STRAIN FIELD AROUND A MOVING DISLOCATION

Dislocations are seen to move with constant velocity $\underset{\sim}{V}_o$. We shall look for solutions corresponding to experimental observations and assume for the phase and the fluid velocity the following expressions :

$$\phi = \phi(\underset{\sim}{r} - \underset{\sim}{V}_o t) \text{ and } V_x = V_x(\underset{\sim}{r} - \underset{\sim}{V}_o t)$$

Dislocations correspond to singularities of the phase ϕ . Following ref /15/ one defines local variables corresponding to physical quantities without singularities. Indeed except in the core of the dislocation, the compression $\partial\phi/\partial x$ and the tilt of the layers $\partial\phi/\partial y$ are well defined quantities.

Let us call $\underset{\sim}{m}$ the vector with components corresponding to the tilt (along $\underset{\sim}{y}$) and the compression (along $\underset{\sim}{x}$).

A dislocation $\varepsilon_d = +1$ at the origin $x=0$, $y=0$ such as on figure 13 is defined by a density $\underset{\sim}{j}$ such that :

$$\underset{\sim}{\nabla} \times \underset{\sim}{m} = \underset{\sim}{j} = \varepsilon_d 2\pi \; \delta x \; \delta y \qquad (16)$$

This corresponds to the relation : $\int_C \underset{\sim}{m} \cdot d\underset{\sim}{l} = +\varepsilon_d 2\pi \qquad (17)$

where the sign + corresponds to the circuit (C) shown on fig. 13 .

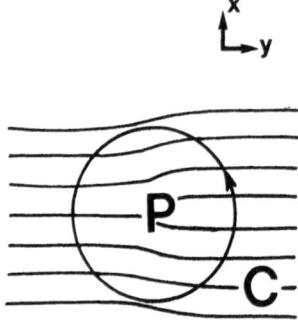

Fig. 13 : Edge dislocation of positive sign (the corresponding dislocation density is a vector along z , perpendicular to the plane of the figure)

Let us now transform eqs 12 and 13 in terms of the new variable $\underset{\sim}{m}$.

Dislocations are moving with velocity $\underset{\sim}{V}_o$ much lower than the sound velocity ($\sim (B/\rho)^{1/2}$). This allows one to neglect the left hand side of eq. 12 and to derive the dynamical equation for $\underset{\sim}{m}$ (with use of relation $\partial\phi/\partial t = -\underset{\sim}{V}_o \cdot \underset{\sim}{\nabla}\phi$) :

$$(\zeta_{\parallel} \, \partial^2/\partial y^2 + \zeta_{\perp} \, \partial^2/\partial x^2) \, \underset{\sim}{V_o} \cdot \underset{\sim}{m} = -q_o^2 \, \{(1+ \lambda_p \zeta_{\parallel}) \, \partial^2/\partial y^2 + \lambda_p \zeta_{\perp} \partial^2/\partial x^2\} \, \underset{\sim}{\nabla} \cdot \underset{\sim}{\phi} \, (\underset{\sim}{m})$$

$$(18)$$

where $\qquad \underset{\sim}{\nabla} \cdot \underset{\sim}{\phi} = -\delta F/\delta \phi$

$$\underset{\sim}{\phi} \, (\underset{\sim}{m}) = B(- \Lambda^2 \Delta_{\perp} \underset{\sim}{m}_{\perp} + m_x \cdot \underset{\sim}{x}) \qquad\qquad \text{for smectics} \qquad\qquad (19)$$

$$\text{and} \quad \underset{\sim}{\phi} \, (\underset{\sim}{m}) = B_{\perp} \, m_y \, \underset{\sim}{y} + B_{\parallel} \, m_x \, \underset{\sim}{x} \quad \text{for the elliptical shear instability} \,\,(20)$$

One obtains easily from eqs (16) and (18) the Fourier transform (noted $\wedge$) components $\hat{m}_y$ and $\hat{m}_x$ of $\underset{\sim}{m}$.

In the smectic case :

$$\hat{m}_x = - \frac{2 \pi i \, (B \, \Lambda^2 \, q_y^3 \, - i \, V_{oy} \, g(q))}{(B(q_x^2 + \Lambda^2 q_y^4) \, - i \, g(q) \, (q_x V_{ox} + q_y \, V_{oy}))} \tag{21}$$

$$\hat{m}_y = \frac{2 \pi i \, (B \, q_x - i \, V_{ox} \, g(q))}{(B \, (q_x^2 + \Lambda^2 q_y^4) \, - i \, g(q) (q_x V_{ox} + q_y \, V_{oy}))} \tag{22}$$

$$\text{and} \qquad g(q) = \frac{(\alpha_4/2) \, q^4 + \eta_o \, q_y^2 \, q_x^2}{q_o^2 \, (q_y^2 \, + \lambda_p \, ((\alpha_4/2) q^4 + \eta_o \, q_y^2 \, q_x^2))} \tag{23}$$

In the shear instability case :

$$\hat{m}_x = - \frac{2\pi \, i \, (B_{\parallel} \, p^2 q_y - i \, V_{oy} \, g(q) \,)}{(B_{\parallel} \, (q_x^2 \, + p^2 \, q_y^2 \,) \, - ig(q) \, (q_x V_{ox} + q_y V_{oy}))} \tag{24}$$

$$\hat{m}_y = \frac{2 \pi i \, (B_{\parallel} q_x - i \, V_{ox} \, g(q))}{(B_{\parallel} (q_x^2 \, + p^2 \, q_y^2 \,) \, - ig(q) \, (q_x V_{ox} + q_y V_{oy}))} \tag{25}$$

$$\text{with} \qquad g(q) = \frac{\zeta_{\parallel} \, q_y^2 + \zeta_{\perp} \, q_x^2}{q_o^2 \, ((1 + \lambda_p \zeta_{\parallel}) \, q_y^2 + \lambda_p \zeta_{\perp} \, q_x^2 \,)} \tag{26}$$

We shall now focus our attention to the elliptical shear instability in order to compare with experimental results. The smectic case is described in ref. /2/. One first notices that the symmetry of the strain field is modified compared to the static one.

The y and x symmetries are respectively broken by V_{oy} and V_{ox}. The far field is significantly modified by the dislocation motion. On the contrary for the short range strain, $q \gg V_o/D$ Max $\{ \xi_{//}^{1} \lambda_p^{-1} , 1 \}$ where

$$D_{//(\perp)} = \lambda_p q_o^2 B_{//(\perp)}$$

One recovers a strain similar to a static one

$$m_x \simeq y/ (y^2 (D_{//} /D_\perp)^{1/2} + x^2 (D_\perp /D_{//})^{1/2}) \tag{27}$$

$$m_y \simeq -x/ (y^2 (D_{//} /D_\perp)^{1/2} + x^2 (D_\perp /D_{//})^{1/2}) \tag{28}$$

Expressions (24), (25) may be easily inverted in the regime of strong permeation where $g(q) \simeq (\lambda_p q_o^2)^{-1}$

$$m_x = -e^{-((V_{oy} y/2D_\perp)+(V_{ox} x/2D_{//}))} \{D_\perp \frac{\partial}{\partial y} K_o (r/ \xi) + V_{oy} K_o (r/ \xi)/2\} (D_\perp D_{//})^{-1/2} \tag{29}$$

$$m_y = e^{-((V_{oy} y/2D_\perp)+(V_{ox} x/2D_{//}))} \{D_{//} \frac{\partial}{\partial x} K_o (r/ \xi) + V_{ox} K_o (r/ \xi)/2\} (D_\perp D_{//})^{-1/2} \tag{30}$$

where $r = (y^2 (D_{//} /D_\perp)^{1/2} + x^2 (D_\perp /D_{//})^{1/2})^{1/2}$

and $\xi = 2(D_\perp D_{//})^{1/2} /(V_{oy}^2 (D_{//} /D_\perp)^{1/2} + (D_\perp /D_{//})^{1/2} V_{ox}^2)^{1/2}$

The dissymmetry of the long range strain $r \gg \xi$

$$m_x = -\sqrt{\frac{\pi}{2}} ((-y/r\xi)+(V_{oy}/2(D_\perp D_{//})^{1/2}))e^{(-\frac{V_{oy} y}{2D_\perp} + \frac{V_{ox} x}{2D_{//}})-r/\xi} / (r/\xi)^{1/2} \tag{31}$$

$$m_y = \sqrt{\frac{\pi}{2}} ((-x /r\xi)+(V_{ox}/2(D_\perp D_{//})^{1/2}))e^{(-\frac{V_{oy} y}{2D_\perp} + \frac{V_{ox} x}{2D_{//}})-r/\xi} / (r/\xi)^{1/2} \tag{32}$$

is well evidenced for pure climb ($V_{oy} \neq 0$, $V_{ox} = 0$) and glide ($V_{oy} = 0$ $V_{ox} \neq 0$) motions. The strain is exponentially decreasing (screening length $\sim D/V$) in the front of the motion for $y > 0$ (or $x > 0$) in the case of pure climb (or glide). In the wake of the motion $y < 0$ (or $x < 0$) the strain decays with a power law $|y|^{-1/2}$ (or $|x|^{-1/2}$).

The distorsion of the pattern around the moving dislocation is enlightened by the phase difference :

$$\Delta\phi = \phi(y = +\infty, x) - \phi(y = -\infty, x) = \int_{-\infty}^{+\infty} m_y(y,x) \, dy$$

which is calculated with use of the Fourier transform

$$\Delta\phi = \lim_{q_y \to 0} \int_{-\infty}^{+\infty} dq_x \, m_y(q_x, q_y) e^{iq_x x} / 2\pi$$

A main result is that, once a glide component in the motion of the dislocation sets in, there is no long range phase shift in regions toward which the motion is directed but, on the contrary, the whole phase difference appears in the wake of the dislocation as shown on fig. 4 a and c.

For all V_{oy} values, $V_{ox} > 0$

$\Delta\phi = 0$ $\quad x > 0$ $\quad$, $\quad \Delta\phi = +2\pi$ $\quad x < 0$

For all V_{oy} values, $V_{ox} < 0$

$\Delta\phi = -2\pi$ $\quad x > 0$ $\quad$, $\quad \Delta\phi = 0$ $\quad x < 0$

On the contrary for pure climb motion

$\Delta\phi = -\pi$ $\quad x > 0$ $\quad$, $\quad \Delta\phi = \pi$ $\quad x < 0$

one recovers the symmetry $\Delta\phi(x) = \Delta\phi(-x)$ as shown on fig. 4b.

The whole distorsion pattern has been calculated in the case of pure glide motion for different values of the screening length $D_{\parallel} / V_{ox}$; the ratio $D_{\perp}/D_{\parallel}$ has been kept constant at the experimental value 0.06 evaluated in the static limit (section 3b)

Theoretical curves (fig. 14) correspond to $D_{\parallel} = 20$, 200, 2000, where the velocity V_{ox} has been fixed to the experimental value $V_{ox} = 0.16$ layers/sec.

The asymmetry between the front and the wake of the motion is well pronounced for small $D_{\parallel}$ (or large V_{ox}). There is a characteristic non

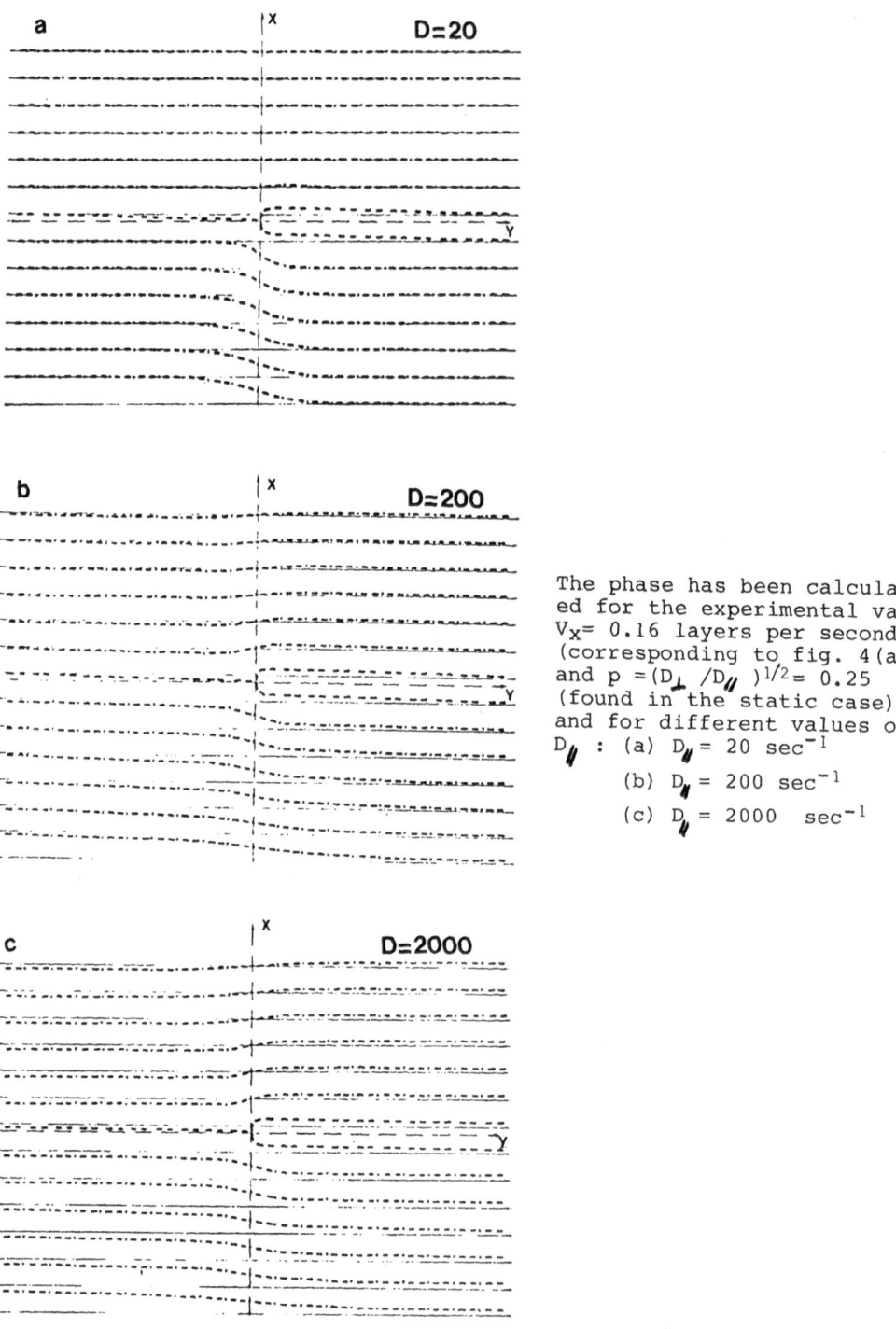

The phase has been calculated for the experimental value $V_x = 0.16$ layers per second (corresponding to fig. 4(a)) and $p = (D_\perp /D_{//})^{1/2} = 0.25$ (found in the static case) and for different values of $D_{//}$: (a) $D_{//} = 20$ sec^{-1}

(b) $D_{//} = 200$ sec^{-1}

(c) $D_{//} = 2000$ sec^{-1}

Fig. 14 : Theoretical roll patterns around a gliding defect.

monotoneous variation of ϕ for $x > 0$ with a minimum for $y < 0$ and a maximum for $y > 0$. Comparison with the experimental distorsions shown on fig. 4a gives a fit of $D_{//}$ ($D_{//} \sim 200$). This parameter leads to a diffusion coefficient (in real units) $D_{//} = 10^{-4}$ cm^2/s. This is in agreement with the theoretical estimate $D_{//} \sim \xi_o^2 / \tau_o^2$ where τ_o and ξ_o are the slowest relaxation time and influence length close to the threshold ($\tau_o \sim Kq^2/\gamma_1$, $q \sim \pi/d_o$.

5. PEACH-KOEHLER FORCE AND EFFECTIVE FRICTION COEFFICIENTS

We want to derive the motion of a dislocation in a sample where some residual stress ϕ^0 is present. One can first estimate the force $\underset{\sim}{P}$ acting on the dislocation under the stress ϕ^0 from the interaction energy :

$$\delta \; F_{int.}^0 \qquad = \int_v \phi^0 \cdot \underset{\sim}{m}^d \; dv \; , \qquad dv = dx \; dy$$

in the following superscript ( )^t, ( )^o, ( )^d will refer to the total, external or dislocation (stress ...), $\underset{\sim}{\nabla}^d = \partial / \partial \underset{\sim}{r}^d$. The force acting on the dislocation (the dislocation is supposed to be located at $\underset{\sim}{r}^d$) :

$$\underset{\sim}{\nabla}^d \; \delta \; F_{int.}^0 \qquad = \int \underset{\sim}{\nabla}^d \; (\phi^0 \cdot \underset{\sim}{m}^d) \; dv \tag{33}$$

can be written, after some algebra, and with the use of $\underset{\sim}{\nabla}^d \underset{\sim}{m}^d = - \underset{\sim}{\nabla} \; \underset{\sim}{m}^d$

$$\underset{\sim}{\nabla}^d \; \delta F_{int.}^0 \quad = - \int_S (\phi^0 \cdot d\underset{\sim}{S}) \underset{\sim}{m}^d + \int \underset{\sim}{m}^d \; (\underset{\sim}{\nabla} \cdot \phi^0) \; dv - \int (\phi^0 \times \underset{\sim}{j}) \; dv \tag{34}$$

where $dS_y = dx$, $dS_x = dy$

This can also be deduced from the total energy :

$$\underset{\sim}{\nabla}^d \; F^t = \underset{\sim}{\nabla}^d \int f(\phi^t , \nabla \phi^t) \; dv = \int (\underset{\sim}{\nabla} \cdot \phi^t) \underset{\sim}{m}^d \; dv - \int_S (\phi^t \cdot d\underset{\sim}{S}) \underset{\sim}{m}^d \tag{35}$$

where f is the energy density
and

$$\phi^t = \phi^0 + \phi^d$$

$$\underset{\sim}{m}^t = \underset{\sim}{m}^0 + \underset{\sim}{m}^d$$

and with use of the identities :

$$\delta F / \delta \phi^t = - \underset{\sim}{\nabla} \cdot \underset{\sim}{\phi}^t$$

and

$$(\partial / \partial x_j^{\ d}) \ \phi^t = -(\partial / \partial x_j) \ \phi^d = - \ m_j^{\ d}$$

After some algebra eq. (35) is written :

$$\underset{\sim}{\nabla}^d F^t = \int (\underset{\sim}{\nabla} \cdot \underset{\sim}{\phi}^0) \underset{\sim}{m}^d dv - \int (\underset{\sim}{\phi}^0 \cdot d\underset{\sim}{S}) \underset{\sim}{m}^d - \int (\underset{\sim}{\phi}^d \cdot \underset{\sim}{\nabla}) \underset{\sim}{m}^d \ dv \qquad (36)$$

writting that $\underset{\sim}{\nabla}^d \delta F^0_{interaction} = \underset{\sim}{\nabla}^d F$ one obtains from eq. (35) and eq. (36) the relation :

$$- \int (\underset{\sim}{\phi}^0 \times \underset{\sim}{j}) \, dv = - \int (\underset{\sim}{\phi}^d \cdot \underset{\sim}{\nabla}) \underset{\sim}{m}^d \ dv \qquad (37)$$

One recognize in the left hand side of this equation the well-known Peach-Kochler force $\underset{\sim}{P}$ /8/,/9/,/10/ :

$$\underset{\sim}{P} = - \int (\underset{\sim}{\phi}^0 \times \underset{\sim}{j}) \ dv \qquad (38)$$

The right hand side is modified with use of the phase dynamic equation (18) for a dislocation moving at a constant velocity V_0 and is written in q space as :

$$\int (\underset{\sim}{\phi}^d \cdot \underset{\sim}{\nabla}) \underset{\sim}{m}^d \ dv = (1/4\pi^2) \int g(q) \ (V_0 \cdot \overset{\wedge}{\underset{\sim}{m}}^d (q)) \ \overset{\wedge}{\underset{\sim}{m}}^d (-q) \ d^2 q \qquad (39)$$

This leads to

$$P_x = - \ \varepsilon_d \ \phi^0_y = \lambda_{xx} \ V_{ox} \qquad (40)$$

$$P_y = + \ \varepsilon_d \ \phi^0_x = \lambda_{yy} \ V_{oy} \qquad (41)$$

$\varepsilon_d = +1 \ (-1)$ for a dislocation of sign $+$ or $(-)$, where for smectics

$$\lambda_{xx} = \int \frac{K_1^2 q_x^6 \ g(q) dq_x \ dq_y}{(K_1 q_y^4 + B_\parallel q_x^2)^2 + (q_x V_{ox} + q_y V_{oy})^2 g^2 (q)} \qquad (42)$$

$$\lambda_{yy} = \int \frac{g(q) \; B_{/\!/}^2 \; q_x^2 \; dq_x \; dq_y}{(B_{/\!/} \; q_x^2 + K_1 q_y^4)^2 + (q_x V_{ox} + q_y V_{oy})^2 \; g^2 \; (q)} \tag{43}$$

and for the shear instability

$$\lambda_{xx} = \int \frac{B_{\perp}^2 \; q_y^2 \; g(q) \; dq_x \; dq_y}{(B_{\perp} \; q_y^2 + B_{/\!/} \; q_x^2)^2 + (q_x V_{ox} + q_y V_{oy})^2 \; g^2 \; (q)} \tag{44}$$

$$\lambda_{yy} = \int \frac{B_{/\!/}^2 \; q_x^2 \; g(q) \; dq_x \; dq_y}{(B_{\perp} \; q_y^2 + B_{/\!/} \; q_x^2)^2 + (q_x V_{ox} + q_y \; V_{oy})^2 \; g^2 \; (q)} \tag{45}$$

Detailed calculations have been given in ref. /2/, we shall only recall the main features focussing our attention on the velocity dependence.

i) Smectics or free-free Rayleigh-Bénard

$$\left\{ \begin{array}{l} V_{oy} \\ V_{ox} \end{array} \right. \quad \text{very small} \qquad \lambda_{yy} \sim \text{constant}$$

$$V_{oy} \gg V_{ox} \qquad\qquad \lambda_{yy} \sim V_{oy}^{-1/3}$$

$$V_{ox} \gg V_{oy} \qquad\qquad \lambda_{yy} \sim V_{ox}^{-1/2}$$

$$\left. \right\} \quad \lambda_{xx} \sim \text{constant}$$

ii) Rigid-Rigid Rayleigh-Bénard

$$V_{oy} \gg V_{ox} \qquad\qquad \lambda_{yy} \sim V_{oy}^{-1/3}$$

$$V_{ox} \gg V_{oy} \qquad\qquad \lambda_{yy} \sim V_{ox}^{-1/2}$$

$$\left. \right\} \quad \lambda_{xx} \sim \text{constant}$$

iii) Elliptical shear

For all V_{oy} and V_{ox} :

$$\lambda_{yy} \sim 2 \log Q + V_{oy}^2 / (V_{oy}^2 + p^2 V_{ox}^2) , \quad \lambda_{xx} \sim 2 \log Q + V_{ox}^2 / (V_{oy}^2 + V_{ox}^2 / p^2)$$

where

$$Q = 2 q_c \; \lambda_p (D_{/\!/} \; D_{\perp} (D_{\perp} + D_{/\!/})) / (D_{/\!/} \; V_{oy}^2 + D_{\perp} V_{ox}^2))^{1/2}$$

and q_c is an ultraviolet cut-off.

Except for the smectics or free-free Rayleigh-Bénard where a pure friction regime exists at very low velocities all these cases present a velocity dependence of λ_{yy}. One recovers for pure glide motion a Siggia-Zippelius law $P_y \sim V_{oy}^{2/3}$ for all velocities in the rigid-rigid case and only for large velocities in the free-free case or in smectics. One finds that λ_{xx} is constant and only depends on a short length scale (q_c^{-1}) for the first two cases. The above simplified presentation hinders a main result (ref. /2/) : in the case of rigid-rigid Rayleigh-Bénard both the permeation and Poiseuille flow are important dissipative processes. For the shear instability one expects at low velocities a relation of the type $P \sim V \log (q_c D/V)$ (the coefficient has been exactly calculated in ref. /2/ in the limit $\lambda_p \zeta > 1$). Eventually note that the friction coefficient λ_{yy} for climb is lowered in the three cases by glide or climb motion. On the contrary there is no effect on glide motion for the two first cases. Although dynamical behavior of dislocations is well described by eq. (40) and (41) in the case of Rayleigh-Bénard instabilities, /16/ our model fails to interpret some experiments observed under elliptical shear. Indeed trajectories such as shown on fig. 5 can never be fitted with the ones deduced theoretically from expressions (40) and (41) (where ϕ^0 is the stress created by one dislocation on the other one). Furthermore experiments on a wedged sample of thickness $\sim x$, which correspond (as will be seen in the next section) to imposing some external stress ϕ_x^0 on the sample, reveal both glide and climb motion. In that case the Peach-Koehler force P_y which results from that stress should only drive a climb motion as predicted by eq. (41). Eventually as quoted in the preceeding section, experiments show that a modification in the rotation sense of the elliptical excitation induces a change in the dislocation velocity. This last point is by no means included in a Smectic A phase dynamics. In the next section, we shall present the smectic C phase dynamics which takes better account of the symmetry of the problem and also introduce the equivalent of the thermomechanical coupling due to the presence of a wedge.

6. SMECTIC C PHASE DYNAMICS IN A WEDGED SAMPLE

The theoretical study /11/ of the elliptical shear instability shows that, at threshold, the direction of the rolls (oy), appears (fig. 15) at an angle α with the axis oy_E of the excitation.

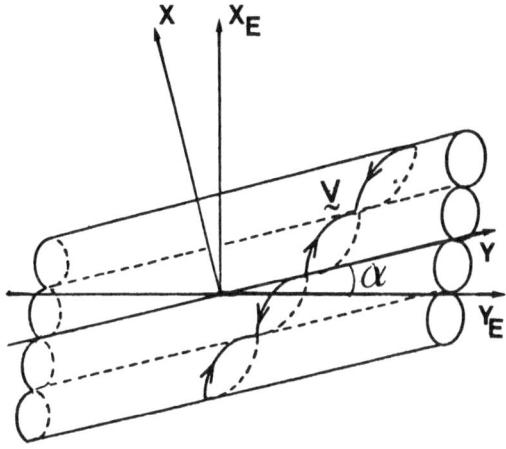

Fig. 15 : Orientation of the direction of the rolls (oy) versus the axis oy_E of the excitation.

Moreover the theoretical analysis shows that the hydrodynamic velocity gets two non zero components V_x, V_y, in the plane oxy (the velocity is not perpendicular to the roll axis). Since a layer is defined by a pair of rolls (of opposite velocity), the direction of the velocity defines a variable similar to the orientation in a smectic C . A change in the rotational sense of the elliptical excitation induces the change $\alpha \rightarrow -\alpha$ but also $(V_y \cdot V_x) \rightarrow -(V_y \cdot V_x)$. This indicates that the smectic term we must add to the elastic free-energy is of the form $E(\partial \phi/\partial x)(\partial \phi/\partial y)$ where $\bar{E}$ changes its sign with the rotational sense of the elliptical shear (in the coordinates oxy the system is invariant under the change $(\bar{E}, y) \rightarrow (-\bar{E}, -y)$). We shall also take into account the wedged structure of the sample. The tilt of the plates $\underset{\sim}{\theta} = (\theta_x, \theta_y)$ plays a role similar to a thermal gradient $\underset{\sim}{\nabla} T$ in smectics or cholesterics. Let us define in reduced units :

$$\delta d = d(\underset{\sim}{r}) - d_o = \underset{\sim}{r} \cdot \underset{\sim}{\theta}$$

The free energy now reads :

$$F = 1/2 \int dxdy \left[B_{//} (\partial \phi/\partial x)^2 + B_{\perp} (\partial \phi/\partial y)^2 + 2C \, \delta d (\partial \phi/\partial x) + 2E (\partial \phi/\partial y)(\partial \phi/\partial x) + \right.$$

$$\left. + 2C' \, \delta d (\partial \phi/\partial y) \right] \qquad (46)$$

The static coupling term $C\delta d(\partial \phi/\partial x)$ simply states that the periodicity of the structure (or the wave vector) depends on the sample thickness. This term is equivalent to the coupling term (in $\underset{\sim}{\nabla}$ T) in smectics or in cholesterics, where it gives the temperature dependence of the structure. The two last terms are characteristic of smectics C . An estimate of the coefficients is obtained by assuming that, at threshold, the periodicity (or wave vector) of the structure only depends on the sample thickness $q = (d_o/d) = 1 - \delta d/d$ (in reduced units). This is a first order approximation in X_E/d, Y_E/d parameters, introduced in ref /11/. Then

$$\frac{B_\perp C - E C'}{B_\perp B_{/\!/} - E^2} = \frac{1}{\pi} \tag{47}$$

An other relation is obtained, using the fact that, at threshold, the angle α defining the direction of the rolls (or equivalently $\partial\phi/\partial y$) only depends on the product (qd) (always at the same approximation level) (ref. /11/). This gives the second relation :

$$C' = E/\pi \tag{48}$$

One obtains from eqs. (47) and (48) the relation between C and $B_{/\!/}$

$$C = B_{/\!/} /\pi \tag{49}$$

The permeation, eq. (13) in presence of a wedge, now reads :

$$\partial\phi/\partial t + V_x = \lambda_p q_o^2 \underset{\sim}{\nabla}.\underset{\sim}{\phi} + q_o^2 \xi_{/\!/} \theta_x + q_o^2 \xi_\perp \theta_y \tag{50}$$

The terms in $\xi_{/\!/}$ and $\xi_\perp$ are respectively characteristic of a smectic A and C. They are equivalent to the thermomechanical coupling in cholesterics (Lehmann effect /17/,/18/) and smectics, and reminiscent of the Soret effect in binary mixtures /1/.

In what follows, we shall focus our attention on a wedge of type I (section 3-d) and take $\theta_x \neq 0$, $\theta_y \neq 0$. The "smectics C Navier-Stokes" equations are :

$$\rho \, \partial V_x/\partial t = -q_o \nabla_x P + q_o \, \delta F/\delta\phi -\zeta_{/\!/} V_x - \nu_{/\!/} V_y \tag{51}$$

$$\rho \, \partial V_y/\partial t = -q_o \nabla_y P -\zeta_\perp V_y -\nu_\perp V_x \tag{52}$$

where we recall that the velocities are as in section 2 averaged over
the sample thickness.

i) Stress due to a wedge

Experimental observations do not reveal any pattern or dust particle
flow, then we shall look for solutions with $V_y = V_x = \partial\phi/\partial t = 0$. Furthermore we use the additional experimental observation that compression or
orientation of the rolls do not depend on y .

The strain field ϕ^0 :

$$\phi^0_y = B_\perp \partial\phi^0/\partial y + E \partial\phi^0/\partial x + C'\delta d \tag{53}$$

$$\phi^0_x = B_/ \partial\phi^0/\partial x + E \partial\phi^0/\partial y + C \delta d \tag{54}$$

must satisfy eq. (50), where for a wedge of type I, $\delta d = \theta \; x$:

$$\lambda_p (\partial\phi^0_x/\partial x) = - \xi_{//} \theta x \tag{55}$$

one obtains the solutions

$$\partial\phi^0/\partial x = \mu + \kappa \theta x \tag{56}$$

$$\partial\phi^0/\partial y = a \tag{57}$$

where a and μ are constants, and $\kappa = -\dfrac{C}{B_{//}} - \dfrac{\xi_{//}}{\lambda_p B_{//}}$

The resulting stress is

$$\phi^0_y = B_\perp a + E \mu + E \kappa \theta x + C'\delta d \tag{58}$$

The two last terms of eq.(58) are simplified with the use of the two relations (48) and (49). Then

$$\phi^0_x = B_{//} \mu + Ea - (\xi_{//}/\lambda_p)\theta x \tag{59}$$

$$\phi^0_y = B_\perp a + E\mu - (E/B_{//})(\xi_{//}/\lambda_p) \theta x \tag{60}$$

The precise form of the stresses depends strongly on experimental conditions. Experiments show that in the region of largest thickness, the system is in fact subcritical : there is no stress on the rolls. For simplicity sake, we suppose the sample rectangular (between $x = + \pi L/a_o$ and $x = - \pi L/a_o$), then the boundary conditions, :

$$\phi^0_x \ (x = +\pi L/a_o) = 0 , \quad \phi^0_y \ (x = +\pi L/a_o) = 0$$

give $\quad a = 0, \quad \mu = (\xi_\parallel / \lambda_p) \, \theta \pi L/a_o$.

The stresses are now for a wedge $+\theta$:

$$\phi^0_x = (\xi_\parallel/\lambda_p) \, \theta((\pi L/a_o) - x) \tag{61}$$

$$\phi^0_y = (\xi_\parallel/\lambda_p) \, (E/B_\parallel) \, \theta \, ((\pi L/a_o) - x) \tag{62}$$

and for a wedge $-\theta$:

$$\phi^0_x = (\xi_\parallel/\lambda_p) \, \theta \, (x + \pi L/a_o) \tag{63}$$

$$\phi^0_y = (E/B_\parallel) \, (\xi_\parallel/\lambda_p) \, \theta \, (x + \pi L/a_o) \tag{64}$$

First notice that the "thermomechanical" coupling induces a stress linearly increasing with x. The stress ϕ^0_y is much lower than ϕ^0_x since one expects $E/B_\parallel \ll 1$. In large samples there will be regions where the critical stress ϕ^c for nucleating dislocations will be reached. One would have a periodic emission of dislocations just as one has a periodic emission of vortices in a Josephson junction. We shall not consider here the question of nucleation but rather a situation in which an isolated dislocation is moving in the stress defined by eqs (61) $\div$ (64).

ii) Motion of an isolated dislocation in a wedge

The procedure is similar to the one defined in sections 4 and 5. Since dynamical equations are linear, the total stress is defined as $\phi^t = \phi^0 + \phi^d$ and the strain field $\underset{\sim}{m}^t = \underset{\sim}{m}^0 + \underset{\sim}{m}^d$, where ( )^0 are the preceeding solutions in the presence of a wedge, and ( )^d refers to solutions for an isolated dislocation in the absence of the wedge (more generally ( )^0 could refer to quantities not directly linked to the dislocation d). In the absence of a wedge the strain field of a dislocation is defined by a dynamical equation similar to eq. (18)

$$h \, \underset{\sim}{V}_{o} \cdot \underset{\sim}{m}^{d} = - \, q_{o}^{2} \, (\partial^{2}/\partial y^{2} + \lambda_{p} h) \, \underset{\sim}{V} \cdot \underset{\sim}{\phi}^{d} \, (\underset{\sim}{m}^{d}) \tag{65}$$

where the differential operator takes into account the smectic C version of Navier-Stokes equations (51), (52)

$$h = \zeta_{/\!/} \partial^{2}/\partial_{y}^{2} + \zeta_{\perp} \partial^{2}/\partial_{x}^{2} + (\nu_{/\!/} + \nu_{\perp}) \, \partial^{2}/\partial_{x} \partial_{y} \tag{66}$$

A straightforward calculation gives the Fourier transform of the strain :

$$\hat{\underset{}{m}}_{x}^{d} \, (q) = - \, \frac{2 \pi \, i \, (B_{\perp} \, q_{y} + E q_{x} - i g \, (q) \, V_{oy})}{B_{/\!/} \, q_{x}^{2} + B_{\perp} \, q_{y}^{2} + 2 E \, q_{x} q_{y} - i g (q) \, (V_{ox} q_{x} + V_{oy} q_{y})} \tag{67}$$

$$\hat{\underset{}{m}}_{y}^{d} \, (q) = \, \frac{2 \pi \, i \, (B_{/\!/} \, q_{x} + E q_{y} - i g \, (q) \, V_{ox})}{B_{/\!/} \, q_{x}^{2} + B_{\perp} \, q_{y}^{2} + 2 E \, q_{x} q_{y} - i g (q) \, (V_{ox} q_{x} + V_{oy} q_{y})} \tag{68}$$

where now

$$g(q) = \frac{\hat{h} \, (q)}{q_{o}^{2} \, (q_{y}^{2} + \lambda_{p} \, \hat{h} \, (q))} \tag{69}$$

$g(q) \to (\lambda_{p} q_{o}^{2})^{-1}$ in the limit of strong permeation, $\hat{h} \, (q)$ is the Fourier transform of the operator (66).

The relation between the stress $\underset{\sim}{\phi}^{0}$ created by the wedge through the "thermomechanical" coupling and the dislocation velocity results from eqs. (38) and (39) where now $\hat{m}_{x}^{d} \, (q)$, $\hat{m}_{y}^{d} \, (q)$ and $g(q)$ are given by eqs (67), (68), (69).

$$P_{x} = - \varepsilon_{d} \phi_{y}^{0} = \lambda_{xy} \, V_{oy} + \lambda_{xx} \, V_{ox} \tag{70}$$

$$P_{y} = + \varepsilon_{d} \phi_{x}^{0} = \lambda_{yy} \, V_{oy} + \lambda_{yx} \, V_{ox} \tag{71}$$

with

$$\lambda_{xx} = \int \frac{(B_{\perp}^{2} \, q_{y}^{2} + E^{2} q_{x}^{2}) \, g(q)}{\mathcal{D} \, (q)} \, dq_{x} \, dq_{y} \tag{72}$$

$$\lambda_{yx} = \lambda_{xy} = - E \int \frac{(B_\perp q_y^2 + B_\parallel q_x^2) \, g(q)}{\mathcal{D}(q)} \, dq_x \, dq_y \tag{73}$$

$$\lambda_{yy} = \int \frac{(B_\parallel^2 q_x^2 + E^2 q_y^2) \, g(q)}{\mathcal{D}(q)} \, dq_x \, dq_y \tag{74}$$

and $\mathcal{D}(q) = (B_\parallel q_x^2 + B_\perp q_y^2 + 2E \, q_x q_y)^2 + g^2(q) \, (q_x V_{ox} + q_y V_{oy})^2$ (75)

Expressions (70) and (71) may be inverted

$$V_{ox} = -\varepsilon_d(\Phi_y^0 \, \lambda_{yy} + \Phi_x^0 \, \lambda_{xy})/(\lambda_{yy} \, \lambda_{xx} - \lambda_{xy}^2) \tag{76}$$

$$V_{oy} = \varepsilon_d(\Phi_x^0 \, \lambda_{xx} + \Phi_y^0 \, \lambda_{yx})/(\lambda_{yy} \, \lambda_{xx} - \lambda_{xy}^2) \tag{77}$$

According to these relations, except for a very particular combination of Φ_x^0 and Φ_y^0 , one will always observe simultaneous glide and climb of the dislocations, with the smectic C symmetry. On the contrary, the smectic A analogy implies E = O and thus λ_{xy} and Φ_y^0 = O : only climb is possible in that case. The experimental relevance of these results is devoted to the conclusion.

7.   CONCLUSION

     The consideration of the rolls systems dynamics, based on a simple smectic A analogy, is successful in describing the strain pattern sur- rounding moving isolated dislocations. The comparison of the calculated patterns with experiments performed on a nematic shear instability allows one to estimate the phase diffusion coefficients. The friction coefficients that one can calculate depend in general on the velocity of the disloca- tion in a non trivial way. Attempts to check these predictions have been performed on the same experiment, mainly with the observation of trajec- tories of pairs of opposite sign defects interacting with each other, and the observation of isolated defect motions in a wedge shape sample. Both, cannot be interpreted in a smectic A analogy. One is thus entitled to wonder if the analogy with a thermodynamical system close to equilibrium makes any sense at all. To that respect, the experiments performed in the wedge shape sample are very revealing : although some experiments

have been performed more than 24 hours after the roll pattern has been set
dislocations are still emitted at the same rate as after 15 or 30 minutes.
Furthermore, their motion involves both climb and glide, even when the
tilt is entirely along the normal to the layers. These observations
raise problems of two different natures. The first, shows that no steady
state (at least in the sense of a stable roll pattern) is reached when
a wedge exists in the sample ; the second reveals that the spatial symmetry of the smectic A is too high. None of them however, invalidates
fundamentally, the idea of taking advantage of the current knowledge
on thermodynamical layered systems, to understand the behavior of
rolls in the large box limit. Indeed, taking a proper account of the
symmetry, simply requires the use of a smectic C analogy ($C_{2\upsilon}$ point
group symmetry) ; furthermore thermodynamical systems submitted to the
action of external fields, need not necessarily evolve toward an equilibrium state (e.g. Josephson junction, Soret effect, etc...) According to
the smectic analogy, the tilt angle $\underset{\sim}{\theta}$ is equivalent to a temperature
gradient $\underset{\sim}{\nabla T}$. Depending on boundary conditions there may exist or not
exist equilibrium solutions in the presence of a temperature gradient : the
thermomechanical coupling has been introduced first by Leslie /17/ to
explain the strong molecular rotation observed at the begining of this
century by Lehmann /18/ in cholesterics submitted to a temperature
gradient. The considerations developed in the sixth section follow
the same line. Do they give a proper account of the experimental observation ?

As already stressed, equations (61, 62) show that, the stress due
to the "thermomechanical coupling" necessarily reaches the critical
value for nucleating dislocations at some point. The observation that
nucleation occurs preferentially in the equatorial plane at the boun-
aries, can be understood if one remarks :

i) Nucleations are always easier at surfaces,

ii) In the equatorial plane, the image interaction is purely repulsive, whereas the x component is attractive elsewhere.

These two facts combined together allow one to understand that the critical stress ϕ^c is first reached at this location. It is now important
to check if the symmetry properties, one can deduce from the equations
of motion, are compatible with experiment or not. The smectic C symmetry
is essentially represented by the elastic constant E, which as we have
seen, is expected to change sign when the sense of rotation of the

ellipse is reversed. From equations (61, 62, 72, 73 and 74) one can deduce that the change :

$$E \rightarrow -E$$

implies :

$$\phi_x^0 \rightarrow \phi_x^0$$

$$\phi_y^0 \rightarrow -\phi_y^0$$

$$\lambda_{xx} \rightarrow \lambda_{xx}$$

$$\lambda_{yy} \rightarrow \lambda_{yy}$$

$$\left.\begin{matrix}\lambda_{xy}\\\lambda_{yx}\end{matrix}\right\} \rightarrow \left\{\begin{matrix}-\lambda_{xy}\\-\lambda_{yx}\end{matrix}\right.$$

Thus from equations (76) and (77), one sees that V_{ox} ought to change sign, but not V_{oy}, which corresponds precisely to experimental observations. An other important symmetry concerns the sign change $\theta \rightarrow -\theta$; as it is clear upon comparing equations (61, 62) and (63, 64) in the central region, changing the sign of θ , changes neither ϕ_x^0 nor ϕ_y^0 , which again is in good agreement with experimental observations.

The only, a priori, puzzling point concerns the experimental non-dependence of V_{oy} on θ , whereas V_{ox} does depend linearly on it .

According to equations (76, 77) and (61, 62) V_{ox} and V_{oy} are linear functions of θ . However ϕ_x^0 is larger than ϕ_y^0 in a ratio ($B_{/\!/}$ /E). It is thus considerably easier to reach the critical stress for nuclea-ting dislocations, on ϕ_x^0 than on ϕ_y^0 . Thus ϕ_x^0 is bound to values lower than ϕ^C , and cannot increase with θ . In the limit $E \ll B_{\perp}$, $B_{/\!/}$, equations (76, 77) can be approximated by (this inequality is suggested by the fact that the strain pattern, as given by the smectic A analogy provides a satisfactory account of experiments):

$$V_{ox} \simeq -\varepsilon_d \phi_y^0 / \lambda_{xx} \tag{78}$$

$$V_{oy} \simeq \varepsilon_d \phi_x^0 / \lambda_{yy} < \varepsilon_d \phi^C / \lambda_{yy} = V_{oy}^c \tag{79}$$

The climb velocity is bound to a value lower than V_{oy}^c , whereas the glide velocity can increase linearly with θ . On the other hand, although ϕ_y^0 is considerably smaller than ϕ_x^0 , V_{ox} can be comparable to or larger than V_{oy} since λ_{xx} is much smaller than λ_{yy} (in a ratio $(B_\perp /B_{//})^2 \sim (0.06)^2$.

There are, thus, good pieces of evidence that the smectic C hydrodynamics, including the thermomechanical coupling gives a fair account of experimental observations. A more complete analysis should include considerations on the dislocation nucleation frequency, on other geometries (tilt along the layers), as well as dislocation interactions. One must also keep in mind that the use of this analogy has to be restricted to the "large box" problem, and that it will fail as soon as the phase approximation breaks down.

Note added in proof :
The smectics equations throughtout this paper do not take into account the renormalization of viscosities due to long wavelengths thermal fluctuations /19/. This is legitimate for non thermodynamical systems.

ACKNOWLEDGEMENTS

We want to thank E.Guyon and P. Manneville for stimulating discussions.

This work has been presented in two communications at the conference on "Cellular structures in instabilities".

REFERENCES

/1/ Martin P.C., Parodi O., Pershan P.S., Phys. Rev. A 6, 2401 (1972)

/2/ Dubois-Violette E., Guazzelli E., Prost J., Phil. Mag. A 48, N°5 727 (1983)
We want to point out that in the present paper, notations are different from those of ref./2/. This change has been made to be in agreement with the notations of this book.

/3/ Newel A., Whitehead J.A., J. Fluid Mech., 38, 279 (1969)

/4/ a- Pomeau Y., Manneville P., J. Phys. Lett. 40, L 609 (1979)
b- Pomeau Y., Manneville P., Zaleski S., Phys. Rev. A 27 2710 (1983)

/5/ a- Siggia E.D., Zippelius A., Phys. Rev. A 24, 1036 (1981)
b- Siggia E.D., Zippelius A., Phys. Rev. Lett. 47, 835 (1981)
c- Cross M.C., Phys. Rev. A, 27, 490 (1983)

/6/ Guazzelli E., Guyon E., Wesfreid J.E., in "Symmetries and Broken Symmetries in Condensed Matter Physics" Ed. N. Boccara IDSET Paris, p. 455 (1981)

/7/ Guazzelli E., Guyon E., Wesfreid J.E., Phil. Mag. A, 48, N°5, 709 (1983)

/8/ Peach M., Koehler J., Phys. Rev. $\underline{80}$ 436 (1950)

/9/ Nabarro F.R. , Theory of Dislocations - Oxford Clarendon Press (1967)

/10/ Friedel J., Dislocations , Pergamon Press (1964)

/11/ Dubois-Violette E., Rothen F., J. Phys. $\underline{39}$, 1040 (1978)

/12/ Pieranski P., Guyon E., Phys. Rev. Lett. $\underline{39}$, 1281 (1977)

/13/ Guazzelli E., "Nematic instability induced by an elliptical shear"
16 mm movie available at SERDDAV 27, rue P. Bert - 94204 IVRY CEDEX
France. The scenario is published in "Non linear phenomena at phase
transitions and instabilities", ed. T. Riste, New York, Plenum
Press, p. 173 (1982)

/14/ Guazzelli E., C.R. Acad. Sci., $\underline{291}$, 139 (1980)

/15/ Pershan P.S., J. Appl. Phys. $\underline{45}$, 1590 (1974)

/16/ Pocheau A., Croquette V., preprint, submitted to J. de Physique

/17/ Leslie F., Molecular Crystals and Liquid Crystals, $\underline{7}$, 407 (1969)

/18/ Lehmann O., Annalen Phys. (4), $\underline{2}$, 649 (1900)

/19/ Mazenko G.F., Ramaswamy S., Toner J., Phys. Rev. Lett. $\underline{49}$, 51 (1982)

DEFECTS AND INTERACTIONS WITH THE STRUCTURES IN EHD CONVECTION IN
NEMATIC LIQUID CRYSTALS.

R. RIBOTTA, A. JOETS

Laboratoire de Physique des Solides
Université de Paris-Sud, Bât. 510
91 405 ORSAY CEDEX (France).

A nematic liquid crystal subjected to an AC electric field shows
sequences of convective structures having a decreasing symmetry, before
chaos. The structures are periodic in space and therefore may contain
defects in the periodicity. These defects interact with each other as
well as with the structure in such a way that they can help to stabilize
or to destabilize such a structure.

1. DEFECTS IN A RECTILINEAR ROLLS STRUCTURE (Williams Domains)

As was presented in the preceding paper (A. Joets, R. Ribotta) the
convective structures have for the lower values of the control parameter translationnal invariance. Thus they are periodic in one direction
with a periodicity equal to two contra-rotative rolls diameters.
A defect will then be a "stacking fault" in the periodicity and hence
will appear as an extra pair of rolls (a pair of disinclinations). The
mechanism by which the defects appear is not yet clear. Usually it is
observed that they result from misfits between adjacent domains in
inhomogeneous samples. These misfits can occur when these domains are
experiencing lateral motion thus creating stresses at some places (a
shear is produced by a dilative stress in the roll). In this process
the defects move inside the structure and it has not yet been clearly
establihed which forces really act on them. Nevertheless it is possible
to separate the dynamical interaction (i.e. inducing motion of structures and defects) from a static interaction. This can be realized when
the density of defects present in the sample is low (typical separation
distance ≈50 rolls) and when the field is applied at a sufficiently low
rate. A possible cause for a misfit between two adjacent (along a perpendicular to the roll axis) convective zones could be when the threshold is different for two adjacent zones. As the wave-vector varies
above threshold there would be a slight discrepancy between the two
zones and therefore a stacking fault may appear.

The deformation field of a typical disinclination pair (here-after called edge-dislocation) is sketched on Fig. 1.

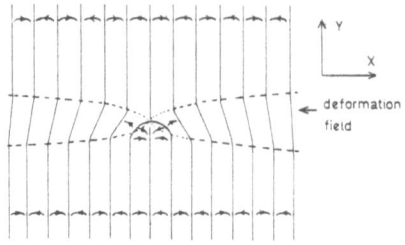

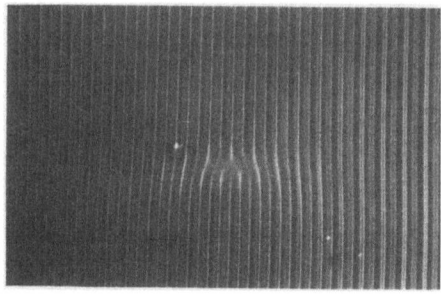

Fig.1 : The elementary defect is an extra pair of rolls.

One can notice that there is much similarity to the deformation field of an edge dislocation in smectic layers [1]. In fact the analogy of convective rolls with layered systems [2] has been developed and used to study the deformation of convective rolls in nematics under mechanical shear [3]. The main result is that the shape of the dislocation can provide an estimate of a parameter which is the ratio of two diffusion coefficients of the phase (the relative position of a roll). The first coefficient $D_\perp$ is associated to the existence of a transverse mode (undulations of the rolls along their axis), the second one $D_{//}$ is associated to a longitudinal mode (compression dilation of the diameter of the rolls). In smectic layers the coefficients are respectively, the curvature elastic modulus K and the normal compressional modulus B. The undulation mode is associated to an energy of order q^4 (q : wave-vector of the deformation) while the compression is associated to an energy of order q^2.

Practically, the aperture of the parabola of the deformation field is proportionnal to the ratio $\lambda = \left(\dfrac{K}{B}\right)^{1/2}$. In other words in quasi-incompressible systems a small value of λ means that curvature is easier than compression, while a large value of λ means that a curvature deformation is easily relaxed by compression of the rolls (variation of their diameters). In the first case the deformation induced by a defect extends over a larger distance than it does for the second case.

The two situations described here are observed in our system and correspond to exciting fields with different frequencies. At low frequencies (from 20 to 40 Hz) the defects extend to larger distances (small value of λ) while at higher frequencies above 70-80 Hz they distort the neighbouring rolls on short distances (large λ) (Fig. 2).

a)

b)

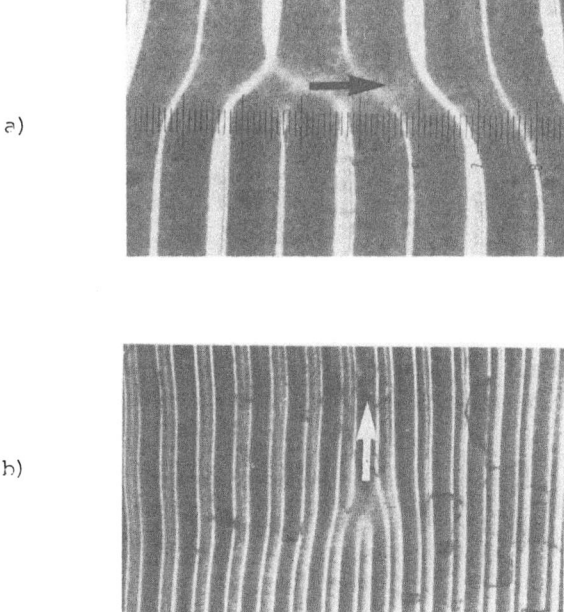

Fig.2 : a) Defect at low frequency : small λ , large deformation
(glide motion).

b) Defect at high-frequency : large λ , small deformation, (climb motion).

Also the motion of these defects is strongly different for small λ and for large λ. At low frequencies the glide is dominant while at high frequencies it is the climb. It can be understood that climb is easier when compression energy is lower than curvature energy. However it is not yet clear which process is analogous here to the permeation (or diffusion). At low frequencies compression is difficult therefore making the climb less easy. In fact one observes that around such a defect there is a strong "shear strain" (tilt of roll associated with local decrease in diameter), and the motion takes place parallel to the direction of the supposed "shear strain", i.e. normal to the roll axis. A shear of the roll would indicate that the "elastic limit" has been reached.

2. DEFECTS IN THE ZIG-ZAG STRUCTURE

The convection takes place in a double set of rectilinear rolls tilted by an angle θ with respect to the original roll axis (say along the $\vec{y}$ direction). Defects can exist which are more generally formed in the zones of bending where the strains are larger. Their shape is similar to that of low frequency defects observed in the rectilinear roll structure. However, they are more unsymmetrical and, close to the core on one side, the deformation is such that the bending of the rolls produces locally a tilt opposite to the general orientation of the rolls (see Fig. 3).

If the defect is on the separation line (flexion line) between the $+\theta$ and $-\theta$ orientation it corresponds to a breaking of that line with a step. For instance on one side of the defect the $+\theta$ domain extends over the $-\theta$ domain. It is observed that the defect will "glide" back and forth on the flexion line so as to increase the area of the $+\theta$ domain at the expense of the $-\theta$ one. There, clearly appears the role of the defect : it helps to stabilize one structure against its symmetrical part. Since the two orientations correspond to the same energy it can be supposed that the motion and the "work" of the defect are a consequence of stresses imposed for instance from the boundaries. That type of defect is reminiscent of the transformation dislocation observed in twinning of crystal lattices [6].

After some time of order (1 to 20 minutes), these defects segregate in order to build grain boundaries (see Fig. 4). These grain boundaries are built first along the $\vec{y}$ direction (Fig. 4.a). Their width increases with the control parameter i.e. the energy fed into the system. Further increase makes new dislocations stack in the bending zones along $\vec{x}$. Now one obtains closed domains of large size typically 2 × 2 mm) (Fig. 4.b). These domains might be of very different size. In that case one observes slow rearrangements by elimination of some small domains inside larger ones.

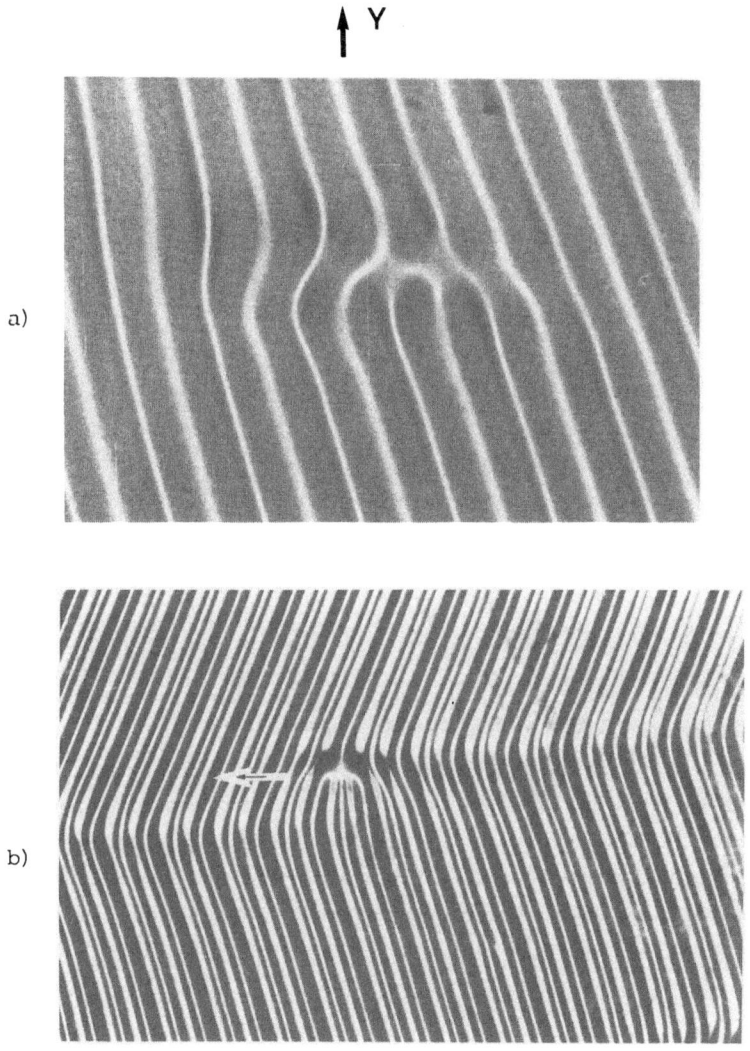

Fig. 3 : a) Defect in a zig-zag structure.
   b) Transformation dislocation at a flexion line

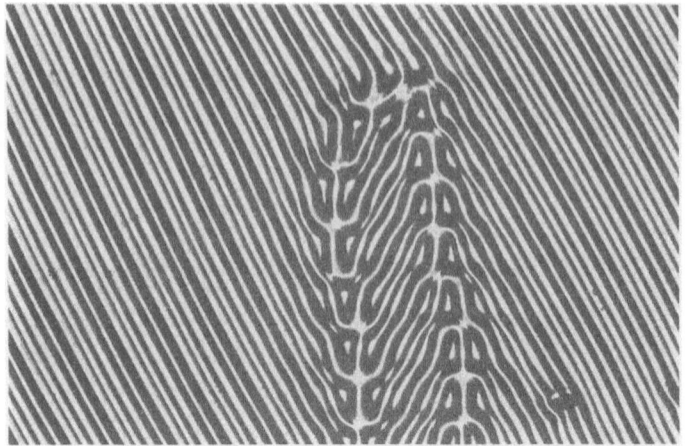

Fig. 4 : Grain boundaries between two domains with opposite
tilt. Rectangles can be seen in the central part. A
varicose deformation exists in the $\vec{x}$ direction close
to the boundaries.

Looking at the grain boundaries one can recognize in their central
part quasi-rectangular structures. Along the $\vec{x}$ direction the rectangles
are replaced by varicose deformation of the tilted rolls. The transition between rectangles and varicose is sudden in space (over 1 or
roll diameters) and this is compatible with a first-order-like transition from varicose to bimodal structure. The varicose deformation on
the contrary vanishes continuously as expected when going along $\vec{x}$
towards the central part of the tilted domain (Fig. 5).

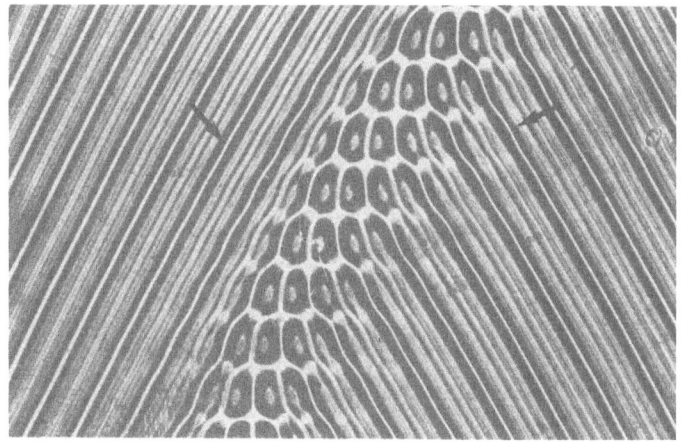

Fig. 5 : As the grain boundary width increases the varicose increases its amplitude.

3. DEFECTS IN THE VARICOSE STRUCTURE

The varicose structure is the result of a modulation of the zig-zag
domain with an oblique wave-vector. The rolls are then periodically
pinched along their axis. A defect will come from a sudden change in
the periodicity and will appear as an extra or a missing period(Fig.6).
This is a typical example of continuous formation of a defect in a
periodic structure. Usually it is observed that such defects are few
in large monodomains.

4. DEFECTS IN THE BIMODAL STRUCTURE

In the rectangle, where there exists a periodicity in both x and
y directions,the situation is apparently quite different from that in
the roll structures. One could think of a defect made of two extra rows
or two extra columns since the periodicity of the structure is twice
the elementary spatial period. In fact one observes that a defect is

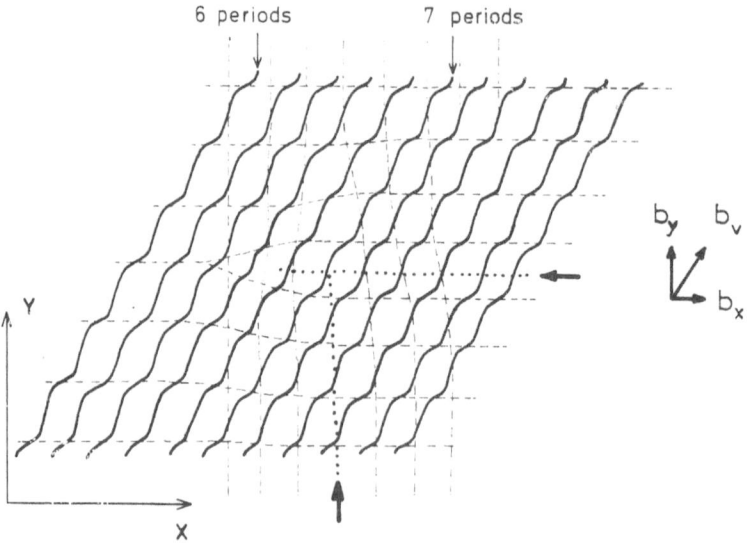

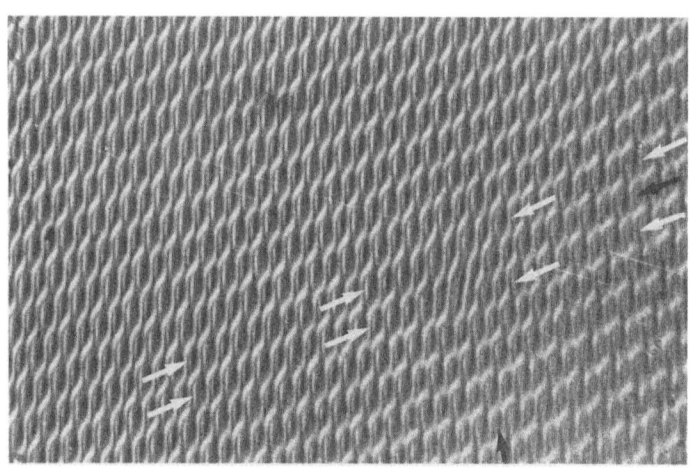

Fig. 6 : Defect in the skewed varicose of "Burgers
vector b_v". The resulting defect in bimodal
will be one extra row(b_y) connected to one
extra column (b_x).

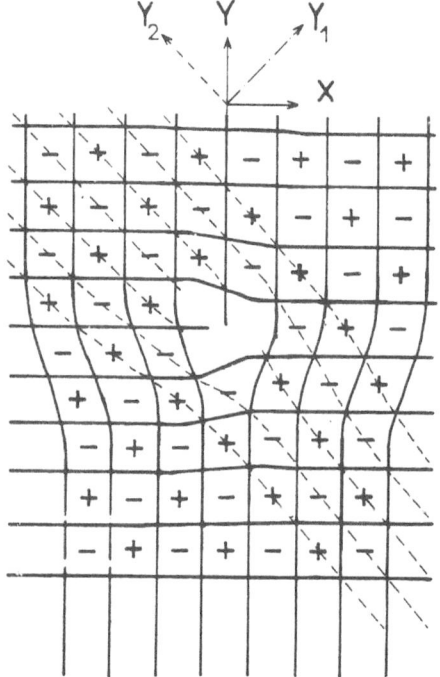

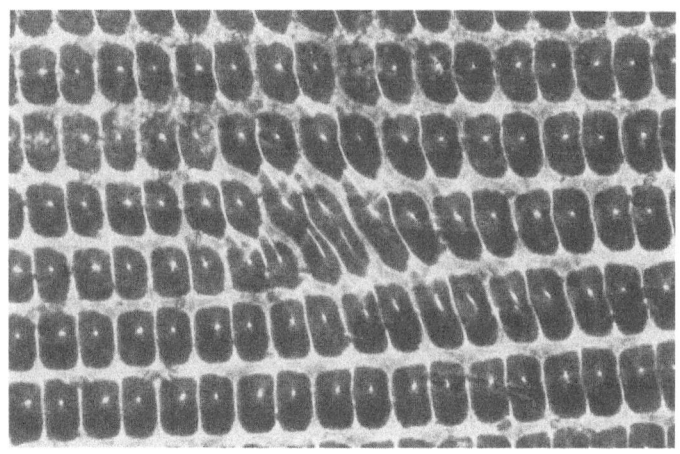

Fig. 7 : Defect in a rectangular bimodal structure.

rather made of one extra now meeting one extra column with a common
core. By continuity of the sense of rotation from one rectangle to the
neighbouring one, one checks that there cannot exist one extra row alone
without an extra column. Moreover that defect is obtained from the
defect in the varicose structure (Fig.7). Thus the Burgers vector of
the varicose defect on the tilted roll can be considered as the sum of
two unitary Burgers vectors along $\vec{x}$ and $\vec{y}$. Such a situation is similar
to that of three dimensional lattices.

5. EVOLUTION OF THE DEFECTS UNDER STRESS

The stresses result from an increase of the control parameter
(here, the voltage V). There are two distinct effects depending on
$d\varepsilon/dt$ the rate in time of the reduced control parameter ε ($\varepsilon=(V^2-V_c^2)/V_c^2$,
where V_c is the critical value at onset).

5.a. Low stress rate

- Elementary "dislocation" : The deformation field changes
strongly as the control parameter is increased. The parabola of the
deformation profile decreases continuously its aperture down to a
line almost in the $\vec{x}$ direction (the gliding direction). That corresponds to a sheared pattern around the dislocation. On the "shear
line" other dislocations can be created by pair which glide apart from
each other after the nucleation (Fig. 8).

- Grain boundaries in the zig-zag structures : The width of
the boundary increases and the varicose deformation which is associated to that boundary extends over larger distances in the $\vec{x}$ direction.
Thus a domain of varicose structure can be directly created from that
boundary. The rectangular bimodal convection is created at the same
time and behind it. This is an example of nucleation from the grain
boundaries (Fig. 5).

- Rectangular bimodal structure : Here, continuous increase of
the control parameter will tend to eliminate the "dislocations" that
may be present. The process which is not yet elucidated, seems to be
connected with the appearance of an oscillatory instability between
zig-zag and bimodal [5].

5.b. High rate of stress

For a stress rate of order 50 Volt/second a large density of
dislocations is suddenly created and the extension of the core(associated to a large shear) takes place with some delay (1 second). These

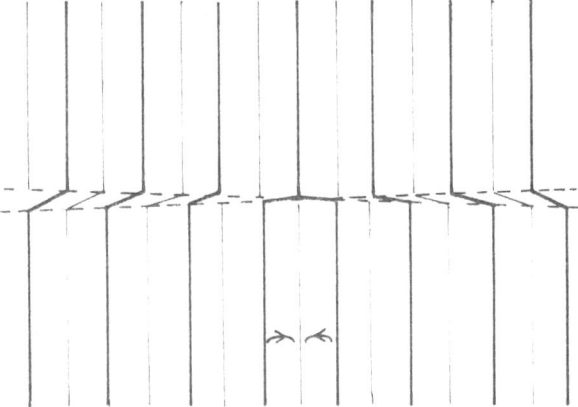

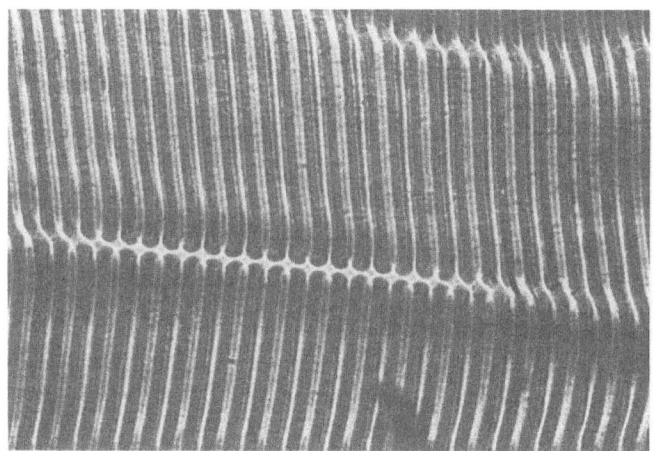

Fig. 8 : Extension of the core (with shear) for a high value of the control parameter.

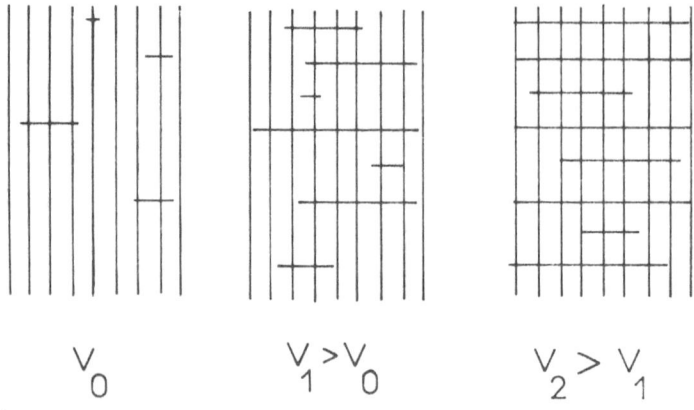

$$V_0 \qquad V_1 > V_0 \qquad V_2 > V_1$$

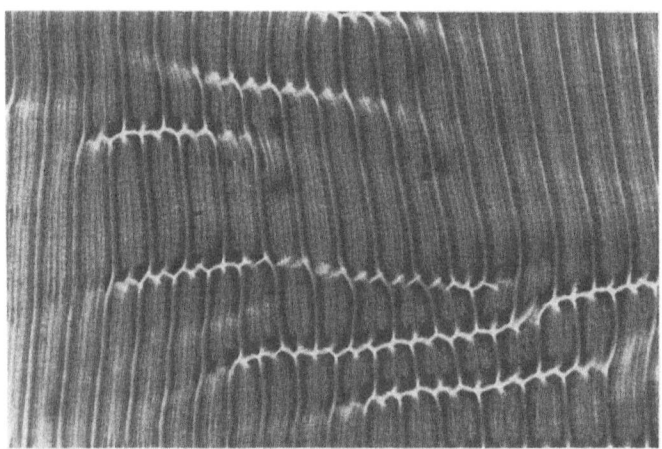

Fig. 9 : Building of rectangular bimodal structure
directly from rectilinear rolls, by periodical stacking of dislocations "sheared" by
application of high rate of stress.

"shear lines" increase in density and while extending tend to pile-up in the $\vec{y}$ direction, with a regular spacing close to the varicose half-period or to the length of a rectangle.

Thus the bimodal structure appears on the whole sample, within times of the order of some 5 to 10 seconds. The defects that may have remained are eliminated, the stress being kept constant for some 10 to 100 seconds. That formation process of a bimodal structure directly from a roll structure through periodic shearing by the defects periodically stacking may be regarded as a martensitic transformation [6] (Fig.9).

Experimentally it is preferable, in order to obtain a bimodal structure, to use that process because a stable state is reached faster since the defects are "eliminated" in the process.

CONCLUSION

Convective structures in nematic liquid crystals subjected to electric field can be developed in extended geometries. They offer a wide set of situations for the study of defects. In particular the two main types of defects show respectively glide and climb motion. These defects analyzed by analogy with the defects in smectic-A layers correspond to a change of magnitude in the ratio of smectic-like elastic moduli (compression, curvature). The shape of the defects allows one to predict the structure that will appear for a higher stress. Therefore defects appear clearly as nuclei for the growing of structures of decreasing symmetry. Indeed we can show that these structures can be obtained from defects either by continuous slow process or through martensitic-like transformations.

REFERENCES

1) P.G. de GENNES, C.R. Acad. Sci. , Paris, $\underline{275}$, 939 (1972)

2) J. SWIFT, HOHENBERG, Phys. Rev. A $\underline{15}$, 319 (1977)
   J. TONER, D.R. NELSON, Phys. Rev. B $\underline{23}$, 316 (1981)

3) E. GUAZZELLI, E. GUYON, J.E. WESFREID in "Symmetries and Broken
   Symmetries in Condensed Matter Physics". Proc. of Colloque Pierre
   Curie, Ed. N. Boccara, IDSET - Paris (1981)

4) J. LAJZEROWICZ, J.J. NIEZ, J. de Phys. Lettres, Paris $\underline{40}$, L-165
   (1979) also in "Solitons and Condensed Matter Physics"
   Ed. A.R. Bishop, T. Schneider, Springer-Verlag (1978)
   J. LAJZEROWICS, Ferroelectrics, $\underline{35}$, 219 (1981)

5) R. RIBOTTA, A. JOETS, to be published

6) J. FRIEDEL in "Les Dislocations" Gauthier-Villats, Paris (1956)
   also J.W. Christian in "The Theory of transformations in metals
   and alloys" Pergamon Press (1965)

QUANTITATIVE STUDY OF THE DISORGANIZATION OF HEXAGONAL CONVECTIVE STRUCTURES

R.OCCELLI[+],E.GUAZZELLI[++],J.PANTALONI[+]

[+]Laboratoire de Dynamique et Thermophysique des Fluides (L.A.72)
Université de Provence,Rue H.Poincaré
13397 MARSEILLE CEDEX 13,FRANCE

[++] E.R.A. 1000:Département de Physique des Systèmes Désordonnés,
Rue H.Poincaré,Université de Provence
13397 MARSEILLE CEDEX 13,FRANCE[*]
[*]The present work has been performed at E.R.A. 1000:E.S.P.C.I.,Laboratoire d'Hydrodynamique et Mécanique Physique,10 Rue Vauquelin,75231 PARIS CEDEX 05.

When we examine pictures of convection in large aspect ratio cells when the initial conditions are not controlled and even at moderate departure from threshold ε ,we find patterns with structural defects such as dislocations.

We are motivated by the influence of these defects on the disorganization of the structure when the control parameter ε is increased.For this purpose we consider rather the geometrical descri-ption of the disorder than the temporal one(the data were taken after a period (15 minutes for the nematic hydrodynamic instability, several hours for the Bénard-Marangoni instability) during which the dislocations density decreases noticeably by pair annihilation or elimination at the edge of the sample and then stabilizes).

The study has been performed for two instabilities well controlled and understood which present hexagonal structure:
-The Bénard-Marangoni instability in a horizontal fluid layer heated from below with a free upper surface/1/
-A nematic hydrodynamic instability induced by a circular shear/2/3/. The choice of such instabilities is motivated by the fact that they can be easily observed (shadograph observation for the Bénard-Marangoni instability,microscopic observation or diffraction for the nematic hydrodynamic instability) within cells of large aspect ratio.

real space　　　　　**reciprocal space**

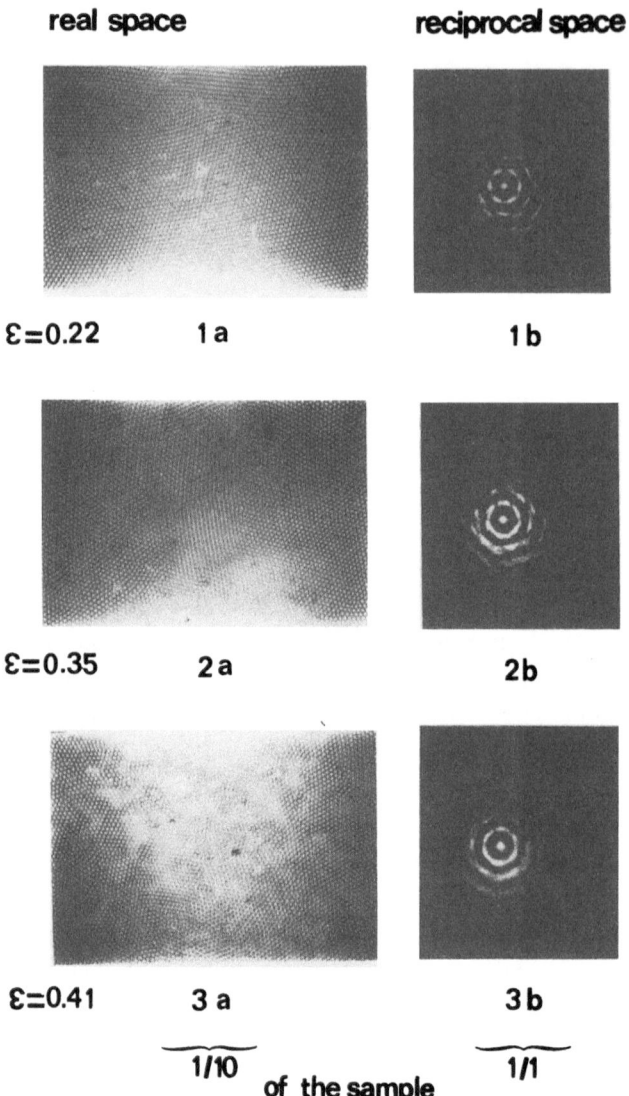

$\mathcal{E}=0.22$　　　**1a**　　　　　　**1b**

$\mathcal{E}=0.35$　　　**2a**　　　　　　**2b**

$\mathcal{E}=0.41$　　　**3a**　　　　　　**3b**

$\overbrace{}$　　　　$\overbrace{}$
1/10　　**of the sample**　　**1/1**

Figure 1

Nematic hydrodynamic instability convective structures: hexagonal
structures (a) and corresponding diffracted images(b).

The hexagonal pattern is shown in fig 1.1.a. The reciprocal image
is made up of spots of finite extent suggesting the possibilty of
an order such is present in a 2D solid(fig 1.1.b). Increasing
$\mathcal{E}$, arcs of diffraction develop around the original points (fig1.2.b)
This feature corresponds in real space to clusters of crystallized
hexagonal patterns slightly disoriented from one another and
separated by grain boundaries(Fig 1.2.a). The occurence of a
finite density of dislocations induces the loss of the
translational order between each crystallite but not completely
that of the orientational order. At very large $\mathcal{E}$,we obtain
the characteristic ring pattern of a 2D "liquid" state(fig 1.3.a
and b)

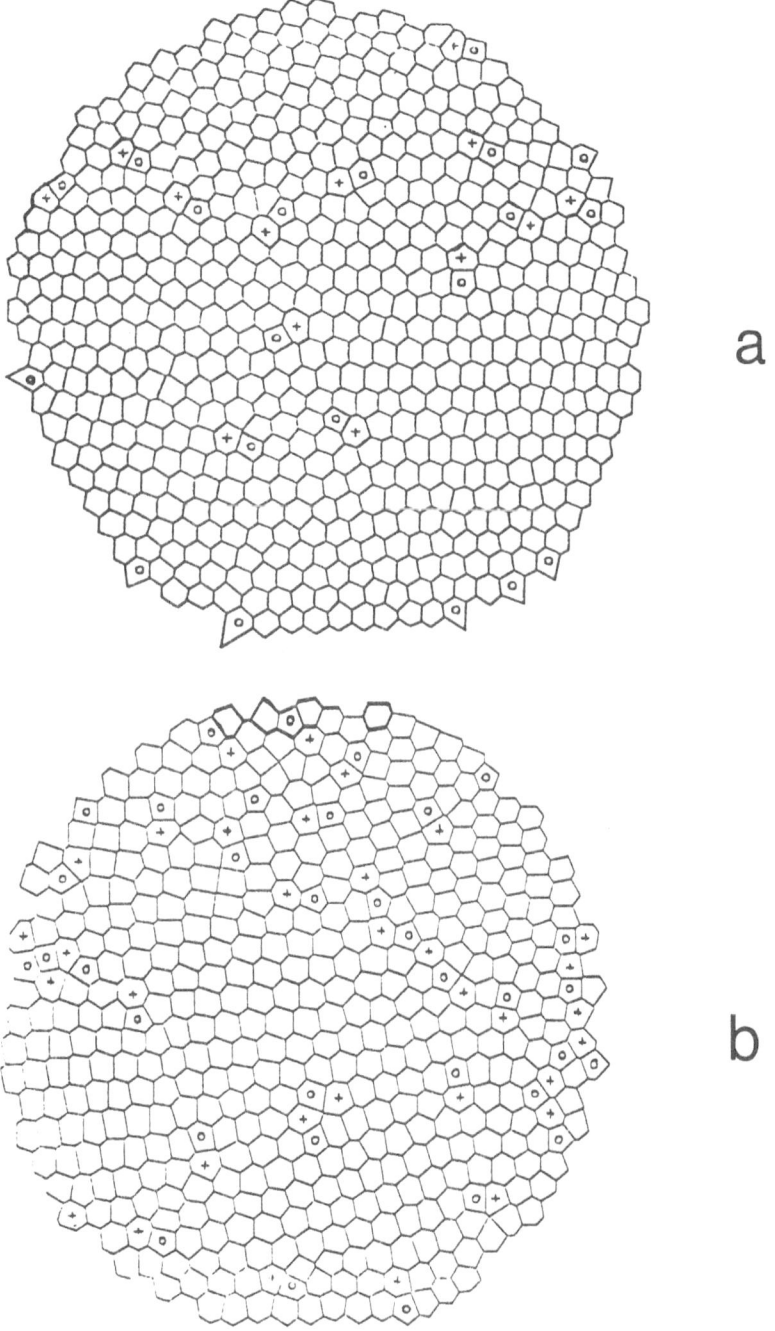

Figure 2
Reconstruction of the hexagonal structure in the Bénard-Marangoni
instability case.
a) ε =-0.03 ;percentage of defects 4.6% (Benard-Marangoni instability
is a subcritical instability)
b) ε =0.68 ; percentage of defects 16.0%

The number of hexagons is over 3000 for the Bénard-Marangoni instability and 40000 for the nematic hydrodynamic instability.

In this paper we present the geometrical methods of characterisation of the disorder and the main results.

The reader should have a fairly complete record of the study in ref/4/.

The spatial chaotic situation is principally induced by the dislocations made of pairs of pentagon-heptagon as presented in figure 1.

Following Dreyfus and Guyon /3/ we use the Theoretical tools introduced in the melting of 2D solid/5/6/ for describing the spatial chaos:

-The number (fig 5) and type (fig 2) of defects versus ε .
-The analysis of the translational G(r) (fig 3) and orientational $G_6(r)$ (fig 4) correlation functions.

a) Near the threshold, the structure exhibits few defects made of dislocations and pairs of dislocations (fig 2a).G(r) decreases algebraically (fig 3a)(quasi long-range order) whereas $G_6(r)$ is constant (fig 4a) as in a 2D solid.In the Bénard-Marangoni instability case the orientational order is weaker(quasi long-ranged).

b) Above the threshold,a dramatic increase of the number of defects is observed (fig 5)(occurrence of grain boundaries);this corresponds to a loss of the translational order (fig 3b): G(r) changes from algebraic to exponential (fig 3b) and $G_6(r)$ which was constant becomes algebraic (fig 4b) as in an intermediate phase (oriented fluid),this also corresponds to the formation of arcs of diffraction in the reciprocal space (fig 1.2b)

c) At higher ε both translational and orientational order are short ranged (fig 3c and 4c) as in 2D liquid.

These results show a qualitative similarity to those observed by Mc Tague and all/6/ in the melting of 2D thermodynamic hard disk system with a r^{-6} repulsive interaction.

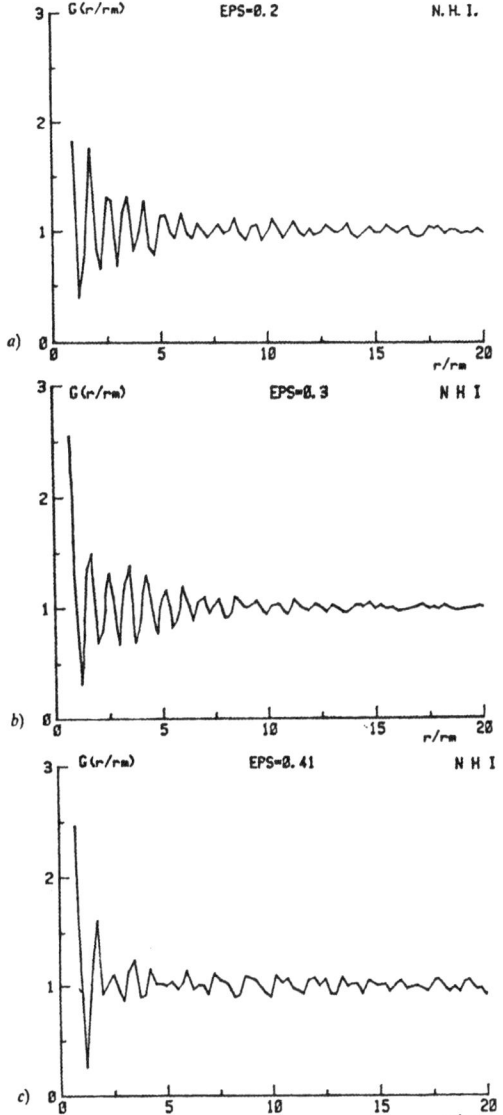

Figure 3
Translational correlation functions for the nematic hydrodynamic instability:
a) ε =0.20 ; b) ε =0.30 ; c) ε =0.41

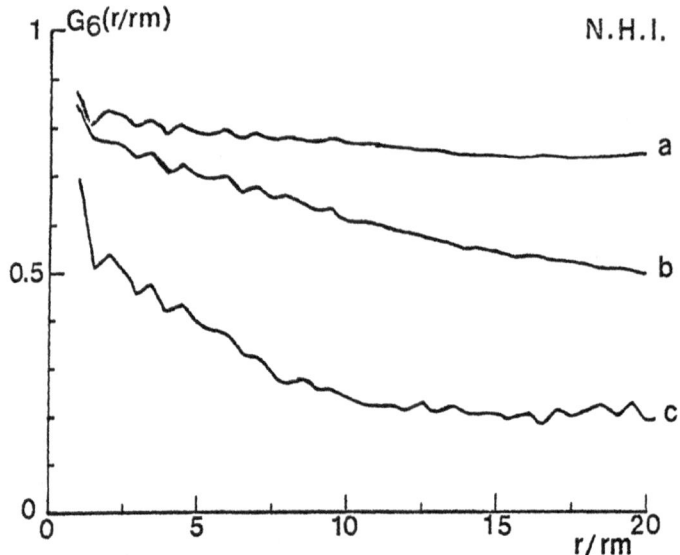

Figure 4: Orientational correlation functions for the nematic hydrodynamic instability. a) $\mathcal{E}$ =0.20 ; b) $\mathcal{E}$ =0.30 ; c) $\mathcal{E}$ =0.41

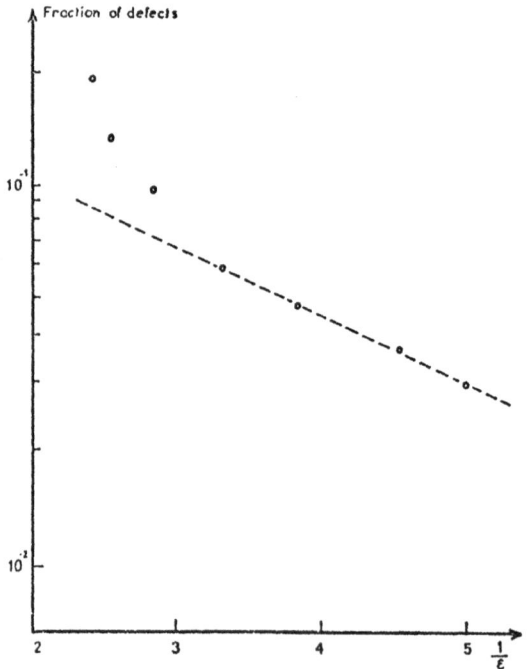

Figure 5 : Ln(fraction of defects) versus $\mathcal{E}^{-1}$ for the nematic hydrodynamic instability.

However our numerical calculations are restricted to fairly small lattices(all runs reported here are performed on a set of cells chosen in the center of the sample:2000 hexagons out of 40000 for the nematic hydrodynamic instability and 800 out of 3000 for the Bénard-Marangoni instability) and do not permit an evaluation of a larger range disorder. A direct evaluation of the structure factor of the 2D pattern can be obtained by a laser beam diffracting on the total area of the cell as in the nematic hydrodynamic instability case as presented in fig 1.

In conclusion this study gives a proper quantitative description of the 2D dissipative structures using the tools used in the theories of 2D melting.But we have at present no understanding of the mechanism that would play the role of the thermal energy in our systems.

<u>REFERENCES</u>

/1/ J.PANTALONI,R.BAILLEUX,J.SALAN,M.G.VELARDE
    J.Non Equilib.Thermodyn. Vol 4(1979) p201

/2/P.PIERANSKI,E.GUYON Phys.Rev.Lett. <u>39</u> 1281(1977)
    E.DUBOIS-VIOLETTE,F.ROTHEN;J. de Phys. <u>10</u> 1039 (1978)

/3/J.M.DREYFUS,E.GUYON;J.dePhys. <u>42</u> 283 (1981)
    see also E.GUAZZELLI,E.GUYON C.R. Heb.Sean.Acad.Sc. <u>292</u> II 141(81)

/4/R.OCCELLI,E.GUAZZELLI,J.PANTALONI;J.de Phys.Lett. <u>44</u> 1567(1983)

/5/J.M.KOSTERLITZ,D.J.THOULESS; J. Phys. C118(1973)
    D.R.NELSON,B.HALPERIN Phys.Rev. <u>1319</u> 2459(1979)

/6/J.P.Mc TAGUE,D.FRENKEL,M.P.ALLEN;Ordering in two dimensions SK.
SINHA Ed. North Holland(1980) p147.

SHEAR MODES IN LOW-PRANDTL THERMAL CONVECTION

J.M.Massaguer, I.Mercader

E.T.S.d'Enginyers de Camins, Canals i Ports, and
Facultat d' Imformatica
Universitat Politecnica de Barcelona
Barcelona, Spain

ABSTRACT

Non-linear, time-dependent numerical simulations of low Prandtl
thermal convection are presented. The model is based on a modal expansion and includes a vertical vorticity mode. Shear instability of the
primary flow -an hexagonal cell- is proved, and the bifurcation is examined. Two new families of solutions emerge from the bifurcation. One
of them, steady, is conjectured to be unstable; the other one, periodic,
is examined as a candidate to explain the high frequencies observed in
mercury near the transition to time-dependence.

1. INTRODUCTION

Bifurcation patterns for high and low Prandtl number convection are
essentially different. In the former, as the Rayleigh number R is increased the system evolves through a hierarchy of well organized structures
increasing in complexity. In the latter, a complicated time-dependence
is reached even at very small supercritical Rayleigh numbers: the system
jumps from a two-dimensional, steady regime to a three-dimensional,time
-dependent motion in a narrow, poorly defined strip of parameter space
(Krishnamurti 1973).The first attempt to explain the differences in behaviour is in Busse(1972, Clever and Busse 1974). They proved that,at
low Prandtl, non-zero vertical vorticity solutions bifurcate from two-
dimensional rolls. These solutions are oscillatory in character and the
instability sets in as a periodic bending of rolls. However, the agreement between theory and experiments is satisfactory only for Prandtl
numbers not much smaller than one, as in the case of the air, whose
Prandtl number is $\sigma = 0.71$. At lower Prandtl numbers the agreement is
poor. The frequencies predicted by the theory for the onset of oscillatory instability in mercury ($\sigma = 0.025$) are significantly different
from the experimental values measured by Krishnamurti.

A lot of work has been done recently to model low-Prandtl thermal
convection. An obvious issue was to explore two-dimensional models in

different geometries(Jones, Moore and Weiss 1976; Proctor 1977; Busse 1981; Clever and Busse 1981). Their results prove, consistently,that for intermediate values of the Rayleigh number, in the low Prandtl regime, the only two-dimensional stable solutions are the so-called inertial: at leading order the pressure gradient balances the advection of momentum. The stability of these inertial solutions against three-dimensional perturbations has been checked only in the axisymmetric case(Jones and Moore 1979) and they have been found unstable , stressing our feeling that all these flows represent unstable balances. Alternatively, several modes have been analyzed using Galerkin expansions, either in three dimensions(McLaughlin and Martin 1975) or using two-dimensional planforms(Toomre,Gough and Spiegel 1977,1982). In these techniques the selection of modes is crucial and the results in the quoted papers, even if consistent at high Prandtl numbers, at low Prandtl numbers show poor agreement with experiments as far as time-dependence is concerned(Krishnamurti 1973).

Difficulties in modelling low Prandtl number convection must be associated, in our opinion, to the enhanced contribution of the shear modes. It is well known that near the onset of convective instability the velocity field is, roughly speaking, independent of the Prandtl number. Thus, low and high Prandtl number mean, respectively, high and low Reynolds regime, with shear instabilities playing a fundamental role in the former but not in the latter. From current knowledge in shear flows(Orzag and Kells 1980; and references therein) we can infer that the most dangerous instabilities might set in as three-dimensional finite amplitude instabilities. However, no such instabilities are present in the previously quoted papers.

Our purpose is to address that issue. The model we present for low Prandtl convection is the most simple we could build displaying finite amplitude instability against three-dimensional perturbations. We have assumed a planform expansion with only one mode, in the same way Toomre et al..(1977) did, but including as a shear-mode a non-zero vertical vorticity component. The structure of the model equation is discussed in part II, the most relevant properties of the steady solutions are presented in part III, and some work in progress on the time dependent problem is presented in part IV. Finally, we make some conjectures about the physical relevance of the model.

2. ESTRUCTURE OF THE MODEL EQUATION

The main purpose of this work is to analyze the role played by the vertical component of the vorticity in low Prandtl convection. We can write the third component of the vorticity equation as

$$\partial_t \omega_z + \underline{v} \cdot \nabla \omega_z = \underline{\omega} \cdot \nabla v_z + \sigma \nabla^2 \omega_z \qquad (1)$$

where v_z and ω_z are, respectively, the vertical components of the velocity $\underline{v}$ and vorticity $\underline{\omega}$ fields; σ is the Prandtl number and other terms retain their usual meanings

It is important to realize that if $\underline{\omega} \cdot \nabla v_z = 0$, then, the flow tends to $\omega_z = 0$ for a large set of boundary conditions, meaning that some geometrical configurations cannot generate vertical vorticity. The converse is even more important for our purpose: if the vertical vorticity is forced to be zero, the geometry of the flow must be taken, consistently, to satisfy the condition $\underline{\omega} \cdot \nabla \, v_z = 0$. These flows must be considered as two-dimensional even if their geometries look complicated, as in Toomre et al.(1982).

The structure of equation (1) can be made apparent with the following decomposition for the velocity field

$$\underline{v} = \nabla \times \nabla \times (\phi \underline{k}) + \nabla \times (\psi \underline{k})$$

where ϕ and ψ are two scalar fields related to the vertical components of the velocity and vorticity fields by the expression

$$v_z = -\nabla_1^2 \phi \qquad ; \qquad \omega_z = -\nabla_1^2 \psi$$

$\underline{k}$ is the unit vertical vector and $\nabla_1 = \partial_{xx}^2 + \partial_{yy}^2$. The equation (1) can be written now as

$$(\partial_t - \sigma \nabla^2) \nabla_1^2 \psi + \Lambda(\phi,\psi) + \frac{\partial(\nabla^2 \psi, \psi)}{\partial(x,y)} = \frac{\partial(\nabla^2 \phi, \nabla_1^2 \phi)}{\partial(x,y)} \qquad (2)$$

where $\Lambda(\phi, \psi)$ is an operator bilinear in ϕ and ψ.

In the small amplitude limit the equation (2) becomes a linear diffusion equation and the solution asymptotes to $\psi = 0$. If a $\psi \neq 0$ solution is stable, the flow is then constrained by the geometrical condition

$$\frac{\partial(\nabla^2 \phi, \nabla_1^2 \phi)}{\partial(x,y)} = 0$$

and the velocity field is forced to be two-dimensional. Any instability associated to the shape of the flow requires this jacobian to be non-zero. As well known examples of shape instability we can mention cross-roll, zig-zag and oscillatory instabilities. However, even if the shape of the primary flow is stable, the flow can be unstable . In the limit of small ψ-amplitude but finite ϕ,equation (2) can be written for a two-dimensional flow ϕ, no matter whether it is plane or not, as

$$(\partial_t - \sigma\nabla^2)\nabla_1^2 \psi + \Lambda(\phi,\psi) = 0 \qquad\qquad (3)$$

This equation might show a bifurcation for a finite amplitude ϵ of ϕ.If we define the Reynolds number Re = ϵ/σ, we can expect the flow to be unstable for a Reynolds number of order one. Since the only driving is provided by the velocity gradients this instability is of shear type. The existence of this instability is tightly linked to the shape of the primary flow. It has been proved by Busse(1972) that in a two-di-mensional plane flow ϕ(i.e. ϕ is a roll) the volume average $<\psi\Lambda(\phi,\psi)>$ is zero and no source of instability can be expected from equation (3) (i.e. : $\omega_z = -\nabla_1^2\psi =0$). On the opposite, for curved flows this average need not to be zero: only curved flows can show this shear instability. This might explain why Jones and Moore(1979) found direct rather than oscillatory instabilities for low Prandtl axisymmetric convection.

The model we explore is a curved flow. It is the one-mode system de-veloped by Toomre et al. (1977) with the addition of a vertical vortici-ty component, expanded in such a way that we complement their system with the modal expansion of equation (3) -see their paper for details-. We expand the vertical velocity v_z, the vertical vorticity ω_z and the temperature field T in the following way

$$v_z = f(x,y)\ W(z,t)$$
$$\omega_z = f(x,y)\ \xi(z,t)$$
$$T = \overline{T}(z,t) + f(x,y)\ \theta(z,t)$$

where the functions $f(x,y)$ are planforms defined by the equation $\nabla_1^2 f = -a^2 f$ and the additional conditions $\overline{f}=0$, $\overline{f^2} =1$, $\overline{f^3} =1/2\ C$ with an overbar meaning horizontal average. The wavenumber a and the coupling constant C especify the model. In the results shown later we have taken the critical value a = 3.117 and C = 0.408, corresponding to hexagonal planforms.

The system to be solved can be written as

$$(\sigma^{-1}\partial_t - \nabla^2)\nabla^2 W = -Ra^2\theta - C\sigma^{-1}(W\nabla^2\partial_z W + 2\partial_z W\nabla^2 W + 3\xi\partial_z\xi) \qquad (4.a)$$
$$(\partial_t - \nabla^2)\theta = -W\partial_z\overline{T} - C(2W\partial_z\theta + \theta\partial_z W) \qquad (4.b)$$
$$(\partial_t - \partial_{zz}^2)\overline{T} = -\partial_z(W\theta) \qquad (4.c)$$
$$(\sigma^{-1}\partial_t - \nabla^2)\xi = C\sigma^{-1}(\xi\partial_z W - W\partial_z\xi) \qquad (4.d)$$

with $\nabla^2 = \partial_{zz} - a^2$. We assume rigid boundary condition on top and bottom for the velocity field

$$W = \partial_z W = \xi = 0 \qquad (z=0,1)$$

and perfectly conducting boundaries

$$\theta = 0 \qquad\qquad (z=0,1)$$
$$\overline{T} = 0 \qquad\qquad (z=1)$$
$$\overline{T} = 1 \qquad\qquad (z=0)$$

The heat flux is measured by the Nusselt number N, defined in the steady regime as

$$N = -\partial_z \overline{T} + W\theta ,$$

for time-dependent flow we define two Nusselt numbers N1, N2 as

$$N1 = -(\partial_z \overline{T})_{z=1} \qquad\qquad N2 = -(\partial_z \overline{T})_{z=0}$$

obviously, in a steady regime N1 = N2 = N.

3. STEADY SOLUTIONS

The system (4) admits two families of steady solutions. One of them defined by the absence of vertical vorticity ($\xi=0$), has been explored by Toomre et al.(1977). We have done a stability analysis of these solutions and found them to be unstable to a direct mode. The instability can be defined by the Reynolds number Re = CW/σ evaluated at the maximum value of W. For Prandtls smaller than 0.1, the Reynolds number at the transition point asymptotes to a critical value of Re = 20.84. A second family of solutions defined by the condition $\xi \neq 0$ bifurcates from the first one. The bifurcation is inverse, as can be seen from figure 1 where the Nusselt number has been plotted against Rayleigh number for mercury (σ = 0.025), and takes place at R = 3750. The nose of the bifurcated branch is located at R = 2970 and, in terms of fluxes, the solutions overlap at R = 4104. As can be seen from the figure, the bifurcated solutions are very inefficient in transporting heat: the Nusselt number at the intersection of both branches is N = 1.038 and increases very slowly if compared with the upper branch. This behaviour can be explained as a result of the flow being largely dominated by the horizontal motion (ξ larger than W). The vertical velocity for these flows is, roughly speaking, independent of the Rayleigh number and so do the heat flux and the work of buoyancy.

In figure 2 we have plotted the Rayleigh values at the end of the noses as a function of the Prandtl number. We have included Krishnamurti's experimental results for transitions in air (σ = 0.71) and mercury. At first sight, the agreement between her's and our results seems poor. We have remarked previously, however, that our results asymptote for σ< 0.1 and we are not expecting agreement for larger values. Thus, if the finite amplitude instability is physical, we expect from figure 2 this instability to be associated to the second transition to time-dependence (transition to high frequencies).

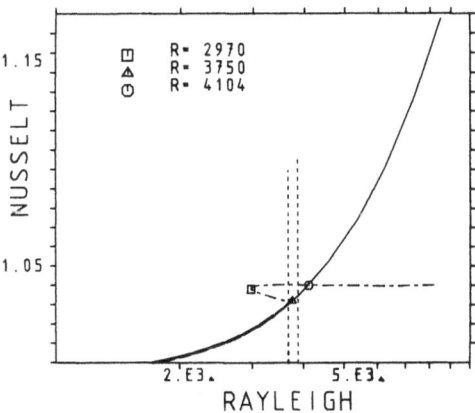

Figure 1.-. Nusselt vs. Rayleigh for mercury(σ = 0.025). Thick solid line: steady ξ= 0 solutions. (Stable). Thin solid line: steady ξ = 0 solutions.(Unstable). Dot-dash line: steady ξ ≠ 0 solutions.(Unstable). Dotted lines: amplitudes of periodic ξ ≠ 0 solutions.

4. TIME-DEPENDENT SOLUTIONS

From the analysis of steady solutions we have realized that the upper limit for small Prandtl number must be located around σ= 0.1. The only set of results available in this range are measurements for convection in mercury and we will concentrate on them. Although no stability analysis of the "ξ = 0 " solutions has yet been done, from our time-dependent results we are confident about all these solutions being unstable: we have time-marched the system (4) from a variety of initial conditions and Rayleigh numbers, and the solutions entered a periodic orbit each time, unless the Rayleigh number was small enough and the flow could be attracted towards a steady, "ξ = 0 " solution. Hence, a finite amplitude instability exists in system (4), with the flow jumping from a steady, two-di-

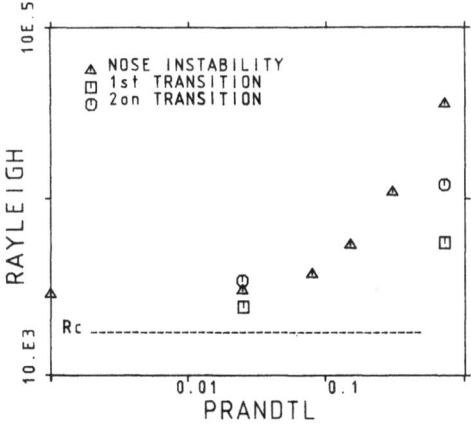

Figure 2.-. Rayleigh values for the noses of the finite amplitude instabilities against Prandtl. Experimental values for transitions are included.

mensional, $\xi = 0$ solution to a periodic, three-dimensional, $\xi \neq 0$ regime.

As a first test for our periodic solutions we have run the case $R= 10^4$, $\sigma = 0.025$ corresponding to mercury. The period found for the oscilations is $\tau= 1.8$ and the average Nusselt number is $N = 1.42$, to be compared with the experimental result $N = 1.57$ -extrapolated by Toomre et al. from Rossby (1969) measurements- and the steady values of $N=1.17$ and $N=1.04$, corresponding, respectively, to the zero and non-zero vertical vorticity solutions. Work is now in progress to explore these periodic solutions but, as a reference, we have plotted in figure 3 the

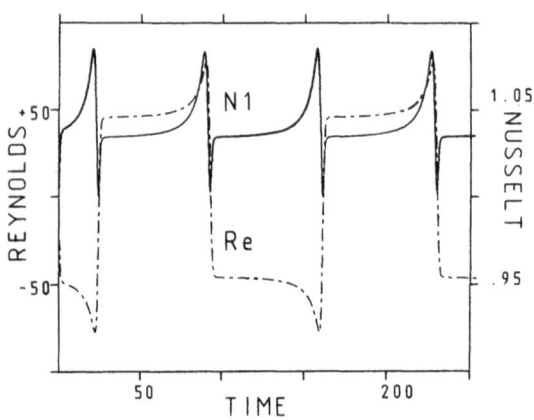

time evolution of Reynolds Re and Nusselt N1 for the case $R=3550$, $\sigma =0.025$ located in the range of high frequencies explored by Krishnamurti. The period found is $\tau=137$. Although this value is larger than the upper limit she reported ($\tau \simeq 100$ in thermal time scales), figure 3 suggests that measurements in fluxes could halve the fundamental periods. If it was so, our periods would be in the experimental range.

Figure 3.-. Time sequence for R=3550, $\sigma = 0.025$. Solid line: Nusselt on top. Dot-dash line: Reynolds at mid layer. Time is measured in thermal difusion units.

A feeling about the amplitude of these periodic motions can be obtained from figure 1. We have plotted there, as a dotted line, the excursion of the Nusselt number N1 for two different Rayleigh numbers (R=3850 and 3650). Their periods are, respectively, $\tau = 20$ and 40.

CONCLUSIONS

From the results previously quoted we can infer that shear type modes play a fundamental role in low Prandtl number convection even at very small Rayleigh numbers -R $\simeq$ 1.7 R_c for mercury in our model- though larger than the critical value for the onset of oscillatory instabilities -R $\simeq$ 1.2 R_c according to Clever and Busse(1974)- ; The absence of oscillatory instability in our model being a consequence, probably, of the curvature of the flow, as suggested by Jones and Moore (1979). The agreement between the periods computed for convection in mercury and Krishnamurti's experimental results, together with the coincidence between the onset of finite amplitude instability and the measured tran-

sition to high frequencies for mercury suggest the following scenario;
The instabilities we have called of shear type cannot appear unless the
primary flow shows some curvature; if the primary flow is plane (rolls,
for instance) an oscillatory instability is needed for bending the primary flow but, a finite amplitude instability of shear type sets in, with
the flow jumping to the metastable branch. In a more realistic model
both, oscillatory and finite amplitude instabilities can be present.
Then, the detailed structure of the patterns in the range of Rayleigh
numbers between both instabilities can be very complicated but, in the
region of shear instabilities, we do not expect the flow to be strongly
dependent on the details of the oscillatory instability, explaining the
agreement between numerical and experimental results.

BIBILOGRAPHY

Busse,F.H. (1972) J.Fluid Mech.52,97.
Busse,F.H. and Clever,R.M. (1981) J.Fluid Mech. 102,75.
Clever,R.M. and Busse,F.H. (1974) J.Fluid Mech. 65,625.
Clever,R.M. and Busse,F.H. (1981) J.Fluid Mech. 102,61.
Jones,C.A.;Moore,D.R. and Weiss,N.O. (1976) J.Fluid Mech. 73,153.
Jones,C.A. and Moore,D.R. (1979) Geophys, and Astrophys. Fluid Dyn. 11,
245.
Krishnamurti,R. (1973) J.Fluid Mech. 60,285.
McLaughlin,J.B. and Martin,P.C. (1975) Phys.Rev. 12,186.
Orzag,S.A. and Kells,L.C. (1980) J.Fluid Mech. 96,159.
Proctor,M.R.E. (1977) J.Fluid Mech. 82,97.
Rossby,H.T. (1969) J.Fluid Mech. 36,309.
Toomre,J.;Gough,D.O. and Spiegel,E.A. (1977) J. Fluid Mech. 79,1
Toomre,J.;Gough,D.O. and Spiegel,E.A. (1982) J.Fluid Mech. 125,99

Note added in proof: As can be realized from the sharpness of
the jumps in Figure 3, the frequency spectrum for the oscilla
tions is very broad,as in Krishnamurti's work.Roughly speaking,
there are significant amplitudes in the range of periods between
150 and 2.

SPATIAL INSTABILITIES AND TEMPORAL CHAOS

S. Fauve, C. Laroche, A. Libchaber, B. Perrin

Ecole Normale Supérieure, Groupe de Physique des Solides

24 rue Lhomond, 75231 Paris Cedex 05

Dynamical system theory has provided a successful point of view about the study of the temporal structure of turbulent flows [1]. However, in meteorology or astrophysics, temporal impredictibility gene- rally occurs concomittantly with a complex spatial structure, and even in controled laboratory experiments the sequence of events leading to chaotic time dependence is often preceeded by the development of sta- tionary spatial patterns. This occurs in thermal convection in a hori- zontal fluid layer heated from below, and one needs to quench a simple convective pattern in order to observe a transition scenario to chaos following the predictions of dynamical system theory [2]. This is usual- ly achieved by the lateral boundaries of the fluid container. However, when the horizontal extension of the layer is large enough compared to its height, a turbulent regime is observed just above the convection onset [3], and may require a stochastic description [4]. Nevertheless the problem of this turbulent state origin has also received determi- nistic explanations, resulting from the analogies between the transi- tion to turbulence through supercritical bifurcations and second order phase transitions, and using the amplitude equation formalism (see the introductory paper to this book). However there is not a clear cut answer, and several explanations have been proposed in this context :
- Non potential effects in the amplitude equations due to coupling with large scale flows [5,6].
- Competition between bulk and boundary effects (unstable selected wavenumber [7] or frustration [8]).
- Motions of the convective pattern defects [9].

We have studied how the turbulent state above the convection onset depends on the relevant experimental parameters (Prandtl number, boun- dary conditions). Our results are reported in section 1 and are connec- ted with the marginal modes at onset in section 2. We show in section 3 how these modes can be controlled experimentally with a horizontal magnetic field, and how the routes to turbulence are affected.

1. TURBULENCE AT CONVECTION ONSET

Qualitative differences in the sequence of events leading to turbulence have been observed for the first time in convection experiments in cylindrical containers of liquid helium by changing the aspect ratio [3,10] . It was shown that an abrupt transition to chaos, involving a low frequency noise occurs at convection onset for large enough aspect ratio (Γ = 57). Visualisation experiments in large aspect ratio containers have been performed in order to make a comparison with helium experiments and to check the different theoretical models (see the papers of V. Croquette and J. Gollub in this book). Since the working fluids, silicon oil and water have larger Prandtl numbers than helium, no turbulent state at the convection onset was reported but two remarkable experimental results were pointed out :
- The existence of multiple stable patterns at convection onset, some of them with defects [11,12].
- The existence of non potential effects, eventually connected with large scale flows [13].
We have studied the time dependence above the convection onset in mercury (a small Prandtl number fluid) with containers of different shapes and aspect ratios. Our results and previous results with helium experiments are reported in table 1.

Shape Γ	Fluid P	Mercury 0.025	Helium $\sim$ 0.6 Libchaber-Maurer [10]	Helium $\sim$ 3 Ahlers-Behringer [11]
C 2		periodic	periodic	—
C 3		chaotic for $R \gtrsim R_c$	periodic	—
C 6		—	chaotic for $R \sim 2R_c$	—
C 57		—	—	chaotic for $R \gtrsim R_c$
R 6×3.3 6×6 10×6		periodic	—	—

Table 1 : Nature of the first time dependent regime for convection experiments with different Prandtl number fluids (P), different container shapes (cylindrical C or rectangular R), and different aspect ratios (Γ = radius/height or lateral lengths/height).

The effect of the Prandtl number on turbulence at the convection onset clearly appears from these results. This is corroborated by the fact that no time dependent regime at onset has been reported for higher

Prandtl number fluids (water, silicon oil, ...). Moreover, besides the
well known aspect ratio effect, the container shape plays an essential
role. In parallelepipedic containers a periodic regime follows a stationary one as the Rayleigh number is increased, whereas for a smaller
cylindrical container a turbulent state is observed at convection onset.
Flow visualisation shows that the flow consists in hot ascending and
cold descending regions of fluid, randomly distributed in space, and
moving on a slow time scale compared to the vertical heat diffusion
characteristic time [14].

2. MARGINAL MODES AT CONVECTION ONSET

In a laterally infinite fluid layer, linear theory predicts the
existence of an infinite number of possible convective modes with an
horizontal wavenumber q_c at convection onset. As noticed by Busse [15]
this degeneracy is twofold. First there is a pattern degeneracy that
corresponds to different planforms (rolls, squares, hexagons, ...), and
that is removed by a non linear stability analysis [16]. The second
degeneracy is connected with the translational and orientational invariance of the problem. The orientational degeneracy is only partly removed when lateral boundaries are taken into account; this depends on
the shape of the container. The equivariance of the convection equations
with respect to a given transformation implies the existence of marginal modes. For instance, it results from the translational invariance
along the x axis, that if $(\vec{v}(x,z), \theta(x,z))$ is a solution of the convection equations, $(\partial_x \vec{v}, \partial_x \theta)$ is a solution of the linearized equations
around $(\vec{v}, \theta)$ with a zero growth rate. These marginal modes are the
first that one expects to become unstable through non linear coupling
with the primary flow. This is the case of Eckhaus, zig-zag and cross-
roll instabilities of convection rolls. However, as long as vertical
vorticity is not taken into account, these modes can only describe a
relaxational behavior and not a chaotic regime, at leading order in the
amplitude equation formalism.

In linear theory, vertical vorticity is not coupled with the vertical velocity and the temperature, and obeys a diffusion equation

$$(\partial_t - P \nabla^2) \omega_z = \text{non linear terms}$$

With stress-free boundary conditions, horizontal velocity modes of
wavenumber k with no z-dependence are possible, and their linear growth
rate is

$$\eta = - P k^2$$

For a finite Prandtl number these modes can be marginally stable ($\eta \to 0$)

if k → 0 (container of infinite horizontal extent). Non linear coupling
with convective modes leads to skewed-varicose and oscillatory instabilities [5,17]. For rigid boundary conditions, horizontal velocity modes
independent of z are not allowed, but the damping of vertical vorticity
modes is small for a small enough Prandtl number. Vertical vorticity
may thus explain the chaotic regime just above the convection onset at
small Prandtl number. Moreover the non linear terms in the vertical
vorticity equation depend on the convective pattern, and therefore on
the shape of the container. For a three-dimensional pattern (it is the
case in a cylindrical container), a subcritical shear instability generates a low frequency time-dependent flow (see the paper of
J.M. Massaguer and I. Mercader in this book).

With small aspect ratio containers, or large Prandtl number fluids,
the linear damping of vertical vorticity increases, and chaos at convection onset is not observed. Vertical vorticity effect is therefore a
possible answer to the problem of turbulence at onset.

3. EFFECT OF A HORIZONTAL MAGNETIC FIELD

We report in this section the study of the different type transitions to turbulence in convection in a cylindrical container of mercury
of aspect ratio $\Gamma = 3$ subjected to a horizontal magnetic field, that
suppresses the rotational invariance. We show that this has the same
qualitative effects as varying the geometry of the fluid container
(shape, aspect ratio). As mentioned in table 1 a turbulent flow is observed just above the convection onset in the absence of magnetic field.
When the field is applied, the convective pattern takes the form of
rolls parallel to the field axis in the bulk of the container, and the
flow becomes stationary for a large enough field amplitude. The main
effect of the field is to inhibit velocity variation along its axis,
but it does not affect the motion of rolls parallel to its axis.
Consequently the critical Rayleigh number for the onset of convection
is unchanged, but the number of marginal modes at onset decreases. This
structural change is manifested in Nusselt number measurements. As
shown in figure 1 the convective heat flux initial slope increases with
the magnetic field amplitude, in agreement with the fact that parallel
rolls are associated with the highest convective heat transport, when
the boundaries have a much higher heat conductivity than the fluid [16].

The different flow regimes are plotted in figure 2, in a two parameter space R, Q; R is the Rayleigh number; the Chandrasekhar number,
Q, proportional to the square of the magnetic field amplitude, measures
the ratio between the anisotropic viscosity due to the field and the

kinematic viscosity. In the diagram of figure 2 each curve represents
a transition where spatial or temporal order is lost. The curve $R_1(Q)$
corresponds to the onset of time dependence, and always involves three-
dimensional patterns in the observed range $0 < Q < 5000$. This is known
for higher Prandtl number fluids and is associated with the skewed-
varicose instability [11], but somewhat surprising when a large magnetic field two-dimensionalize the flow. The time dependent regime occurs
after a Hopf bifurcation for $Q \gtrsim 300$ and after a saddle node bifurcation for $Q \gtrsim 2000$. Two kinds of periodic state result, that are rather
similar to those recently reported in experiments with liquid helium
[18,19]. In the range $300 \lesssim Q \lesssim 2000$, damped transient oscillations and
chaotic time dependence are observed at bifurcation from stationary to
time-dependent regime.

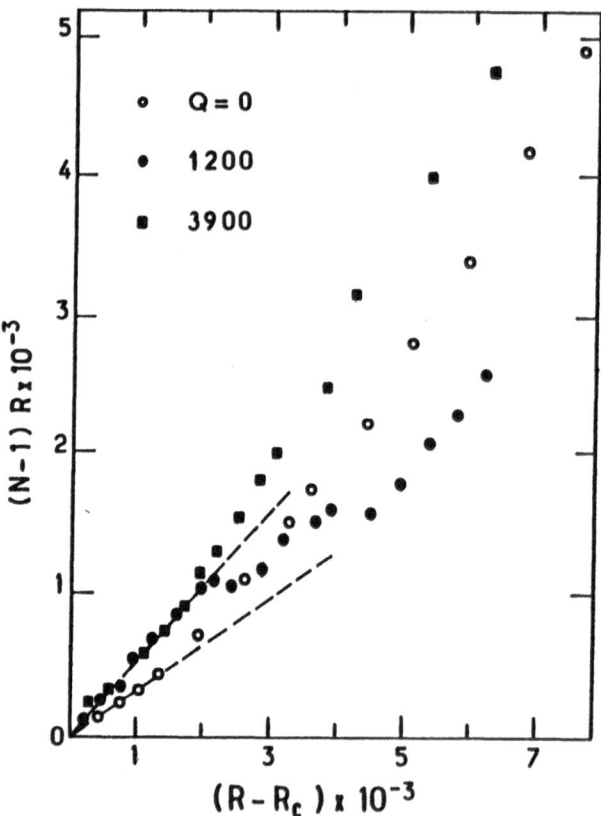

Figure 1 : Convective heat flux characterized by the Nusselt number N,
versus the Rayleigh number for different Chandrasekhar numbers Q. Note
the existence of negative slopes for Q = 1200.

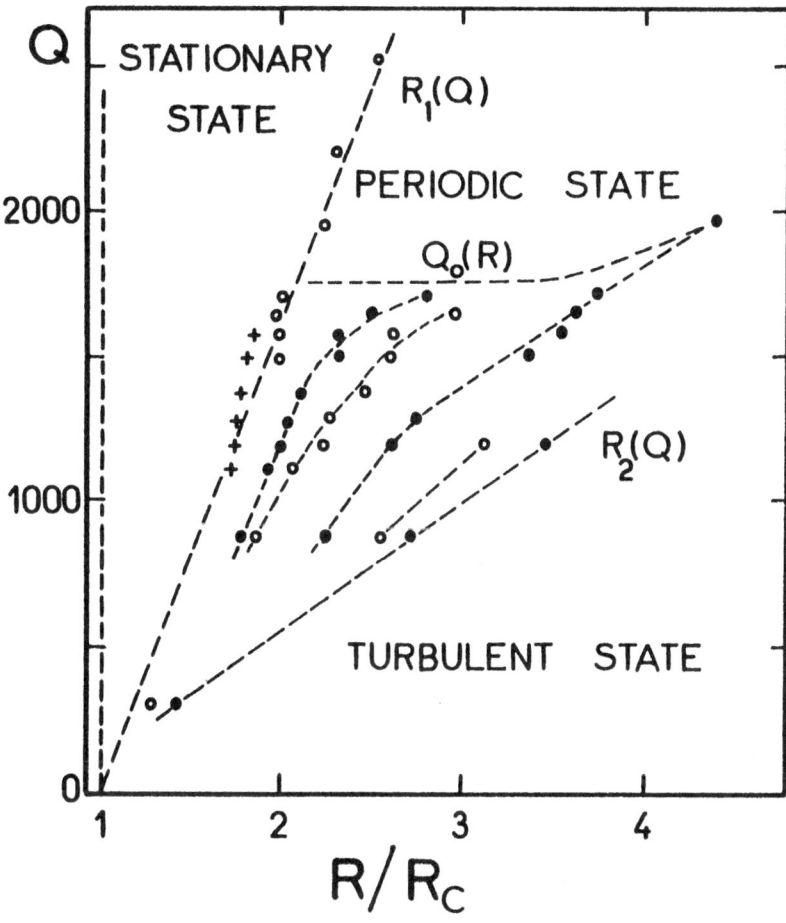

Figure 2 : Nature of the time dependent states as a function of the Rayleigh and Chandrasekhar numbers. The dotted curves indicate a transition to a different regime : damped transient oscillation (+), sustained oscillation (o), chaotic time dependence (•).

As the Rayleigh number is increased, several bands of periodic states are found within the chaotic region. These complex transitions are connected with the existence of negative slopes in the convective heat flux versus Rayleigh number curve (see figure 1, $Q = 1200$), that indicates structural changes of the flow. This behavior is not observed for $Q > Q_0(R)$ in the diagramm of figure 2, where the time dependence involves the interaction of only a few spatial modes. In this parameter range the transition to chaos has the same characteristics as in low dimensional dynamical systems, now well understood in the framework of dynamical system theory. The next step is to consider the more realistic case, where a lot of spatial modes may be excited. The experiment presented here is a first attempt in this way.

REFERENCES

1. J.P. Eckmann, Rev. Mod. Phys. 53, 643 (1981).
2. A. Libchaber, S. Fauve, C. Laroche, Physica 7D, 73 (1983).
3. G. Ahlers and R.P. Behringer, Phys. Rev. Letters 40, 712 (1978).
4. H.S. Greenside, G. Ahlers, P.C. Hohenberg, R.W. Walden, Physica 5D, 322 (1982).
5. E.D. Siggia, A. Zippelius, Phys. Rev. Letters 47, 835 (1981).
6. M.C. Cross, Phys. Rev. A27, 490 (1983).
7. Y. Pomeau, P. Manneville, Phys. Letters 75A, 296 (1980).
8. Y. Pomeau, P. Manneville, J. de Physique 42, 1067 (1981).
9. E.D. Siggia, A. Zippelius, Phys. Rev. A24, 1036 (1981).
10. A. Libchaber, J. Maurer, J. de Physique Lettres 39, 369 (1978).
11. J.P. Gollub, J.F. Steinman, Phys. Rev. Letters 47, 505 (1981).
12. V. Croquette, M. Mory, F. Schosseler, J. de Physique 44, 293 (1983).
13. A. Pocheau, V. Croquette, J. de Physique (to appear Jan. 1984).
14. S. Fauve, C. Laroche, A. Libchaber, B. Perrin, to be published.
15. F.H. Busse, Hydrodynamic Instability and the Transition to Turbulence, Editors : H.L. Swinney and J.P. Gollub, Springer Verlag (1981).
16. A. Schlüter, D. Lortz, F. Busse, J. Fluid Mech. 23, 129 (1965).
17. F.H. Busse, J. Fluid Mech. 52, 97 (1972).
18. R.P. Behringer, H. Gao, J.N. Shaumeyer, Phys. Rev. Letters 50, 1199 (1983).
19. R.W. Walden, Phys. Rev. A27, 1255 (1983).

TEMPORAL AND SPATIAL ASPECTS OF THE ONSET OF CHAOS IN A TAYLOR INSTABILITY

SUBJECTED TO A MAGNETIC FIELD

par P. TABELING, C. TRAKAS

Laboratoire de Génie Electrique de Paris
Plateau du Moulon
91190 GIF-SUR-YVETTE - FRANCE -
Laboratoire associé au C.N.R.S. aux Universités
PARIS VI et PARIS XI et à l'E.S.E.

1. INTRODUCTION

   We present herein some results obtained on a Taylor experiment in the presence
of an external magnetic field. We shall describe three distinct transitions to chaos,
obtained for different values of the magnetic field. The spatial structures of the
instability modes which we have observed crucially depend on the external magnetic
field. Our purpose is to show some typical sequences of events leading to turbulence,
and to exhibit their relation with the flow structures.

2. DESCRIPTION OF THE EXPERIMENTAL ARRANGEMENT

   The experimental arrangement is shown in Figure 1. The mercury is confined between two concentric copper cylinders. The dimensions of the mercury cell are : 40 mm
for inner radius R_1, 41.17 mm for outer radius R_2, and 26.7 mm for height L. The
fluid is driven by the interaction of an electric current I directed radially and
an external magnetic field B_o parallel to the common axis of the cylinders. The temperature of the flow is maintained to constant value $20 \pm 0.01°C$.

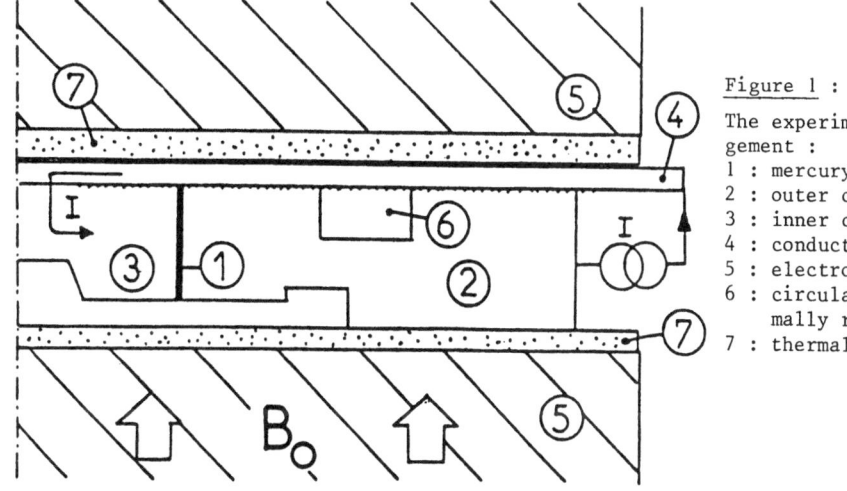

Figure 1 :
The experimental arran-
gement :
1 : mercury ;
2 : outer cylinder ;
3 : inner cylinder ;
4 : conducting plate ;
5 : electromagnet pole ;
6 : circulation of ther-
mally regulated water
7 : thermal insulators.

The detection system is composed of 80 electrodes located along a generant of the outer cylinder, and in electrical contact with the fluid. This system allows for characterizing spatially the observed modes of instability. Moreover, it is possible to measure the mean rotation of the liquid metal by measuring the voltage drop between the two copper cylindrical pieces (1).

3. THE OBSERVED REGIMES

The experiments are performed at fixed values of the external magnetic field ; the control parameter is electrical current I. At low values of current I, the flow is laminar ; no voltage fluctuation (with an amplitude larger than the instrumental noise) is visible at the output of the detectors. Above a critical value I_o (which depends on the external magnetic field), voltage fluctuations, with significant amplitudes, appear. Depending on the value of the external magnetic field, we have observed either periodic or chaotic signals. Figure 2 represents a synthesis of the various regimes which have been obtained experimentally (2) : the ordinate is the critical value of the Taylor number at the instability point ; the abscissa is Chandrasekhar number Q which characterizes the intensity of the external magnetic field ; this number is defined by $Q = \sigma B_o^2 d^2/\eta$, where σ, η are respectively the conductivity and the viscosity of the fluid and $d = R_2 - R_1$ is the gap width. The curve drawn in continuous line on Figure 2 represents the critical points (Q, T_o) obtained experimentally, while that drawn in dashed line represents the critical points, obtained numerically, and corresponding to the onset of stationary modes (Taylor cells) (3).

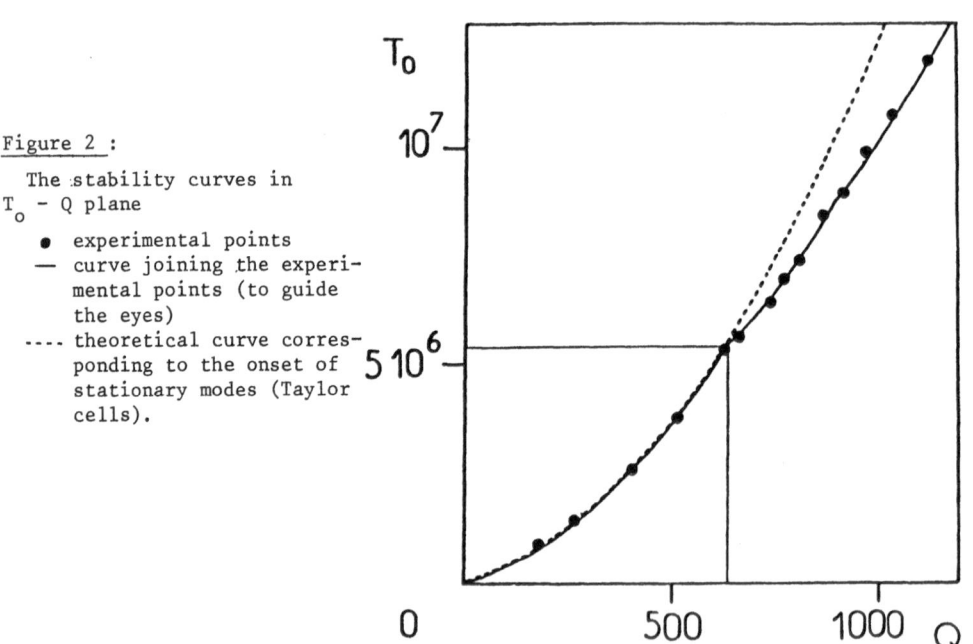

Figure 2 :

The stability curves in T_o - Q plane

- • experimental points
- — curve joining the experimental points (to guide the eyes)
- ···· theoretical curve corresponding to the onset of stationary modes (Taylor cells).

Figure 2 displays two ranges : the range of "large" values of the magnetic field
(Q > 620) and that of "low" values of the magnetic field (Q < 620). In the first
range, the experimental curve clearly lies below that corresponding to the onset
of stationary modes. We shall show that in such a range, the laminar flow is unstable
against oscillatory disturbances in form of spiraling vortices. In this case, the
transition to turbulence corresponds to a temporal chaos.

In the second range (Q < 620), the experimental curve is very close to that corresponding to the onset of stationary modes. We shall show that in this case, turbulence
corresponds a spatio-temporal chaos.

The point located at the frontier between the two ranges defines a kind of critical
value for the magnetic field ($B_o \simeq 0.82T$). At this point, the codimension of the system is equal to 2. However, the dynamics of the system in the vicinity of this point
has not been studied because of the absence of spatial order for $B_o < 0.82T$ (see §6).

4. TEMPORAL CHAOS AFTER THREE SUCCESSIVE BIFURCATIONS

The events which we describe herein have been observed at large values of the magnetic field ($B_o > 1.07T$, which corresponds to Q > 1000). Figure 3 represents a
Fourier analysis of the fluctuating tension measured at the output of a detector, for
$B_o = 1.15T$, and for a small value of discrepancy to threshold $\varepsilon = I/I_o -1$ (here, I_o is
equal to 24.586A). We obtain a monoperiodic regime, which can be characterized by its
amplitude A_1 and its phase ϕ_1. These two quantities are a priori functions of space
variable Z (measured along a generant) and angular coordinate θ. Figure 4 shows the
variations of amplitude A_1 and phase ϕ_1 with coordinate Z. A_1 is a slowly varying
function of Z ; it does not possess mirror symetry $Z \rightarrow L - Z$; ϕ_1 can be written in
the form :

$$\phi_1 = 2 \pi Z/\lambda_1 + m \theta + \phi_{10}$$

where λ_1 is a wavelength and m is an integer (we have measured m = 4). For $B_o = 1.15T$,
λ_1 is equal to 3.3 mm. Wavelength λ_1 is an increasing function of the magnetic field,
as shown in Figure 5.

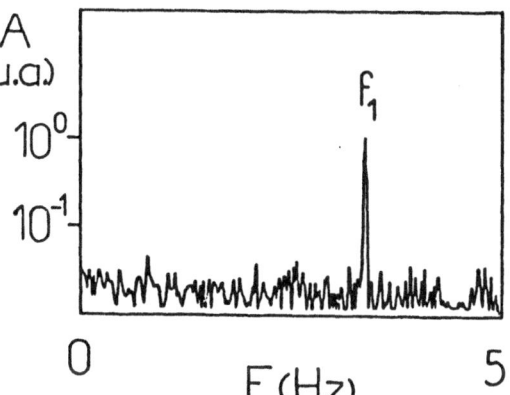

Figure 3 :

At large values of the magnetic field,
the first event observed at the output
of the detectors is a monoperiodic oscillation of the tension ; the figure
represents the Fourier analysis of such
a signal for $\varepsilon = 3.10^{-3}$ and $B_o = 1.15T$

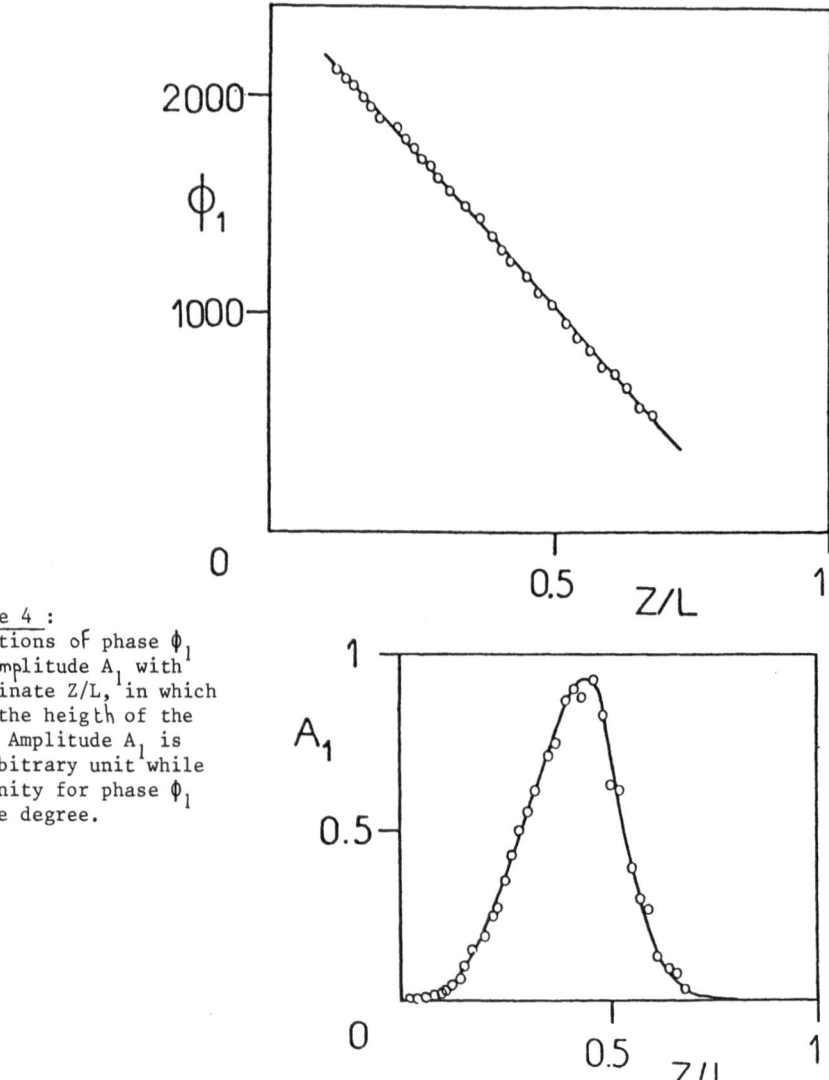

Figure 4 :
Variations of phase ϕ_1
and amplitude A_1 with
coordinate Z/L, in which
L is the heigth of the
duct. Amplitude A_1 is
in arbitrary unit while
the unity for phase ϕ_1
is the degree.

The spatial structure of such modes of instability is in form of spiraling vortices. As the critical discrepancy is increased, a new oscillator f_1' appears (see
Figure 6). By similar arguments as those used above, it is possible to show that the
spatial structure of f_1' is in form of spiraling vortices similar with the preceding
ones, but with a helicity of opposite sign. The nodal lines, obtained numerically by
using our measurements, are represented in Figure 7. We obtain a two-dimensionnal
structure, which is composed of two periodic structures with distinct orientations.

At still larger values of the critical discrepancy, a third frequency f_2 appears
(see Figure 8). It consists of an amplitude modulation of the previous two-frequencies
regime, as shown in Figure 8. The spatial structure of this new regime is very similar
to that of the two-frequencies regime.

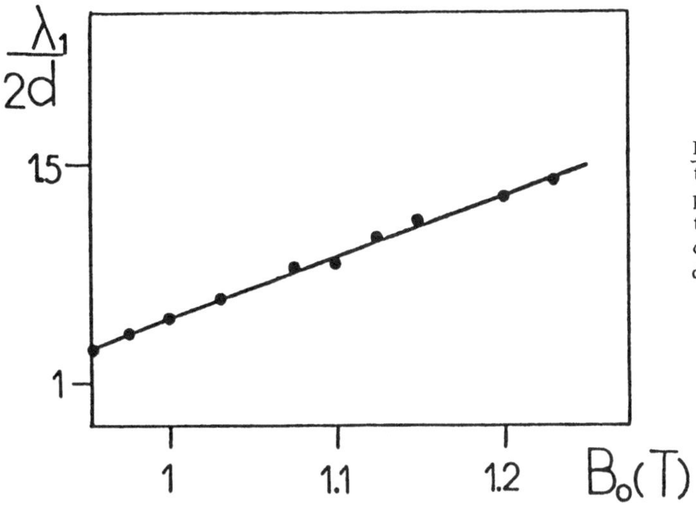

$\dfrac{\lambda_1}{2d}$

15—

1—

1   1.1   1.2 $B_o(T)$

Figure 5 : Variations of the wavelength of the periodic structure with the magnetic field. The continuous line has been drawn to guide the eyes.

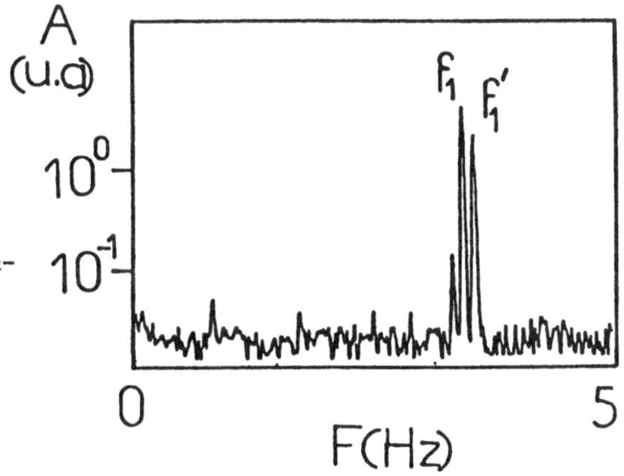

A (u.d)

10^0—

10^{-1}—

f_1 f_1'

0   F(Hz)   5

Figure 6 : Spectral analysis of the voltage drop at the output of a detector, for B_o = 1.15T and ε = 10^{-2}. The regime is biperiodic.

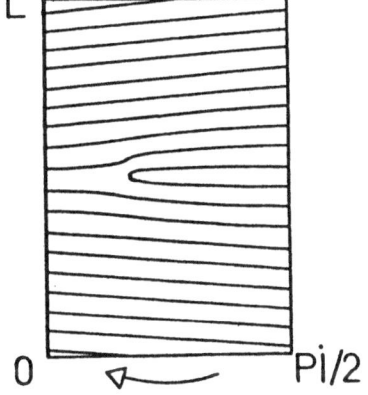

L

0   Pi/2

Figure 7 : The nodal lines of the flow, calculated by using the experimental results, for ε = 10^{-2} and B_o = 1.15T (two frequencies regime f_1, f_1'). As time is increased the nodal lines are advected by the mean flow, while the central defect slowly moves between the two periodic structures of distinct orientations.

Figure 8 : Three frequencies f_1, f'_1, f_2. We have not observed any frequency locking. This regime is stable in time. The values of the parameters which correspond to the figure are B_o = 1.15T and $\varepsilon = 2.10^{-2}$.

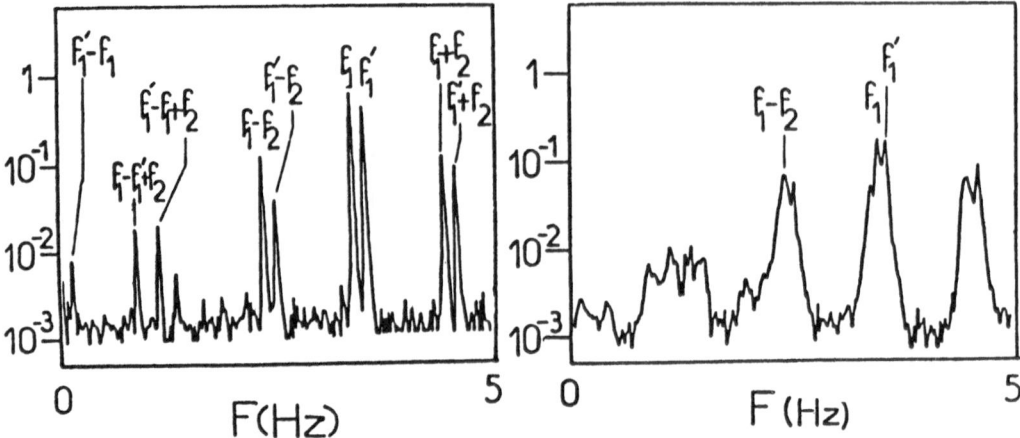

Figure 9 : Onset of turbulence. This figure represents an averaged Fourier spectrum, obtained for B_o = 1.15T and $\varepsilon = 2.4\ 10^{-2}$.

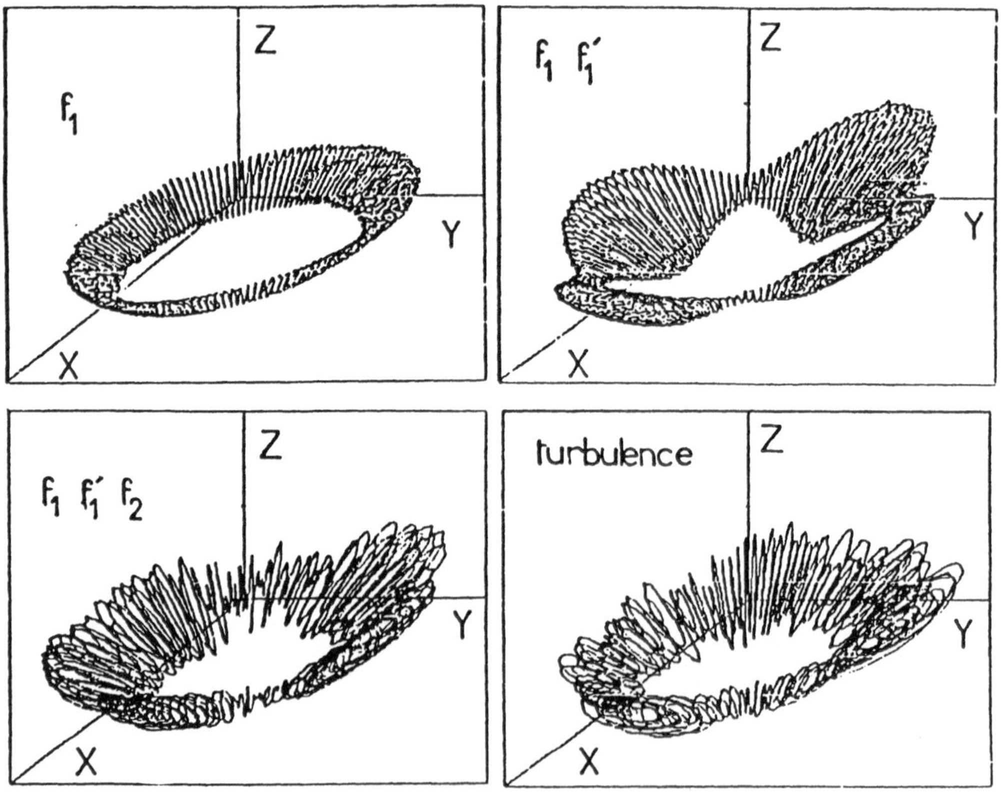

Figure 10

The transition to turbulence occurs after a new increase in the critical discrepancy (see Figure 9). The scenario (in the "coarse" sense) associated with such a transition is in agreement with the general scheme proposed by RUELLE & TAKENS (4).

The four successive bifurcations which we have observed at large values of the magnetic field are represented on Figure 10 : from signal $v(t)$, obtained at the output of a detector, we have constructed the trajectories defined by $X = (X_o + v(t))$ $\cos \Omega_g t$, $Y = (X_o + v(t)) \sin \Omega_g t$ and $Z = v(t+\tau) + Z_o$, where X_o, Z_o, τ, Ω_g are arbitrary.

To each dynamical situation, we can associate a coherent spatial structure of the flow. It is possible to interpret the onset of turbulence in terms of dynamical systems with a small number of degrees of freedom.

5. CHAOTIC DYNAMICS OF RELAXATION BETWEEN TWO STATES OF FLOW

At intermediate values of the external magnetic $(0.82T < B_o < 1.07T)$, the first event observed at the output of the detectors is a chaotic signal. Figure 11 shows a typical recording of voltage fluctuations for $B_o = 0.92T$ and $\epsilon = 10^{-3}$. The regime is quasi periodic (see the Fourier spectrum of Figure 11, performed on a 50s. sample). The spectrum exhibits two uncommensurate frequencies, f_1 and f_2. The spatial structures associated with each oscillator are in form of spiraling vortices, similar to those obtained in the range of large magnetic fields (see §4). The wavelengths of the two modes of instability are very close to each other, while the values of the azimuthal wave-numbers are significantly different. We have measured $m = 4$ for mode f_1 and $m = 5$ for mode f_2. The chaotic character of the regime is related to the fact that amplitudes A_1 and A_2 of the spectral peaks slowly fluctuate (see Figure 12). The amplitudes of the spectral peaks turn out to be stochastic quantities. This behaviour contrasts with the great stability in time of the phase variables associated with each oscillator f_1 and f_2. Spatially, the system indefinitely relaxes between two periodic structures in form of spiraling vortices. The turbulent state corresponds, in this case, to a temporal chaos.

6. SPATIO-TEMPORAL CHAOS

At low values of the external magnetic field $(Q < 620)$, the flow regime is turbulent just above the instability point. Figure 13 represents the direct recording of the voltage fluctuations at the output of a detector, for $B_o = 0.82T$ and $\epsilon \simeq 10^{-3}$. The signal involves time scales of the same order of magnitude as the mean rotation of the flow and time scales much larger. The Fourier analysis of a sample of the signal shows that turbulent energy is concentrated around a limited number of discrete frequencies. In contrast with the preceding range of values of B_o (see §5), the phase variable, associated with signal $v(t)$, fluctuates erratically. The spatial structure associated with such a regime is disordered. It is difficult , without any visua-

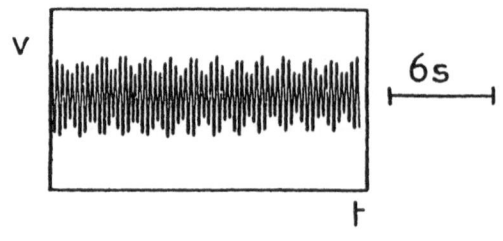

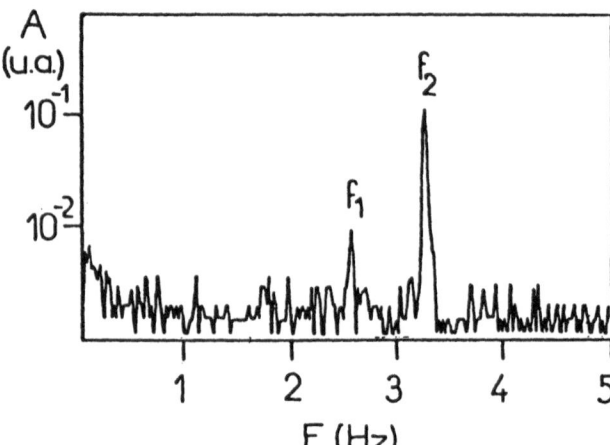

Figure 11 : Direct recording
of the tension at the output
of a detector, and the corresponding Fourier analysis
B_o = 0.92T and ε= 10^{-3}.

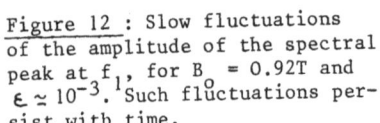

Figure 12 : Slow fluctuations
of the amplitude of the spectral
peak at f_1, for B_o = 0.92T and
$\varepsilon \simeq 10^{-3}$. Such fluctuations per-
sist with time.

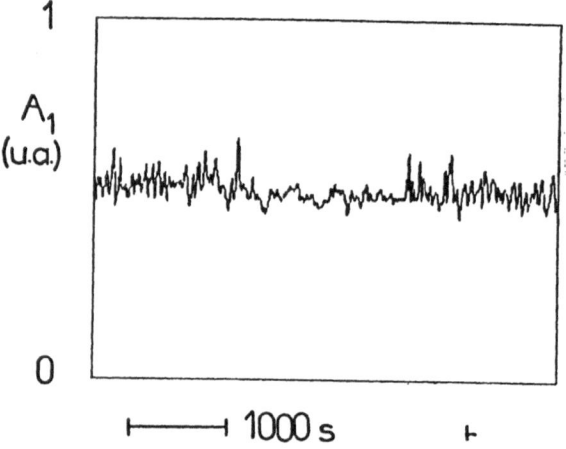

lization technique, to characterize such a structure ; one can reasonably think
that it consists of azimuthal modes with a great number of defects (owing to our
experimental observations, the dynamics of such defects appears to be very complicated ; it involves many dist i nct feotures , such as bursts, overdamped oscillations, long lived transients, etc...).

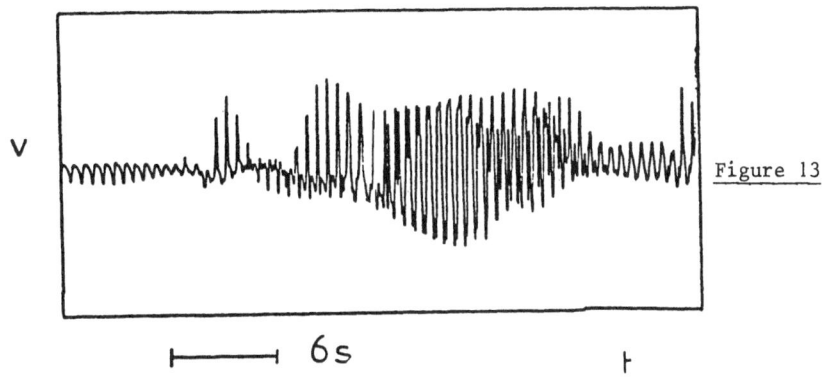

V

⊢——⊣ 6s   ⊦

Figure 13

7. CONCLUSION

We have shown a few examples of regimes of flow observed in a Taylor experiment subjected to an external magnetic induction. This type of experiment allows for showing the links existing between the dynamics of a particular non-linear system and its spatial features. It would be obviously important to be able to specify, more quantitatively, the characteristics of such links.

REFERENCES

(1) This type of information has been widely used in previous experiments (see for instance Phys. Fluids, 24 (1981), 406). It is possible, by using this information, to define a kind of effective viscosity of the system, and study its variations with control parameter I. This study allows for exhibiting the first threshold of the flow.

(2) Figure 2 is also representative of the values of the thresholds obtained in a preceding experiment (Phys. Rev. Lett. A49, (1982), 460).

(3) The threshold values of the Taylor number corresponding to the onset of stationary cells can be found in the Chandrasekhar's monograph ("Hydrodynamic and Hydromagnetic Stability", Oxford University, University Press, Oxford, 1961) and also in J. Fluid. Mech., 112, (1981), 329.

(4) RUELLE, D et TAKENS, F. Commun. Math. Phys., 20, (1971), 167.

ELECTRO-HYDRO-DYNAMICAL CONVECTIVE STRUCTURES AND TRANSITIONS TO CHAOS IN A LIQUID CRYSTAL

A. Joets, R. Ribotta

Laboratoire de Physique des Solides- Bât.510
Université de Paris-Sud
91 405 ORSAY CEDEX (France)

Convection in anisotropic fluids such as liquid crystals may provide a new model for the study of transitions to chaos. In effect, the
anisotropy that comes here from the orientational ordering may, for
instance, raise the degeneracy for the first convective structure and,
as will be suggested here, appears as the possible mechanism for selection of the different structures before chaos. Until now, intensive
study of convective structures in liquid crystals had been restricted
to the first convection which produces a set of parallel rolls. For
instance, a nematic liquid crystal subjected to an electric field develops the so-called Williams Domains [1]. Attempts to investigate the
transitions to chaos in that system have led to yet unclear observations [2]. We present here a typical sequence of transitions to chaos
in a nematic subjected to an AC electric field which shows evolution
of the convection from a roll structure to a bimodal rectangular structure and followed by a chaotic structure initiated by defects .

1. THE EXPERIMENTAL SET-UP

The sample is a layer of nematic sandwiched between glass plates
coated with semi-transparent indium-oxyde electrodes. The molecular
alignment of the nematic denoted by the unit vector $\vec{n}$ (the director)
is parallel to the plates along to $\vec{x}$ direction. Such an anchoring is
obtained by rubbing the plates, previously coated with a polymer (poli-
amid) along x . The electric field is applied across the sample
along $\vec{z}$. The nematic (Merck Phase IV and V) is of negative dielectric
anisotropy $\varepsilon_a = \varepsilon_{//} - \varepsilon_\perp = -0,2$. In the experiment the control parameter
of the system is the voltage V, while the frequency f of the field is
additional parameter. The thickness of the cell is of 50 μm and its
lateral dimensions are typically 2×3 cm. Then at the first convective
threshold about 600 rolls are present. Due to the large birefrengence
of the nematic any change in the molecular orientation will produce
a change in the wavefront of a light passing through the sample.

Thus periodical bending of $\vec{n}$ along $\vec{x}$, as is the case for the first
convective structure produces focal lines which correspond to upwards
and downwards motions. That flow can be roughly visualized by introducing small glass spheres (3-5 µm in diamter). The particular feature
of a liquid crystal is that the director $\vec{n}$ and the velocity field $\vec{v}$
are directly coupled. The observations are made directly under polari-
zing microscope and measurements are made on photographs.

2. THE FIRST CONVECTIVE STRUCTURE - WILLIAMS DOMAINS

For a fixed frequency of the AC electric field (typically f = 60Hz)
the voltage V is continuously increased from zero. The liquid crystal
is at first in a rest state and the sample appears homogeneously transparent. Then at a well defined value $V = V_R$ (typically 6 Volt), bright
lines appear along $\vec{y}$ and one can check that the system is under convection. The rolls are aligned along $\vec{y}$, i.e. they are perpendicular to the
molecular direction $\vec{n}$. It is the so-called Williams Domains structure
in the conduction regime (Fig. 1).

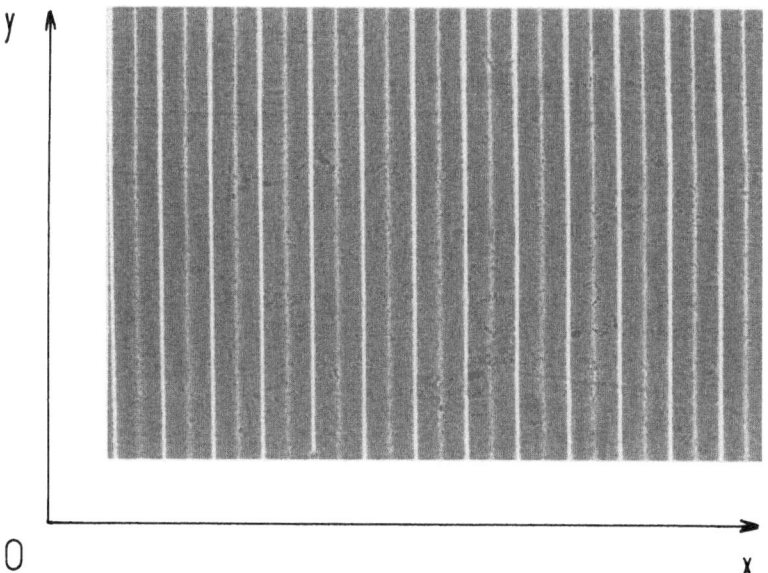

Fig. 1 : Williams Domains

The transition from the rest state to the Williams Domains has been fully analyzed and recognized as a second order-like transition [3]. The macroscopic variable which would stand for the order parameter is the angle ϕ of $\vec{n}$ over $\vec{x}$ in the xOz plane. If now the frequency is varied one finds a transition line starting from a low frequency to a cut-off frequency f_c. In fact as we will see in the next paragraph that line changes its nature for the lower frequency part, contrary to what was started up to now.

Let us recall briefly the basic mechanism for the first convection [4]. Considering the molecular alignment in the xOz plane, let us start from a fluctuation of bend of the form $\phi = \phi_o \sin qx$ (Fig. 2).

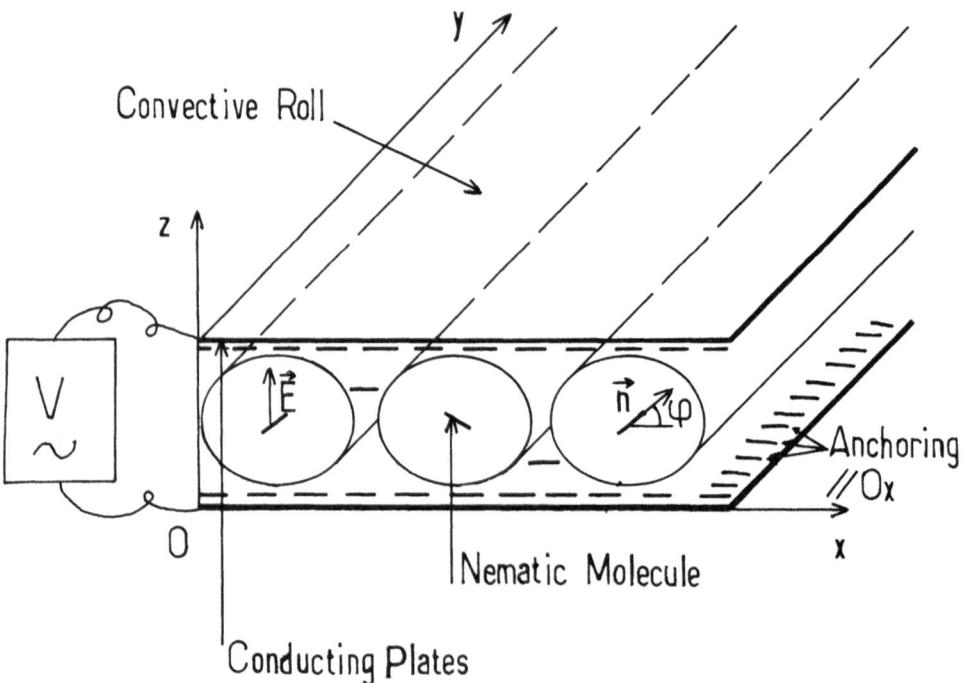

Fig. 2 : Schematic of a sample.

The ionic charges always present in the sample tend to focus on the high curvature zones by a motion along the director (positive conduction anisotropy). The charges accumulate with alternative sign along $\vec{x}$. That produces a periodic transverse component of the local electric field which produces a destabilizing torque on the molecules tending to increase the molecular bending. On the other hand the drag of charges produces a flow which, because of the coupling with the director $\vec{n}$, acts also as a destabilizing mechanism. The restabilizing mechanism is the restoring elastic torque. The convection sets in when the balance of the destabilizing plus stabilizing torques vanishes. The threshold is expressed in V^2 rather than in V [5]. The second order-like character transition has clearly been demonstrated both experimentally and theoretically [3].

3. THE OBLIQUE ROLL STRUCTURE : ZIG-ZAG

Setting the frequency f at some intermediate value (around 60 Hz) the voltage V is gradually increased from zero. For the first threshold V_R, the convection sets in and we obtain the already explained structure : the Williams Domain. Increasing the voltage further, after a second threshold V_z produces an undulation of the rolls along the y axis (Fig. 3).

The undulation is static, starting from zero amplitude with a finite wavelength Λ of order 5 to 10 roll-diameters. The amplitude increases then with the voltage. The wavelength increases smoothly, than at some voltage $V > V_z$ the deformation suddenly increases sharply. Beyond that voltage the defects appear in the high-curvature regions. Finally, domains of oblique straight rolls are formed tilted by an angle $\pm\theta$, symmetrically to the $\vec{y}$ direction (Fig. 4).

In fact, the zig-zag structure was already observed but has neither been analyzed nor recognized as the first state of a new instability[6]. Experimentally we measure above the threshold V_z the maximum tilt angle θ, as a function of the reduced voltage $E_1 = (V^2 - V_z^2)/V_z$. We find that $\theta \sim \theta_0 E_1^{0.43\pm0.07}$ as a result typical of a direct bifurcation.

We propose the following scheme to explain this buckling of the rolls. Increasing the voltage, the maximum angle of bending of the molecules ϕ_0 reaches a value for which the director alignment becomes unstable against transverse fluctuations in the xOz plane. Due to a restoring elastic torque imposed by the molecular anchoring on the plates, the roll is undulated along $\vec{y}$ and that than produces a modulation of the

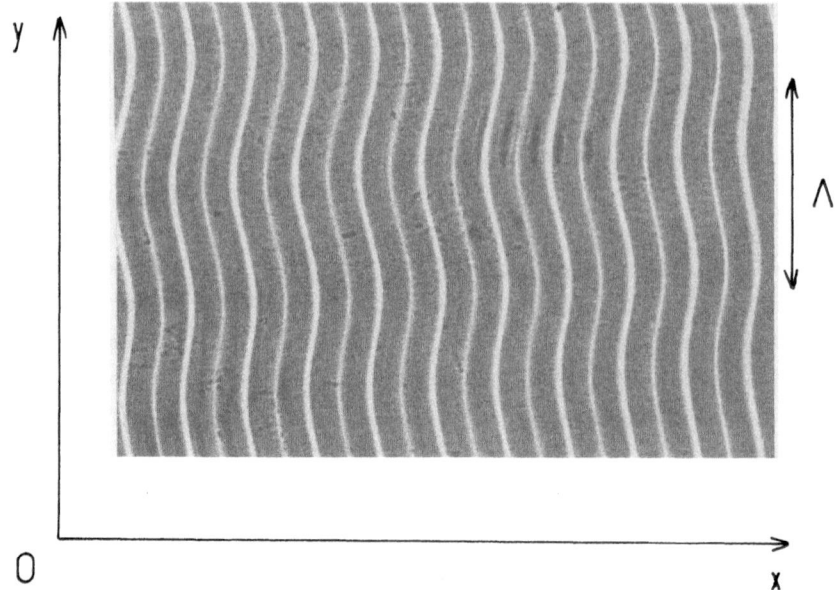

Fig. 3 : Undulations

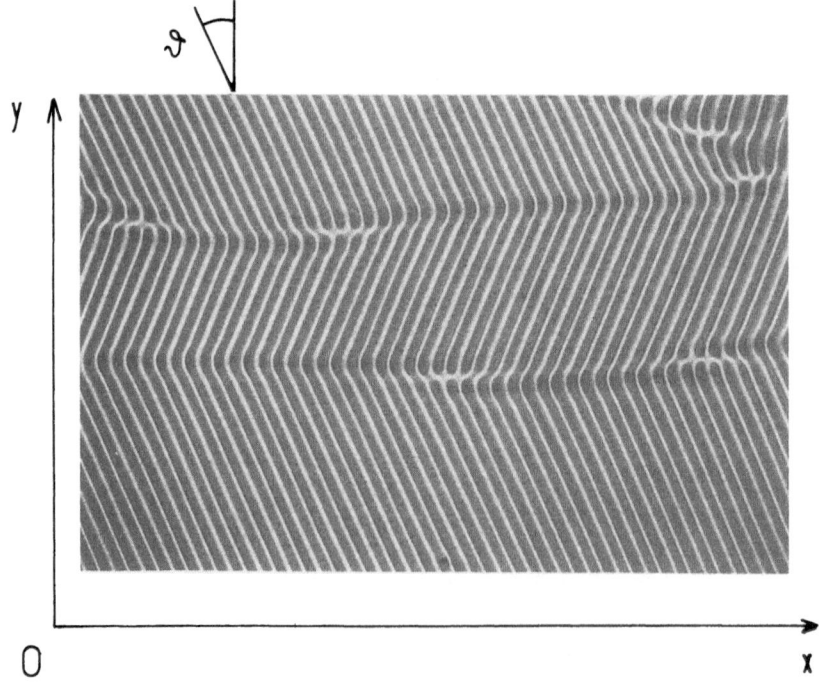

Fig. 4 : Zig-zag

density of ionic charges along $\vec{y}$. The result is a small periodic transverse component, added to the local electric field, in the xOy plane,
which will act to increase the molecular splay. The coupling of molecular orientation and velocity field results in a vertical vorticity component which in return increases the ionic charges drag and therefore
the transverse electric field. Then the bending of the roll is amplified and the result is an undulation controlled by the splay with wavevector along $\vec{y}$.

Now setting the frequency f of the AC field at a low value (f ≃ 20Hz)
one obtains a discontinuous transition (first-order-like) from the rest
state, directly to a well developed zig-zag structure. The Williams
Domain structure does not exist at this frequency. By measuring the
threshold voltages for the different structures as a function of the
frequency, on obtains a diagram of transition lines delimiting the stability domains (Fig. 5).

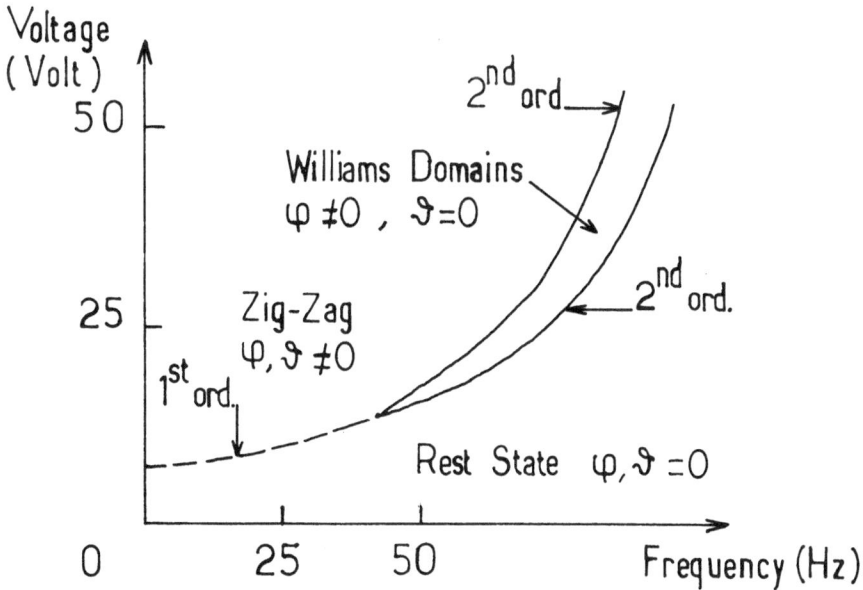

Fig. 5 : The first transition lines to the zig-zag

The first new observation is the existence of a triple point, close to a multicritical point; the experimental uncertainty does not allow one to distinguish the relative position of the two points. One notices a close analogy with the thermodynamic phase diagram for the Nematic-Smectic A-Smectic C transitions in a liquid crystals [8].

4. THE SKEWED VARICOSE

We have thus obtained oblique roll structure tilted by $\pm\theta$. Now, increasing the voltage again at the same fixed frequency of 60 Hz produces, after some well defined threshold V_v (typically 7 volt) a static modulation along the roll axis with a period slightly larger than the roll diameter. That modulation increases in amplitude as the voltage is increased and it is noticed that its phase is slipping from one roll to its neighbour. The result is a periodical pinching along the roll axis, corresponding to a deformation of wave-vector $\vec{q}_v$ oblique with respect to the roll axis and of a direction tending to get close to $\vec{Ox}$. (See Fig.6). This is the skewed varicose [7].

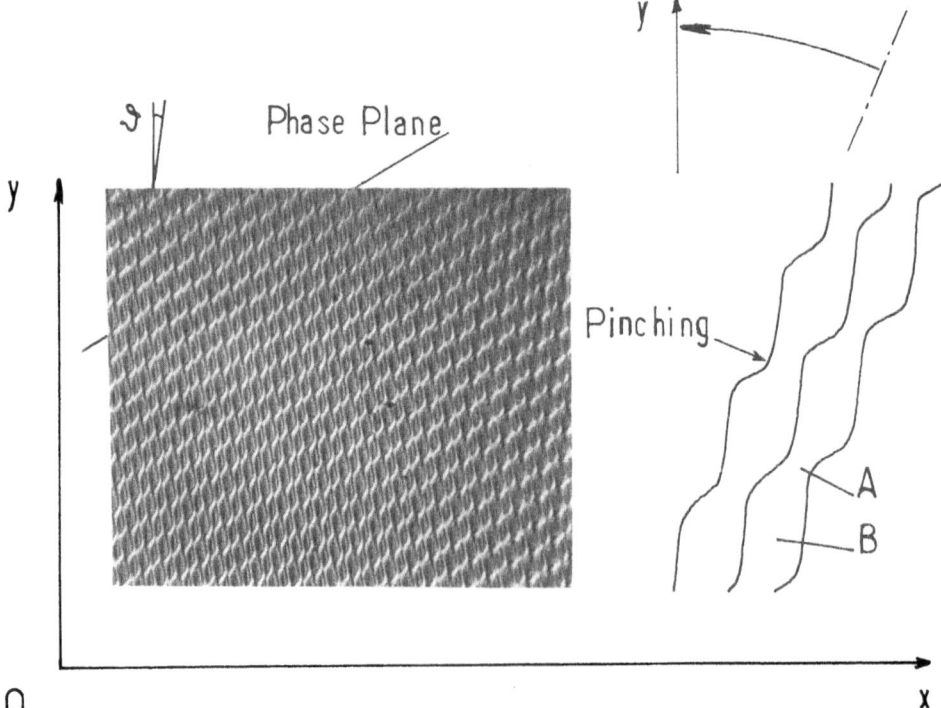

Fig. 6 : Varicose

The amplitude of the modulation increases continuously with the applied voltage V, as for a direct bifurcation. Finally, the effect of the varicose is to impose (see Fig.6) a period shear in the pinched zones A which allows the other non deformed part B of the roll to come back to an alignment along $O\vec{y}$ (see Fig.6).

The skewed varicose was observed in Rayleigh-Bénard, not as the final state of an instability but rather as a transient deformation [9]. In our experiment, we believe that the coupling of the orientation with the velocity field plays an essential role in the selection mechanism as has been suggested for the zig-zag structure. A model remains to be developed. It can only be suggested here that twist deformation of molecular alignement could come into play as the dominant elastic mode. When varying the frequency we obtain a new transition line for the thresholds for that instability : $V_v(f)$ [7].

5. THE TWO DIMENSIONAL STRUCTURE

As the voltage is further increased in a "varicosed" structure one suddenly obtains a two-dimensional structure : there appears then a double periodicity in the flow. One is along $\vec{x}$, the initial periodicity is therefore recovered. The other one is along $\vec{y}$. Thus the elementary structure is a rectangle. By the motion of small glass spheres introduced in the sample it is demonstrated for the first time that one gets a bimodal structure as depicted hereafter. Each cell has a motion opposite to its direct neighbours (Fig.7).

The fact that inside the same rectangular cell one has two independent rotationnal flows $\vec{\omega}_1$ and $\vec{\omega}_2$, implies an important dissipation due to strong shears. For higher frequencies the rectangular structure cannot be obtained easily from a "varicosed" structure due to a high density of dislocations which move quasi erratically. Before formation of rectangles the rolls undergo such large shears that they break and oscillate lateraly from $+\theta$ to $-\theta$ around $\vec{y}$ (Fig.8). The interaction with defects is then very complex (see next paper).

It is only for a higher and well-defined voltage V_B that stabilisation on the bimodal structure is obtained. The transition is first-order-like and the oscillations have the character of an oscillatory often observed in some systems around an inverse bifurcation [10, 7].

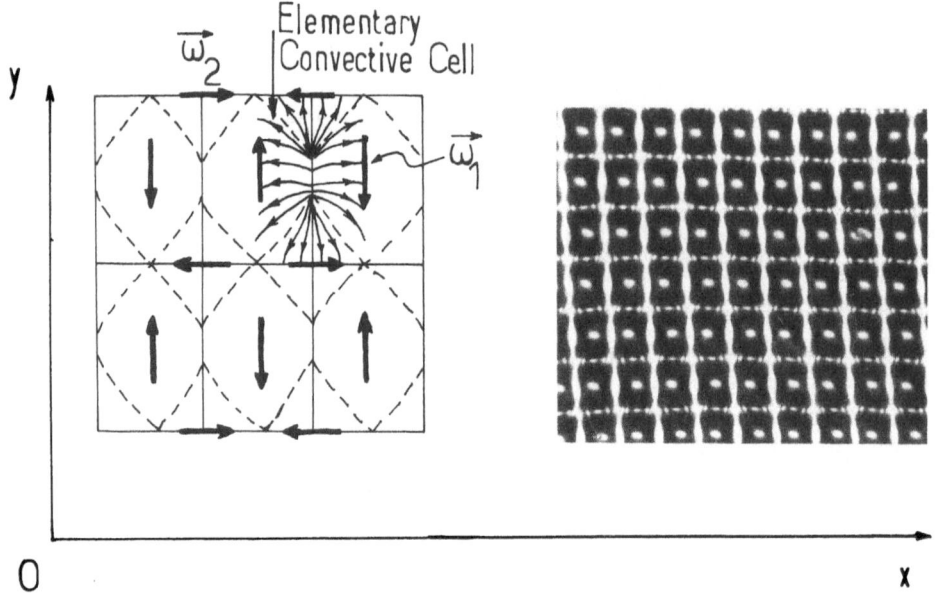

Fig. 7 : Rectangular structure

Fig. 8 : Oscillation varicose-rectangles

6. LOSS OF ORIENTATIONAL ORDERING : CHAOTIC STRUCTURES

Increasing the voltage further beyond the formation of rectangular bimodal structure one observes a zone in the V-f diagram where the
bimodal is stable. Then almost suddenly there appears a first mode in which
the symmetry around the mid-plane xOy is broken.

a. The quasi-ordered "chaos"

That mode appears first as oscillatory instability of rectangles towards a zig-zag mode, around 30 volts i.e. the rectangles open
successively along each diagonal in order to form rolls tilted by $\pm\theta$
around $\vec{Oy}$ as for the oscillating instability which preceded the bimodal. However here the rolls thus formed are shorter in length and
there appear inside each of them one disclination loop which is a
closed singular line for the orientation of the director $\vec{n}$ [11]. These
loops are of a size comparable to the portion of the roll formed in the
oscillatory process. The whole bimodal structure breaks into domains
oscillating incoherently. This is an example of melting associated
with formation of disclination loops (Fig.9). The alignment is therefore so much perturbed on small scales that macroscopic properties
such as optical birefringence are no longer constant on scales of order
some light wavelengths. The sample becomes almost turbid although some
ordering remains inside. In that mode there is now a loss of symmetry
with respect to the mid-plane parallel to xOy. This is the so-called
Dynamic Scattering Mode 1 [11].

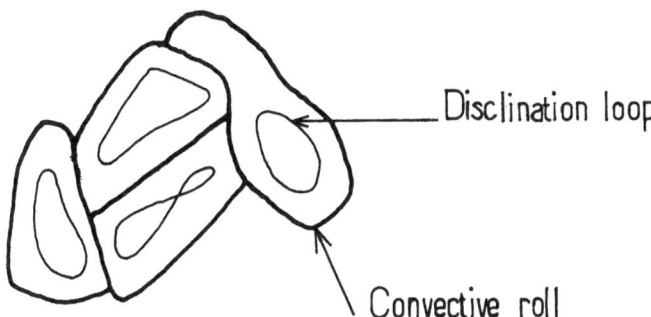

Fig. 9 : Dynamic Scattering Mode 1

b. The full chaos

For higher value of the voltage (typically V ∿ 60 Volt) inside
the quasi-ordered chaotic structure there appears suddenly a large density of much smaller   disclination loops which are nucleated at some
places in the sample. These loops of about 2-5 μm in diameter are
almost circular in shape can overlap and move erratically around their
mean position. The nucleated structure propagates inside the sample at
an average velocity of some 200 μm/s (comparable to the rotation velocity inside a roll) and in an anisotropic way : faster along the
initial alignement direction $\vec{x}$ than along $\vec{y}$. In these areas light scattering experiments show a correlation not larger than 1-3 μm which is
of the order of a light wavelength. In that respect it is called full chaos
since strong de correlation in alignment would mean de correlation
in velocity field on the same scales. In fact that structure scatters
strongly the light and the sample is highly turbid. This is the so-
called Dynamic Scattering Mode 2 which was first suggested to be used
in displays [12] and in fact started the renewal of interest for liquid
crystals.

7. STRUCTURES FOR LOWER AND HIGHER FREQUENCIES

We briefly present observations made at low frequencies close to
D.C. and at high frequencies close to the cut-off.

a. Low frequency range : f < 1 Hz

In this area the situation is less clear, in part because uni-
polar charge injection might be present [13]. Increasing the voltage
from zero, one first observes distorbion in the alignment which takes
the form of hexagons, while convection has not been detected. In fact
that symmetry is consistent with the angle of 30° found for the zig-zag
at low frequencies. For higher voltages it becomes difficult to distinguish the different structures mainly because their rise and decay
times are comparable to the period of the AC fields. However some
structures can be observed in a transient regime for low voltage values.

b. High frequency range : f close to f_c

Close to the cut-off frequency which indicates the limit where
the relaxation time for the changes is close to the period of the
applied frequency, an other behaviour for the structures can be expected. At threshold for convection the rolls aligned along $\vec{y}$ area are not
fixed in space but rather move along $\vec{x}$. In fact it is the deformation

which varies sinusoidally in time. The motion can take place in either direction $\pm\vec{x}$. That effect would originate from a cross-over of the two mechanisms : charge conduction and curvature oscillation with a phase lag [7].

For higher voltage the sequence observed for intermediate frequencies is still present. However the density of dislocations becomes very high, and their motion (mainly climb) is fast (typical velocity > 500 μm/s). These structures can be now mixed with the chevron-like structure which takes place at a slightly higher voltage. Some further work is under way in order to characterize these structures.

CONCLUSION

Electro-Hydrodynamical convective instabilities in a nematic liquid crystal seem now well recognized and show a sequence of structures from the rest state to the chaos which can be obtained spontaneously and reversibly in an extended sample. These structures of decreasing symmetry make the system go from a one mode convection to a bimodal flow. Further on, the chaos is obtained through formation of disclination loops. It has also been measured a settling time for the instabilities of typical values of the order of one minute for voltages in excess of 5 % over the threshold.

Up to now the mechanisms for these structure are not completely elucidated. However our first model suggests that the orientational ordering plays a dominant role in the selection mechanism for the structures. In "small boxes" where the lateral boundaries become important new situations are found that can be explained from the behaviour in extended geometry.

Thus it appeats that such a system that can provide structure in short times with total reproducibility, might be an interesting model for further studies in the transitions to chaos.

REFERENCES

1. R. WILLIAMS , J. Chem. Phys. $\underline{39}$, 384 (1963)

2. S. KAI, H. HIRAKAWA, Mol. Cryst. Liq. Cryst. $\underline{40}$, 261 (1977)

3. W. SMITH, Y. GALERNE, S.T. LAGERWALL, E. DUBOIS-VIOLETTE, G. DURAND
   J. Phys., C-1, $\underline{36}$, 237 (1975)

4. W. HELFRICH, J. Chem. Phys. $\underline{51}$, 4092 (1969)

5. ORSAY LIQUID CRYSTAL GROUP, Mol. Cryst. Liq. Cryst. $\underline{12}$, 251 (1977)

6. C. HILSUM, F.C. SAUNDERS, Mol. Cryst. Liq. Cryst., $\underline{64}$, 25 (1980)

7. R. RIBOTTA, A. JOETS, to be published

8. J.H. CHEN, T.C. LUBENSKY, Phys. Rev. A $\underline{14}$, 1202 (1976)

9. F.H. BUSSE, R.M. CLEVER, J. Fluid Mech. $\underline{91}$, 319 (1979)

10. H.N.W. LEKKERKERKER, J. Phys. Lettres, $\underline{38}$ (1977) p. L 277
    E. GUYON, P. PIERANSKI, J. SALAN, C.R. Acad. Sc. $\underline{287\ B}$, 41 (1978)

11. P.G. de GENNES, The Physics of Liquid Crystals, Clarendon Press,
    Oxford, 1974

12. G.H. HEILMEYER, L.A. ZANONI, L. BARTON, Proc. IEEE, $\underline{56}$, 1162 (1968)

13. N. FELICI, Rev. Gen. Elect. $\underline{78}$, 717 (1969)
    P. ATTEN, J.C. LACROIX, J. Mec. $\underline{18}$, 469 (1979)

ROTATING DISK FLOWS, TRANSITION TO TURBULENCE

M.P. Chauve, G. Tavera

Institut de Mécanique Statistique de la Turbulence, L.A. N° 130
12, Avenue Général Leclerc
13 003 MARSEILLE, FRANCE.

1. INTRODUCTION

The problem of the rotating coaxial disks is usually considered with regard to two geometries noted G 1 and G 2.

Geometry 1 : Gl, is defined by the flow field above an infinite flat disk which rotates around an axis perpendicular to its plane with an uniform angular velocity ω .

Geometry 2 : G2, is defined by the flow field between two infinite parallel flat disks respectively rotating at varius angular velocity (ω_1, ω_2).

We shall discuss further the equivalence of G1 and G2 to their laboratory transposition (flows between finite disks).

The resolution of Navier-Stokes equations in the case G1, for an axisymetrical stationary flow was done by KARMAN (1921) and COCHRAN (1934), who postulated that the axial velocity is independent of the radial coordinate. Later, GREGORY STUART and WALKER (1955) worked on the stability of this flow. BATCHELOR (1951) showed that KARMAN solution was applicable when the fluid is enclosed between two rotating infinite disks (G2).

Since, the theoretical and experimental researches have been investigated mainly for the stability analysis in the case of G1 and for the identification of different possible axisymetrical stationary flows in the case of G2.

In this work, for the case G2, we describe an experimental study of the incompressible flow field between two finite parallel disks, one stationary and the other rotating at constant speed.

2. GENERALITIES

In the ideal cases G1 and G2, two adimensional parameters R_r and R_ℓ are naturally introduced in the steady state problem.

In the case of G1, the equivalence of the actions of the viscosity and centrifugal forces, drive to estimate the thickness δ of the layer of fluid carried by the disk as $\delta = (\nu/\omega)^{1/2}$ (SCHLICHTING 1979). Consequently the adimensional radius R_r is defined by $R_r = r/\delta = (r^2\omega/\nu)^{1/2}$.

In the case of G2 the comparison between the height ℓ of fluid and the thickness δ is expressed through the parameter R_ℓ (adimensional height) as follows : $R_\ell = \ell/\delta = (\ell^2\omega/\nu)^{1/2}$. The starting rotation stage can be expressed through the parameter $\tau(t)$, ratio between the laps of viscous diffusion and the rate of rotation as follows :
$$\tau(t) = \ell^2\omega(t)/\nu = R_\ell^2(t), \quad \text{(FLORENT, N.N. DINH and V.N. DINH, 1973)}.$$

In other respects the equivalence of experimentation with regard to G1 and G2 find expression in two more parameters p_1 and p_2, so that :

$p_1 = a/\delta$, maximal of $R_{r=a}$ linked to experimental apparatus if a is the radius of rotating disk in case G1,

$p_2 = \ell/a$, kind of form parameter.

a. Already known characteristic phenomena

G1 : On the disk an axisymetrical similar flow (KARMAN, 1921) spreads at the central area $R_r < 300$. Its stability swings ($R_r = 300$) to swirling instability with spirally shaped vortex. The horizontal extension of the area where these kind of instabilities leads stays for $300 < R_r < 500$ and the height stands around δ. For $500 < R_r < 550$ more complex flows are lying and show still at that point a certain spatial organisation which suddenly disappears when $R_r > 550$: here begins a turbulent flow. (MALIK, WILKINSON, ORSZAG, 1981).

G2 : The study of only axisymetrical flows was really worked specially about a stationary case to which a degeneration of the solution (might be non single for given R_ℓ) is shown by the numerical solution of Navier-Stokes equations (HOLODNIOK, KUBICEK, HLAVACEK, 1981). This non single solution ($R_\ell > 15$) goes sometimes with non single types of possible flows (ROBERTS, SHIPMAN, 1976 - MELLOR, CHAPPLE, STOKES, 1968), (possible cells, the number of which seems to be between 1 and 5, Fig.1, are limited by parallel plans to those of the disks and on which the axial velocity vanishes). But the one cell flow

seems to be peculiar. So for experiments associated with numerical degenerate situations as to type of flows, only these kind of results are
seen (a boundary layer spreads on each disk and in the middle area the
whole fluid rotates as a rigid body). Its rate is a third or so of the
rotating disk, Fig.2, (MELLOR, CHAPPLE, STOKES, 1968 and OLIVIERA,
BOUSGARBIES, PECHEUX, 1982).

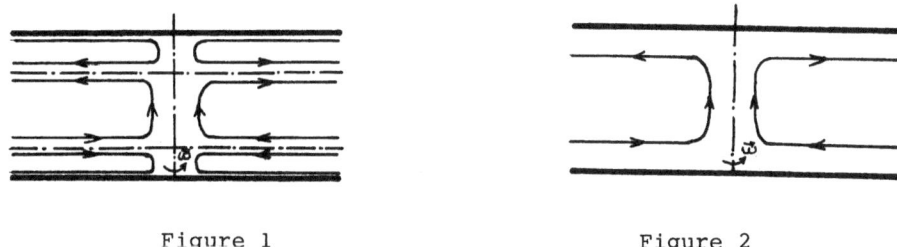

Figure 1   Figure 2

Even if it seems easy to think that there is a connection between
the degeneration of solution of the stationary problems and the lack
of care for the generative history of the flow (how to set the disk to
rotate, same question ("?") than COLES'one, 1965, about flows between
concentric rotating cylinders), it has not been possible to be aware
of its nature.

An other important aspect of G2 is about idealistic geometry
compared to experimental possibilities (as the parameter p_2). For instance, under certain conditions ($p_2 = 0.1$) zones of recirculation
through the fluid are seen by the edge of the disk, as also central a-
reas similar to those have been computed (OLIVIERA and al., 1982 -
ADAMS, SZERI, 1982), in the case of the ideal geometry G2.

All the foregoing studies are implicitly referring to low e-
nough values of R_r where no swirling instability is showing. Our experimental "tool" allowed us to consider experimental configurations
where not only such phenomena can be obvious but also a zone of transition to turbulence.

b. Notes

    - Equations of the problems relating to geometries 1 and 2

    Cylindrical coordonates system :

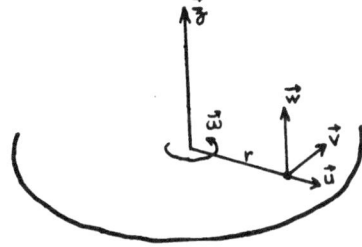

Let $\eta = z/\delta$, $v = \omega\, r\, g(\eta)$, $\omega = -2(\,\omega\nu\,)^{1/2}\, h(\eta)$, $u = \omega\, r h'(\eta)$, $P/\rho = \omega\nu\, p(\eta) + (1/2)\, \lambda\, \omega^2 r^2$, and reporting in the momentum equation, the continuity equation being identifically satisfied, the system of ordinary differential equations is arrived at as follows :

$$\begin{cases} h''' + 2\, hh' - h'^2 = \lambda - g^2, \\ g'' + 2\, hg' - 2\, h'g = 0, \\ p' = -2(h'' + 2\, hh'). \end{cases}$$

- Solutions.

The previous similarity solutions change the problem of equations with partial derivates, into boundary problems of differential equations. So, it is possible to calculate the flows relating to G1 and G2 in writing the respective boundary conditions. Although it is easily done in G1 case, the specific complexity of the numerical problem (ie. degeneration) makes the G2 case a lot more difficult.

Fig. 3 shows a scheme of computed stream lines in G1 case.

Figure 3

In G2, the experimental evidence of the peculiar fact taken by one cell flows, naturally leads us to investigate the stability of the set of the axisymetrical numerical possible solutions (especially towards perturbations in the set of solutions from one to another). However, as far as we are aware, it has not been carried out yet.

On the other hand, in the case of G1 the singleness of the solution allows us to study the stability of axisymetrical flow for the benefit of a non axisymetrical flow. In 1955, STUART started this study which has been recently completed with KOBAYASHI, KOHAMA, TAKAMADATE, 1980, and MALIK and al., 1981). The critical value of parameter $R_r \simeq 295$, the number of spirally shapes swirling instabilities - 21 -, and the quasi-missing phase velocity can be deduced from it.

With regard to G2, it will be stated later how similar phenomena can be shown up in our study.

3. EXPERIMENTAL SET UP AND DEVICE

The disk is made with duraluminium; it has a 450 mm diameter and
is 30 mm thick. During visualisations the disk is covered with a black
anti reflective layer. The disk is fixed on a hollow arbar from the
bottom of which goes a set of tubes to supply with dye every injector
distributed on the whole surface of the disk without preventing the
rotation. At the other extremity an optical code maker measures the
rate of rotation ($0,05 < \omega/2\pi < 4$) with a relative accuracy of
0.003, while an electronic stabilizer connected with the engine, controls
the regular rotation of the axis. To get easily an axisymetrical flow,
the disk is set in a circular box (800 mm diameter), made with plexiglass, the top and the bottom of which can move separately for chosen
distances. The first box is enclosed inside an other plexiglass one
($900 \times 760 \times 60$ mm^3). The whole set up is put on a metallical stand
where stays the motor part (Fig.4).

The scale model is filled with strained demineralized and U.V.
treated water : moreover the bath is kept at a constant temperature.
Visualisations are realised with rhodorsil (a white emulsion of oil
and silicone); its density has been corrected up to a value sensibly
greater than the one of the water.

With a central injector, the dye can be spread on the whole
surface of the disk with or without any rotation. For measurements
with hot film probe the scale-model is surmounted with an apparatus
making the probe displacement in the three directions of the space. In the
perpendicular direction to the disk the shiftings are realised with a
0.01 mm accuracy; the hot film probe is a classical fiber-film probe
55R11, ("nickel film deposited on 70 µm diameter quartz fiber, overall length 3 mm, sensitive film length 1.25 mm, copper and gold plated
at the ends. Film is protected by a quartz coating approximatively
2 µm thickness"). The probe body has a 3 mm diameter; this probe is
connected to a DISA 55M01 Constant Temperature Anemometer.

The main difficulty in this set up is to get a perfect paral-
lelism with the disk and the top of the box. Actually the adjustement
is realised with 3 chocks put on the disk.

Figure 4

4. RESULTS

a. Geometry G1

The experimental realisation of this geometry is in fact a geometry G2 with a great R_ℓ parameter. The foregoing results are done for a minimal value of R_ℓ = 50.

For $R_{r=a}$ = 180, the photos 1,2 and 3 below, show the diffusion of the dye from the central injector to the edge of the disk. It stays close to the rotating plan and makes obvious the stream lines. The inclination of the spirally shaped stream lines shown in the photos is in accordance with the calculation (-40°,SCHLICHTING, 1979).

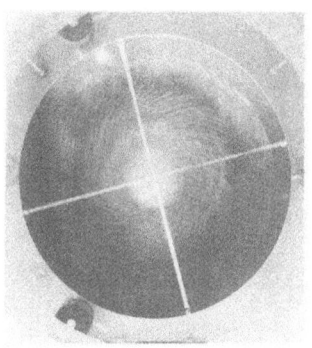 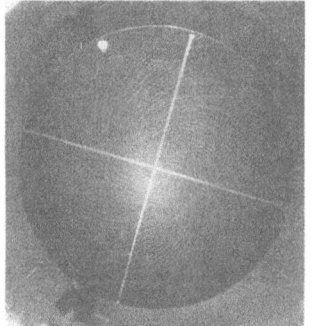 

Photo 1   Photo 2   Photo 3

Photo 4 shows the STUARTS'S swirling instabilities and the zone of transition to turbulence. The critical value of R_r from this photo is $(R_r)_c = 350$, there are 27 spirals and a zone of transition about $(R_r)_t = 495$.

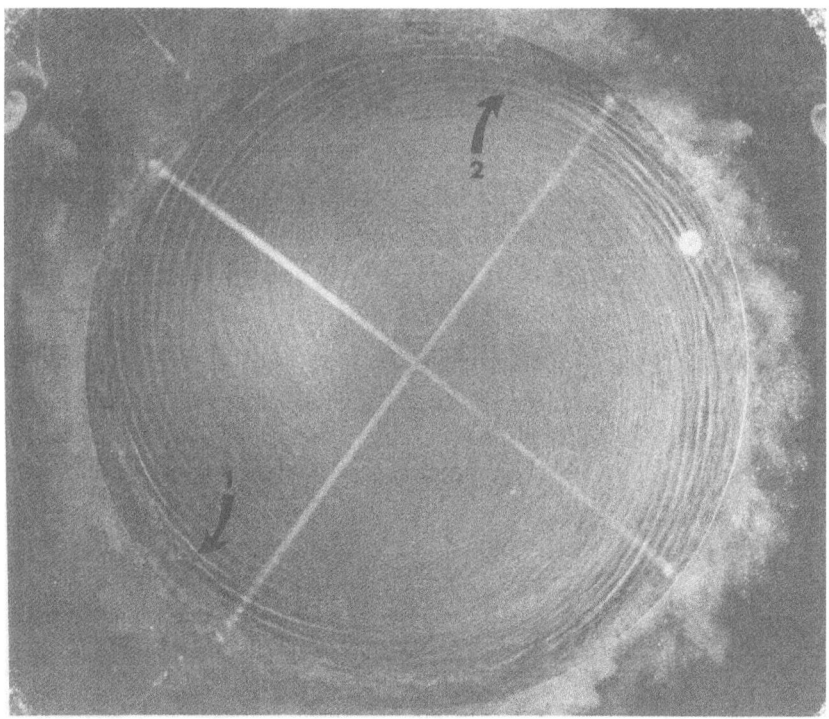

Photo 4

Some significant details (pointed with two arrows on the photo) are relevant to the destabilization of STUARTS'S spirals.

The segmentation of the swirls can be seen (arrow 1) going to
a flow breaking into more and more little structures but still identi-
fiable ones. Sometimes a less usual set of bifurcations (arrow 2) occur. It means an anomaly of the spatial repartition of the spirals
which leads to their dislocation.

In other respects some experimentations with higher angular
velocities show the displacement of the instabilities area to the center of the disk and allow the verification of the independency of the
results with regard to the edge effects.

It looks interesting to compare these visualisations with the
results got with a fiber-film probe.

Fig.5 shows at the start of the rotation, the signal from the
probe brought into the instability area and far from the disk as
d = 0.8 mm, about $\hat{\delta}$, ($\hat{\delta}$ = $\sqrt{2\pi\ \nu/\omega}$ = 1 mm).

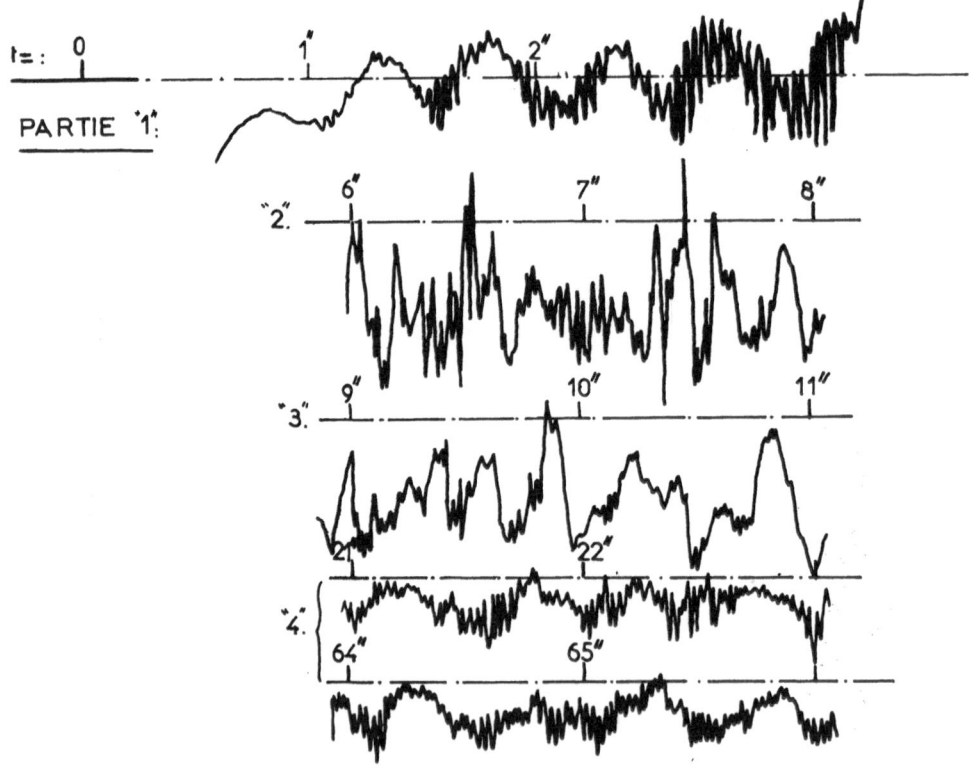

Figure 5

- part 1: the time of the viscous diffusion from the disk to
the probe is around 0.6 seconds; this time defines the zone (part.1).

- part 2: owing to the facts that the dimensions of the instabilities are $\hat{\delta}$ and that the surface of the disk shows punctual defaults
in keeping the same plan (about 0.3 mm), the time of viscous diffusion
in this zone is about 3 or 4 seconds.

- part 3: a chaotic aspect of the signal is observed. It relates
to the competition between different ways of flows which are selecting
themselves to end at part 4 into an asymptotic state. Note that the
time of viscous diffusion for a height of fluid of 20 mm is about 7 minutes : it is a lot greater than the time which defines part 3; this
difference obviously comes from the increase of the diffusion relating
to the chaotic flow.

- part 4: studying at the asymptotic state, the norm of the instability velocity is lower than the one in part 1. That can probably
be explained by the difference of shears in both cases.

The stationary state being reached a spectral analysis was done of the azimuthal component in terms of R_r and in terms of the distance to the disk for given R_r in typical points of the phenomena. The signal is previously strained with an analog pass band filter in the range of 3 to 150 Hz.

Fig.6 gives the results for $R_r = 361(S1)$, $R_r = 448(S2)$,
$R_r = 514(S3)$, $R_r = 599(S4)$ and $d = 0.6$ mm.

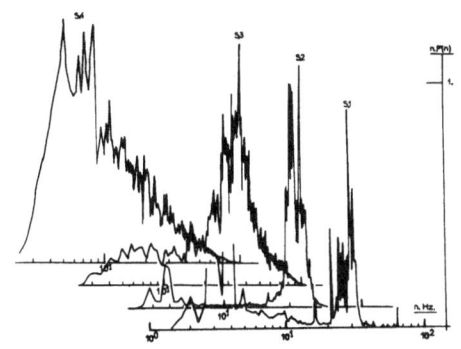   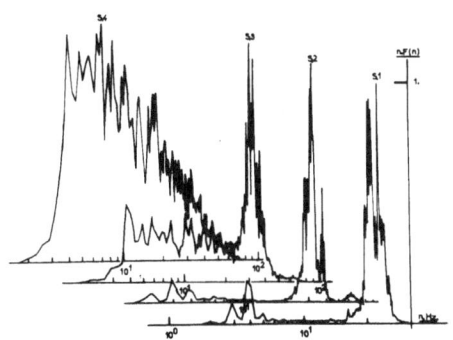

Figure 6   Figure 7

S1 : two spectral zones with the same relative strength are
seen; the one near 4 Hz and the other near 30 Hz. The first one comes
from the punctual errors of the disk for keeping the same plan, the second one from the STUARTS'S instabilities. The influence of the proxi-
mate wall of the box explains the equivalence of levels between the two zones.

S2 : such like equivalence is not found in S2 because of the main part here taken by instabilities.

S3 : the point of measurement stands in a zone usually called zone of transition, which goes with a widening spectral zone of instabilities (these cannot be distinguished anymore in photo 4). The spectral gap in S1 and S2 tends to fill up even if the level of zone of instabilities stays dominant.

S4 : this is not found again in S4, where the global aspect has the characteristic look of a turbulent zone. The integral scale of which seems related to the errors of the disk about the plan even if here the measure point stands outside the rotating disk.

Note that the determination of $(R_r)_c$ with a fiber-film probe gives a value of 276 totaly consistent with MALIK'S (287) analysis (MALIK and al.), whereas the value from visualizations was too high.

Fig.7 visualizes the results got with d = 0.6 mm (S1), 0.8 mm (S2), 1 mm (S3), 1.4 mm (S4) and R_r = 448. The points of measurement stand in the horizontal area where the swirling instabilities are present.

On a global view, the remarks expressed about Fig.6 can be transposed here; and also Fig.7 allows us to estimate the thickness of the zone of instabilities which seems to be about $\hat{\delta}$ ($\hat{\delta}$ = 1 mm). On the top of this area stays a turbulent flow; its spectral characteristics are equivalent to the one obtained outside of the disk in the turbulent zone (see Fig.6; S4). This spectral repartition is also found in the whole fluid layer as to 0.6 mm < d < 20 mm and R_r = 448.

b. Geometry G2

The experimental results of geometry G2 are in photo 5 and photo 6 for $p_2 = 9.10^{-3}$ and respectively at R_ℓ = 3.5 and R_ℓ = 4.

In both cases, the critical R_r is $(R_r)_c$ = 268. The here mentioned values are such that the height of fluid ℓ is lower than $\hat{\delta}$ (ℓ = 2 mm). Therefore it must be noted that the gap to G1 is quite large and that R_c keeps roughly the same value anyway. However the inclination of the observed spirals is very low, a lot lower than STUARTS'S spirals (11°), until they practically blend into circles.

For $p_2 \simeq 4.10^{-3}$ and $R_\ell \simeq 1$, we observed such like circle instabilities near the center of the disk when STUARTS'S type of instabilities were showing at the periphery of the disk : these two set being disjoined.

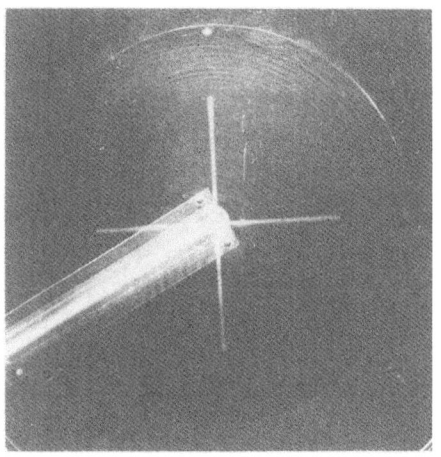 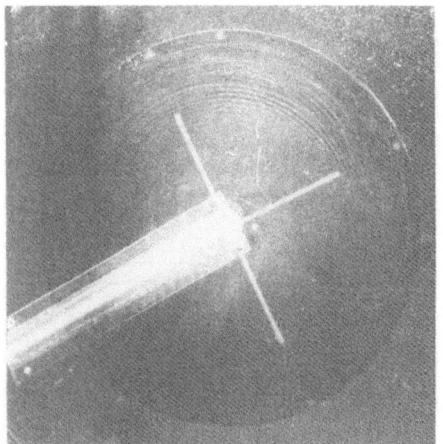

Photo 5   Photo 6

Hot film has been used to measure in a case of numerical degeneration (R_ℓ = 635, HOLODNIOK and al., 1981) the evolution of azimuthal velocity shown in Fig.8 for R_r = 150.

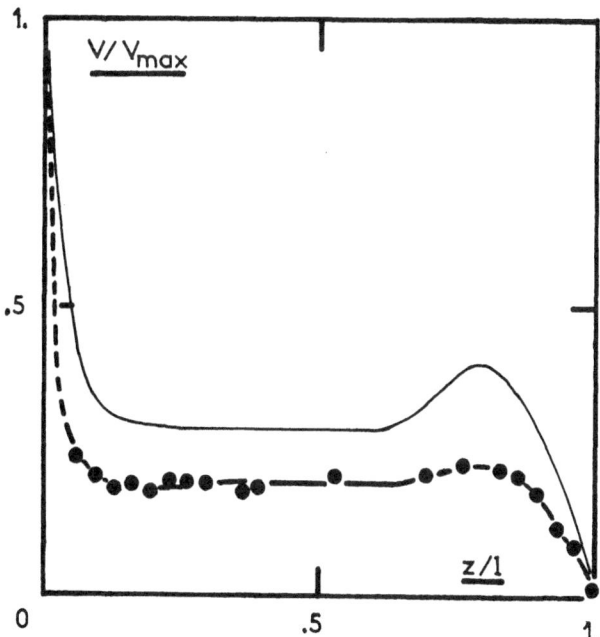

Figure 8 :   ● measured, ───── calculated (HOLODNIOK and al.)

Among numerical solutions reached by HOLODNIOK and al., it is possible to identify one of them in our experimental profile. That corresponds to a one-cell flow whose peculiar role, we have already dwelt on.

In these experimental conditions every attempt to bring out obvious accuracy of other solutions was unavailing. Note that only one cell flow could have been realised with other values of parameter p_2 and R_ℓ.

5. CONCLUSIONS

From this work it comes out that it was difficult to define in a more precise way the conditions allowing the analysis of stability in geometries G1 and G2 not without disturbing by the finited dimensions of experiment. So, it is necessary to sharpen our knowledge about possible flows even with no instabilities.

In other aspects, there is great difficulty to define the magnitude of the value regarding the acceptable geometrical errors (for example, how precise is the rotating disk in keeping the same plan) which allows the conclusions relating to the development of instabilities to carry a good enough general validity.

ACKNOWLEDGEMENT

The authors thanks   C.E.A. (D.E.S.I.C.P./D.G.I.) which studied and built the basic structure of the experimental set up.

REFERENCES

ADAMS, M.L., SZERI, A.Z., 1982, J. Appl. Mech. 49, 1.

BATCHELOR, G.K., 1951, Quart. J. Mech. Appl. Math. 4, 29.

COCHRAN, W.G., 1934, Proc. Camb. Phil. Soc. 30, 365.

COLES, D., 1965, J. Fluid Mech., 21, 385.

FLORENT, P., DINH, N.N., DINH, V.N., 1973, J. Mec., 12, 555.

GREGORY, N., STUART, J.T., WALKER, W.S., 1955, Phil., Trans. A248, 155.

HOLODNIOK, M., KUBICEK, M., HLAVACEK, V., 1981, J. Fluid Mech.,108,227.

KARMAN, T., Von, 1921, Z. angew. Math. Mech. 1, 233.

KOBAYASHI, R., KOHAMA, Y., TAKAMADATE, C., 1980, Acta Mech. 35, 71.

MALIK, R.M., WILKINSON, S.P., ORSZAG, S.A., 1981, A.I.A.A. 19, 1131.

MELLOR, G.L., CHAPPLE, P.J., STOKES, V.K., 1968, J. Fluid Mech., 31,95.

OLIVIERA, L., BOUSGARBIES, J.L., PECHEUX, J., 1982, C.R.A.S., 294 II, 1163.

ROBERTS, S.M., SHIPMAN, J.S., 1976, J. Fluid Mech., 73, 53.

RUELLE, D., TAKENS, F., 1971, Com. Math. Phys. 20, 167 et 23, 343.

SCHLICHTING, H., 1979, Boundary - Layer Theory, Mc G. H. Book Company.

MORPHOLOGICAL INSTABILITIES IN THE SOLIDIFICATION FRONT OF BINARY
MIXTURES

B. Caroli[*], C. Caroli, S. de Cheveigné, C. Guthmann, B. Roulet

Groupe de Physique des Solides de l'Ecole Normale Supérieure
Université Paris VII, Tour 23 - 2, place Jussieu - 75251 PARIS CEDEX 05

1. INTRODUCTION

The presence of a small concentration of dissolved impurities in materials with atomically rough solid-liquid interfaces has been observed
to cause cellular deformations of the solid-liquid interface under directional solidification conditions (i.e. a sample is drawn at constant
velocity V in a fixed temperature gradient G established around the
melting temperature T_M of the material, in such a manner as to progressively solidify the sample. See Fig. 1)[1].

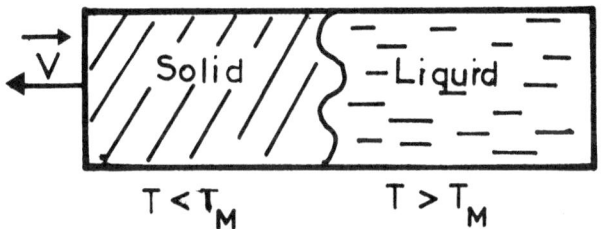

Figure 1 : Directional solidification : the
sample is pulled at constant velocity V in a
fixed temperature gradient established around
melting temperature T_M.

This is the case, for example, in many metals and the phenomenon has
been known to metallurgists - as a nuisance - for a long time[2]. The
physical interpretation of this phenomenon was given in 1964 by Mullins
and Sekerka[3], who were able to determine the conditions of appearance
of this instability. Their model has, since, been refined by several
authors[4,5]. In the past few years, morphological instabilities in
crystal growth have been the object of renewed interest in the context
of recent developments in the study of dynamic instabilities, as described in this book.

* UER Sciences Exactes et Naturelles - Université de Picardie
  33, rue Saint Leu - 80000 AMIENS

2. SOLIDIFICATION FRONT INSTABILITIES

Solidification at finite speed requires transport of heat and of solute away from (or towards) the solid-liquid interface :
- heat since the (first order) phase transition produces latent heat,
- solute since the equilibrium concentration at a given temperature is
not the same in the solid and the liquid (Fig. 2)(except in the case of
eutectic solidification, which gives rise to different types of morphology[1]).

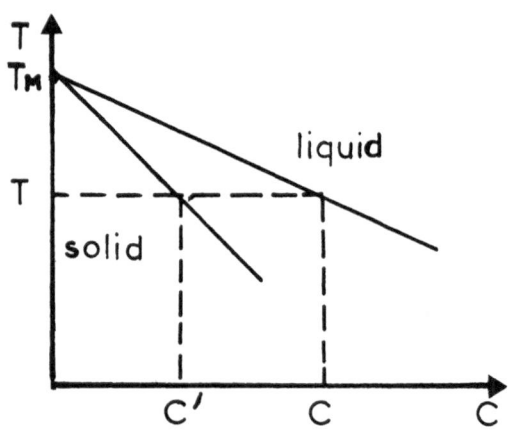

<u>Figure 2</u> : Low concentration part of the phase diagram
of a binary mixture.

Now, heat diffusion is far more rapid than concentration diffusion
which requires mass transport : heat diffusion coefficients vary typically between 10^{-3} cm^2/s in organic materials to 10^{-1} cm^2/s in metals,
while concentration diffusion coefficients in the liquid are of the order of 10^{-5} cm^2/s.

Quite obviously, the slower of the two competing dynamics is the
controlling one. One can thus assume that the latent heat is evacuated
instantaneously, so that the solidification dynamics are primarily governed by solute transport towards or away from the interface.

In directional solidification experiments (Fig. 1), at zero pulling
velocity, the external thermal conditions impose a planar solid front
and planar isotherms. This geometry is preserved at low pulling speeds.

In the case where the equilibrium concentration is higher in the liquid than in the solid (case represented in Fig. 2), the solute accumu-
lates ahead of the solidification front, forming a concentration gra-

dient. Solute diffusion tends to <u>destabilize</u> the planar front : if the solid bulges, the concentration gradient is increased in front of the bulge, and the solute current is enhanced. The solute excess being better evacuated, the growth of the bulge is favored.

This destabilizing effect of the concentration gradient is counter-balanced by the <u>stabilizing effects</u> of the surface tensions which oppo-ses any lengthening of the interface and of the heat gradient : the point is growing into a hotter region, where solidification will even-tually become impossible.

The net result is the appearance of a periodic cellular deformation of the interface when the constraint - for example pulling speed at fixed average solute concentration and fixed temperature gradient - reaches a certain threshold. The wavelength is typically in the 10-100μ range. There is an accumulation of the solute in the grooves between the cells. In thin samples the deformation is no doubt one-dimensional (Fig. 9). In the massive samples studied in metallurgical experiments, both hexagonal and elongated cells are observed[6].

It is to be noted that such effects can in principle occur, not only during solidification but also during fusion, or other phase transforma-tions with a concentration gap between phases[1,5].

3. QUALITATIVE DESCRIPTION OF THE THEORETICAL APPROACH

Various authors have studied the problem of one-dimensional deforma-tions of the solid-liquid interface during directional solidification. We shall only present here a schematic outline of the methods used, referring the reader to a review paper by Langer[1] and to references 3, 4, 5 and 7 for more details.

The dynamics of the system are described by :
(i) diffusion equations for heat and concentration, separately in each phase,
(ii) thermal and concentration boundary conditions at the ends of the (supposedly infinite) sample,
(iii) interface conditions (which play a central role) expressing :
- heat balance
- solute concentration balance (between the mass diffusion currents and the solute rejection rate)
- local thermodynamic equilibrium at the interface which entails conti-nuity of temperature, local concentration equilibrium and the Gibbs-Thomson expression for the temperature of the interface in the presence of solute and of an interface curvature (i.e. the interface is <u>not</u> at

T_M, the melting temperature of the pure material).

In this treatment, various assumptions are made :
- that heat diffusion is far more rapid than concentration diffusion;
- that the mixture is sufficiently dilute to allow the approximation
of the solidus and liquidus curves by their tangents at zero concentration (as in Fig. 2);
- that the densities of the solid and liquid phases are not very
different so that advection in the liquid can be neglected;
- that Rayleigh and solutal convections are negligible.

These assumptions can be improved upon (see for example ref. 5), at
least to some extent.

Much more crucial is the additional assumption of local thermodynamic equilibrium : it implies that the attachment kinetics of the atoms
or molecules joining the solid phase are extremely fast, which can only
be true if the interface is rough on an atomic scale. This restricts
the applicability of the model to metals and a few organic materials,
among which those which solidify as plastic crystals (the cohesive energy in the solid phase being very small, entropy effects are dominant,
which favors atomic roughness of the interface).

It is at the moment completely unclear how to approach theoretically
the same questions in the case of atomically smooth interfaces or
whether facetted cellular fronts exist.

The analysis of the above-described model proceeds in the following
way (which parallels, for example, the treatments of the Rayleigh-Benard
instability) :
  (1) the low pulling speed solution with a planar front is easily
calculated;
  (2) the next step is to study the linear stability of this solution
against a small harmonic deformation. The result is the bifurcation
curve shown on Fig. 3, which separates, in the space of the external
parameters (V (pulling speed) and G (external temperature gradient)) a
region where the planar front is stable against small deformations, and
a region, under the curve, where the front cannot be planar. The scale
of the curve is estimated for organic compounds. It depends on the material (fusion temperature, solid-liquid interface tension) and on
the characteristics of the mixture (average concentration, geometry of
the phase diagram).

These calculations also predict the value of the wavelength of the
first unstable mode.

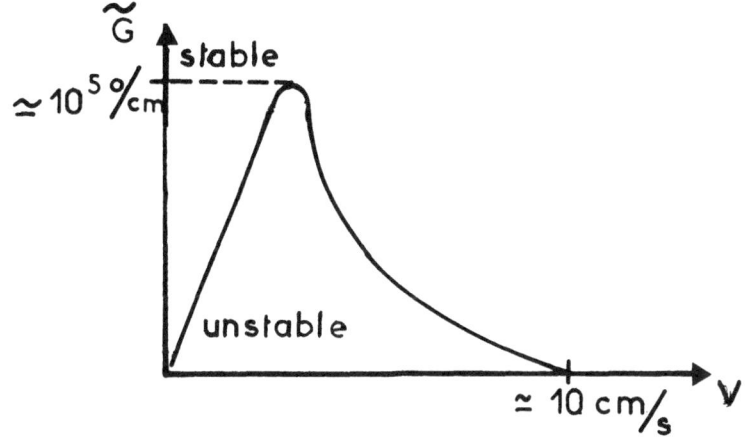

Figure 3 : Bifurcation curve : the plane of the external
parameters (temperature gradient G versus pulling speed
V) is divided into a region where the planar front is
stable against small deformations and one where it cannot
remain planar.

(3) In a second stage, a non-linear expansion of the equations close
to the bifurcation curve allows a prediction about the nature of the
bifurcation. It can be "normal", that is, the out of equilibrium ana-
logue of a second-order phase transition, corresponding to the appearance of a cellular deformation with an amplitude growing continuously
from zero at the bifurcation point. It can also, depending on the values of the various parameters, be of the "sub-critical" type, i.e.
analogous to a first-order transition. If such is the case, the above
mentioned expansions are not, in general, sufficient to predict the
nature of the new configuration of the front.

For example, it is found that, in the vicinity of the origin in
Fig. 3 (i.e. in the small gradient case), the bifurcation can be normal
only in the case of a thin sample, such that heat is essentially trans-
ported by the sample-holding plates. If not, the bifurcation should be
subcritical, that is, one should observe a hysteresis between the
thresholds at increasing and decreasing pulling speeds.

Far from the bifurcation curve, that is in the greater part of the
unstable region of Fig. 3, the behavior cannot be predicted by such
methods.

Let us finally mention that numerical studies of this problem, the
validity of which in principle extends farther from the bifurcation,
have been undertaken recently[9].

4. EXPERIMENTAL STUDIES OF DIRECTIONAL SOLIDIFICATION

As we said, the appearance of front instabilities during the directional solidification of impure metals and dilute alloys has often been observed[2]. But due to the opacity of these materials the samples have to be annealed, then cut for study. A transparent sample, on the other hand, allows observation of the interface <u>during</u> solidification. In most cases transparent organic materials or inorganic salts present facetted growth, but there are a few exceptions where an atomically rough interface allows the appearance of rounded cellular structures. One such case is tetrabromomethane, CBr_4, suggested by Jackson[10]. He observed the type of cellular deformations shown in Fig. 10 but no systematic study of the stability conditions of the interface has, to our knowledge, been undertaken.

The experimental setup is shown in Figure 4. The samples are made of two 20x60 mm x 150 μ glass microscope slides, sealed together with a silicone elastomer so as to form a cell of calibrated thickness. The cells are filled by capillarity with the material in the liquid phase.

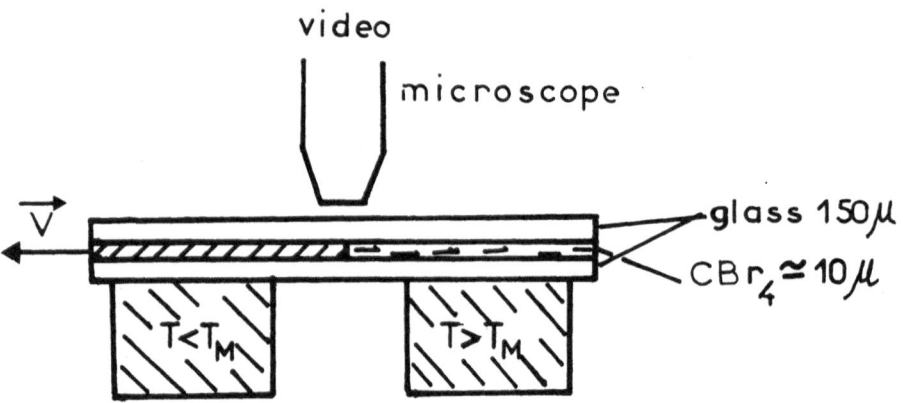

Figure 4 : The experimental setup.

The sample is pressed onto two copper blocks at temperatures respectively above and below the melting temperature of the material to be studied. In the cell, between the blocks, the gradient is constant. The sample is then pulled at fixed speed V across the blocks, and the solid-liquid interface is observed with an optical microscope.

Figures 5 to 10 show the aspect of the interface at various pulling speeds.

Figures 5 to 10 : Morphology of the solidification front at various pulling speeds. Sample thickness : 10 μ. Temperature gradient : 75°/cm.

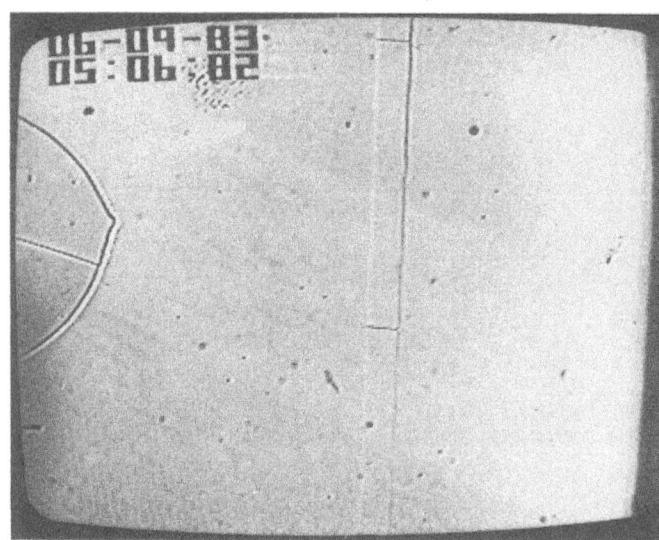

Figure 5 :
V = 6 μ/s
(the bubble to the right should be ignored)

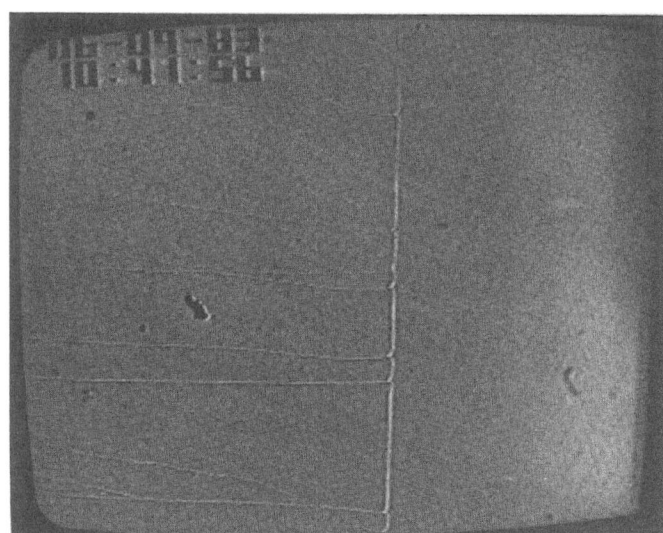

Figure 6 :
V = 15 μ/s

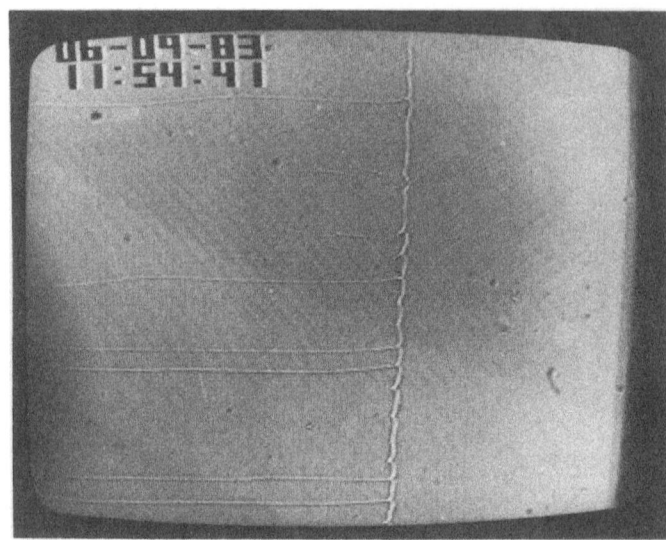

Figure 7 :
V = 16 μ/s

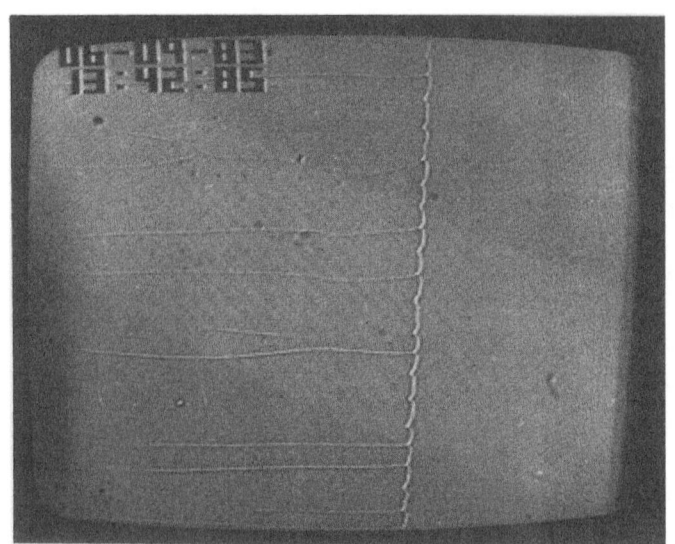

Figure 8 :
V = 17.5 μ/s

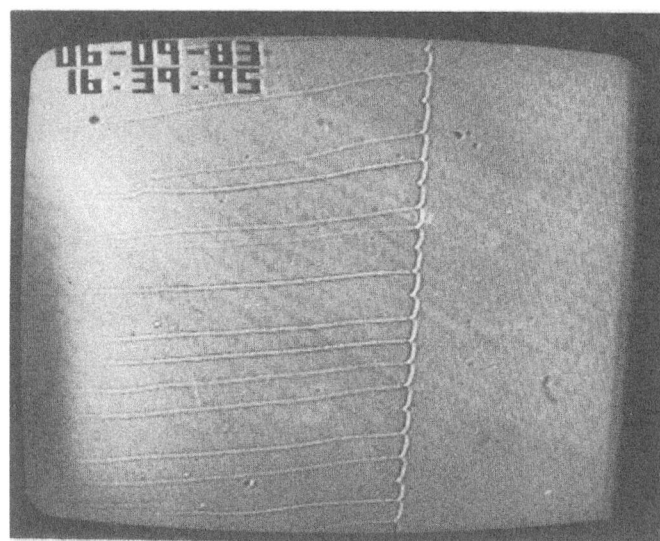

Figure 9 :
V = 27 µ/s

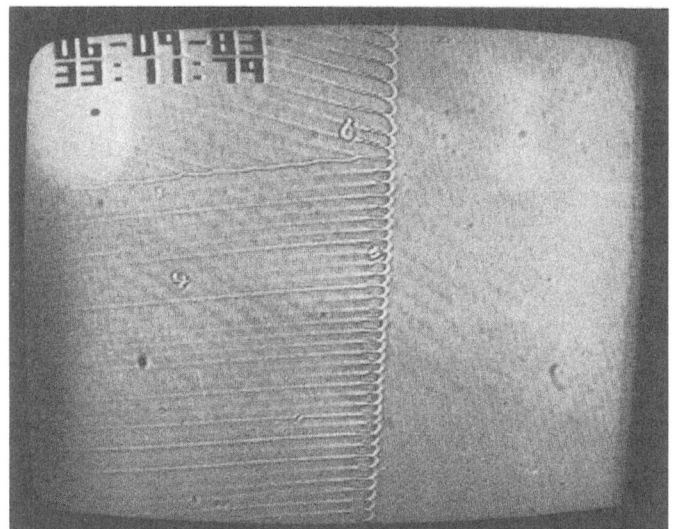

Figure 10 :
V = 150 µ/s

To explore the marginal stability curve, the pulling speed was increased in stages, at different fixed temperature gradients. Below the
threshold speed, transient deformations of the interface are observed.
The threshold is defined as the pulling speed at which the deformation
becomes permanent (at the observation time scale of a few minutes). The
results are shown on Fig. 11 for a 25 μ thick sample, containing roughly 1% impurities (checks were made to ensure that the results were not
affected by any decomposition of the material during the experiment).

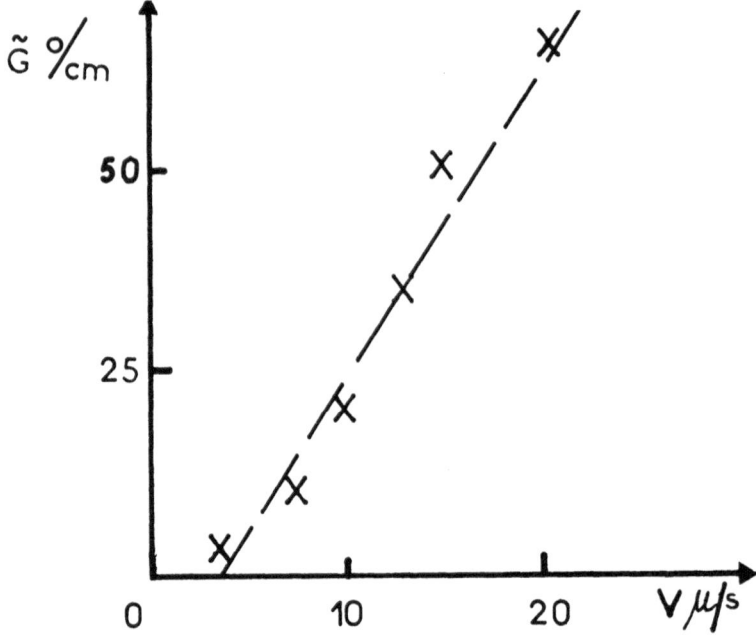

Figure 11 : Experimental stability curve.

It must be noted that only a small region close to the origin of the
stability curve of Fig. 3 can be explored experimentally. The slope at
origin is expressed by[6] :

$$\frac{G}{V} = \frac{(K_S/K_L) + 1}{2} \frac{C_\infty}{\alpha} \frac{T_M}{D_L}$$

where K_S and K_L are the thermal conductivities in the liquid and in the
solid $(K_S/K_L \simeq 2)$;

C_∞ is the solute concentration in the liquid, far ahead of the
front (here, $C_\infty \simeq 1\%$) ;

T_M is the melting temperature $(\simeq 360$ K) ;

D_L is the solute diffusion coefficient in the liquid (of the order of 10^{-5} cm^2/s) ;

α is a function of the slopes of the solidus and liquidus curves and of T_M (usually of the order of 10^2 - 10^3 with concentrations expressed in atomic %).

Note the uncertitude, particularly concerning D_L and α : unfortunately, CBr_4, pure and in mixtures, is not sufficiently well characterized to allow better than a qualitative estimation of the slope : $\frac{G}{V} \approx 5 \; 10^4$- $5 \; 10^5 \; ^\circ s^{-1} \; cm^{-2}$. Experimentally, the slope is found to be $3.8 \; 10^4 \; ^\circ s^{-1} \; cm^{-2}$, a satisfactory agreement, considering the imprecision of the above estimate.

Once the cellular pattern is well established, one can determine a wavelength of the order of 30 μ (the theoretical prediction for the wavelength of the first unstable mode is, with the same lack of precision as above, of about 60 μ[4]). But the problem of the selection of the wavenumber rapidly appears : in Fig. 10 for example, two wavelengths (20 and 30 μ) coexist.

Various phenomena warrant further study. At higher pulling speeds one often finds that the cells grow at an angle (Fig. 10). This is probably because attachment kinetics can no longer be neglected : some faces grow slightly faster than others. Then, at about ten times the threshold speed, the growth begins to appear dendritic.

Besides the grooves between cells, one also observes "lines" in the solid phase (Fig. 6 for example), reminiscent of grain boundaries, but this interpretation has yet to be confirmed.

CONCLUSION

To conclude with respect to the present experiments, it is clear that much remains to be done to gain a better understanding of the various phenomena observed. Such studies will certainly provide a wealth of information on the dynamics of morphologic instabilities.

More generally, in the context of this book, we should like to point out some parallels which can be drawn between the convective instability and the solid-liquid interface one described here. First, the basic physical phenomena are similar : a system, pushed far from equilibrium (by an inverse temperature gradient in the Rayleigh-Benard problem, by an accumulation of solute in the present case) produces a macroscopically ordered, space-periodic response improving the corresponding transport : convection to carry heat, solute evacuation in the grooves between the cells observed here.

The theoretical treatments use, in both cases, the same methods of analysis of dynamic instabilities : linear stability to find instability thresholds, perturbation expansions in the vicinity of bifurcations. The amplitude equations valid close to the thresholds are identical (more exactly the equation for the cellular problem is the same as that of the rigid/free boundary or as the non-Boussinesq Rayleigh problems) except for the absence in solidification of the complementary equation for the motion of vertical vorticity. The difficulties and open problems are, consequently the same : wavenumber selection, stability of the structures, etc...

Less well studied up to now than hydrodynamic instabilities, but providing an easily observable example of out-of-equilibrium structures and, in thin samples, a good approximation of a two dimensional system, solidification front instabilities will no doubt prove complementary in bringing further information about these general questions.

REFERENCES

1. J.S. Langer, Rev. Mod. Phys. 52 (1980) 1, and references therein.
2. J. Friedel, Dislocations, chapter VII, Pergamon Press, Oxford (1964)
3. W.W. Mullins, R.F. Sekerka, J. Appl. Phys. 35 (1964) 444
4. S.R. Coriell, R.F. Sekerka, J. Cryst. Growth 34 (1976) 157-163, and references therein
5. D.T.J. Hurle, E. Jakeman,A.A. Wheeler, J. Cryst. Growth 58 (1982) 163-179 and references therein
6. B. Caroli, C. Caroli, B. Roulet, J. Physique 43 (1982) 1767-1780
7. See for example : L.R. Morris, W.C. Winegard, J. Cryst. Growth 5 (1969) 361-375
8. D.J. Wollkind, L.A. Segel, Phil. Trans. Roy. Soc. 268 (1970) 351-380
9. M. Kerzberg, Phys. Rev. B28 (1983) 247
10. K.A. Jackson, in Crystal Growth : A Tutorial Approach, ed. W. Bardsley, D.T.J. Hurle, J.B. Mullins, North Holland, Amsterdam (1979).

CELLULAR STRUCTURES ON PREMIXED FLAMES IN A UNIFORM LAMINAR FLOW

J. Quinard, G. Searby and L. Boyer

Département de Combustion
Laboratoire de Dynamique et Thermophysique des Fluides
Université de Provence, centre St Jérôme
13397 MARSEILLE CEDEX 13

1. INTRODUCTION

Combustion is a complex phenomenon which involves a multistep chemical reaction with diffusive and convective transport of heat and mass and is generally associated with a strongly turbulent flow. In many situations, this leads to the appearance of large organized structures, first recognized by Smithells [1] and which were not well understood even in the case of a laminar flow.

We present here experimental results concerned with the stability limits of a premixed flame in a laminar and uniform flow. In this case, the flame is a subsonic chemical wave, propagating in a meta-stable mixture, and controlled by the diffusive transport mechanism of mass and energy. The chemical reaction involves many elementary steps (typically 100) and various species, but a global description can be obtained as an overall exothermal reaction :

$$\nu_F \, F + \nu_O \, O \rightarrow \quad \text{Products} + Q$$

where ν_F and ν_O are the stochiometric coefficients of the fuel and the oxydant respectively and Q is the heat released. The reaction rate τ_r^{-1} is given by an Arrhenius law with a very large activation energy E :

$$\frac{1}{\tau_r} \approx \frac{\nu M}{Y_u} \, Y_F^{n_F} \, Y_O^{n_O} \, \exp(-\frac{E}{kT_b}) \tag{1}$$

h is Boltzman's constant

Y_u, ν, M are respectively the initial mass fractions, the stochiometric coefficient and the molecular mass of the limiting component whose concentration controls the reaction ; the exponents n_O and n_F are generally different from the stochiometric coefficients because of the multistep reaction and T_b is the temperature of combustion which is determined by the overall energy balance.

$$C_p(T_b - T_o) = \frac{Y_u Q}{\nu M}$$

T_o is the temperature of the fresh mixture, C_p a mean value of the
specific heat. The reaction rate is thus a strongly increasing function of the temperature, producing a selfacceleration of the combustion process which is finally saturated by the consumption of the
reactants. During the time τ_r, the heat released $\frac{Y_u Q}{\nu M}$ diffuses on a
length $d = \sqrt{D_{th}\tau_r}$, where D_{th} is the thermal diffusivity of the mixture
and thus the flame front propagates with a velocity

$$V_f = \sqrt{D_{th}/\tau_r}$$

Due to the large activation energy

$$\beta = \frac{E}{kT_b^2}(T_b - T_u) >> 1$$

The reactions take place in a very thin layer(fig.1) of thickness d/β,
d being the thickness of the preheated zone, $d = \sqrt{D_{th}\tau_r}$. It is thus
possible to obtain a description of the structure of a planar flame
using an asymptotic expansion in which $\beta \rightarrow \infty$ /2/. However a description
of the dynamical properties of a wrinkled flame is less simple because
hydrodynamics and diffusion are coupled by the modification of the gas
flow velocity through the flame thickness.

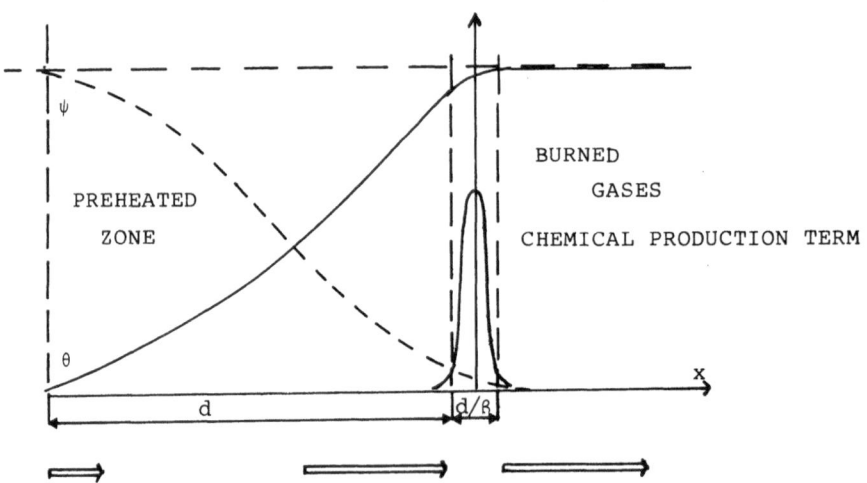

Fig.1 - Structure of the flame front
$\Longrightarrow$ Velocity vector of gas in the frame of reference of the flame
front.
$$\theta = \frac{T-T_u}{T_b-T_u} \quad : \text{ reduced temperature}$$
ψ = reduced concentration of reactants;

2. INSTABILITY MECHANISM

For a planar flame front propagating downwards the onset of instability is revealed by the appearance of structures which are stationary at the threshold and which become self turbulizing as the system
becomes more unstable. Fundamentally, there are principally three distinct types of phenomena that govern the intrinsic stability properties
of premixed flames : hydrodynamic effects, diffusive transport effects
and buoyancy effects.

a. Hydrodynamic effects

When buoyancy effects are neglected and if the entire flame is
simply considered as a surface of discontinuity in density and temperature propagating locally normal to itself at a constant speed V_f with
respect to the upstream gas, then the continuity of the tangential
component of the gas velocity across a tilted flame element requires
the streamlines to be deflected towards the local normal to the surface.
This causes a tube of streamlines to be expanded at the front if the
surface is convex toward the unburnt gas (see figure 2) and thus the
flame encounters a locally decreased fluid velocity and tends to move
even farther upstream. This is a strong destabilizing mechanism whose
dispersion relation is given by /3/ :

$$(2-\gamma)\sigma^2 + 2V_f k\sigma - V_f^2 k^2 /\gamma(1-\gamma)=0$$

where $\gamma = \dfrac{T_b-T_o}{T_b}$ is the expansion ratio, σ the amplification rate and k
the wavenumber of the perturbation written in the normal mode decom-
position

$$a = Ae^{\sigma t} e^{iky}$$

This yields $\sigma = \sigma_1(\gamma)V_f k$ where σ_1 is a positive scalar. It can be seen
that the growth rate is proportionnal to the wavenumber of the dis-
turbance.

b. Diffusive thermal instability

Just ahead of a wrinkled flame front, local gradients of temperature and of concentration induce transverse diffusion of heat and
mass. Through relation (1), it is clear that this will modify the
local flame speed. A bulge directed towards the burnt gases is fed
with a hotter and leaner mixture of fresh gases than a bulge directed
toward the unburnt gases, see fig.2. The effects of transverse heat
fluxes is to modify the local flame velocity in such a way that an
initial perturbation will be flattened out. In other words transverse
heat fluxes are stabilizing, whereas the opposite is true for mass

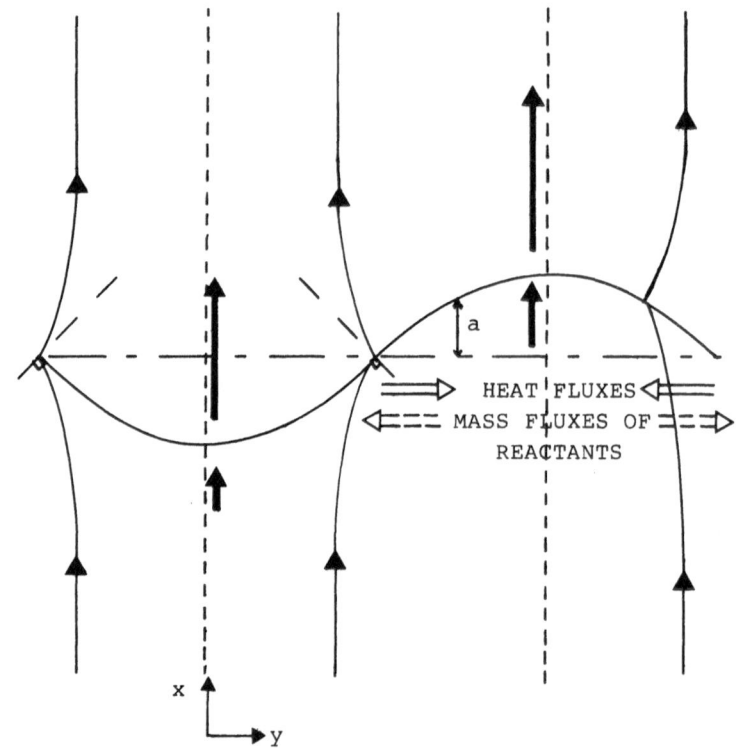

Fig.2 - Flowfield and fluxes for a wrinkled front
        a : deviation from the mean position of the front
        ➡ velocity of gas, ➡ Streamlines

diffusion of the reacting species. The net effect of the diffusion processes will depend on the Lewis number Le which is the ratio of the
thermal diffusivity coefficient to the molecular diffusion coefficient
of the species controlling the reaction rate :

$$Le = \frac{D_{th}}{D_{mol}}$$

The strength of this effect is proportional to the mean radius of curvature of the front and in the absence of hydrodynamical effects the
local equation for the evolution of the front reduces to the diffusion equation.

$$\frac{\partial a}{\partial t} = V_f \mathcal{L} \nabla^2 a \qquad (2)$$

where a represents the local position of the front relative to its
mean position. $\mathcal{L}$ is called the Markstein length and depends on the
diffusive properties of the reacting mixture

$$\frac{\mathcal{L}}{d} = 1 + \frac{\beta}{2} \left(1 - \frac{1}{Le}\right)$$

From (2), it is seen that diffusion will modify the dispersion relation by adding a k^2 term which will stabilize the front at sufficiently short wavelengths.

c. Buoyancy forces

When the flame is propagating downwards, the light burnt gases are located above the unburnt gases. Such an interface is known to sustain gravity waves with a dispersion relation given by $\sigma^2 \alpha -gk$ /4/, g is the acceleration of the gravity. When coupled with the hydrodynamical effects, it turns out that gravity is a stabilizing mechanism with constant strength and is the only mechanism which is effective at large wavelengths ($k \longrightarrow 0$) and the relative strength increases at low flame velocities. Taking advantage of the difference of scale between the flame thickness d and the wavelength Λ of the wrinkles of the front (typically 1cm) an analytical study of the front dynamics of premixed flames has been recently carried out including the coupling of these three fundamental effects /5/. It leads to a rather lengthly dispersion relation which fortunately reduces to a simpler expression for the marginal stability condition of usual hydrocarbons flames :

$$\frac{\gamma}{1-\gamma} k \{ \frac{1}{2} k_c - k(\delta - \frac{k}{kn})\} > 0$$

$$\text{gravity} \quad \text{hydrodynamics} \quad \text{diffusion}$$

with

$$k_c = 2(1-\gamma)gd/v_f^2$$

$$\delta = 1 + \frac{gd}{v_f^2} (1-\gamma)(\frac{\mathscr{L}}{d} - \frac{1}{\gamma} \ln (\frac{1}{1-\gamma}))$$

$$k_n = \{1 + \frac{2+\gamma}{\gamma} \frac{\mathscr{L}}{d} - \frac{2}{\gamma} \ln(\frac{1}{1-\gamma})\}^{-1}$$

The hydrodynamical instability is stabilized by the gravity effect for small wavenumbers whilst perturbations at large wavenumber are stabilized by diffusive effects which are almost always stabilizing when advective cross transport ahead of the flame front is also taken into account. For a given stable condition, the instability threshold can be accessed either by increasing the flame velocity, which increases the strength of the hydrodynamic effect, or by decreasing the effective Markstein length via a modification of the diffusive properties of the gas mixture. However, a main result of this analysis is that the instability appears with a critical length Λ_c which is related only to the expansion ratio and the critical flame velocity /5/ :

$$\Lambda_c = \frac{\pi v_f^2}{(1-\gamma)g} \tag{3}$$

3. APPARATUS

The burner is presented schematically on figure 3. The experimental conditions are as close as possible to the ideal situation of a uniform and laminar flow field on both sides of a stable planar flame front. The mass flow of each gas (fuel, air , nitrogen) is regulated by two cascaded pressure regulators followed by a sonic throat. The residual fluctuations are less than 1‰ . The gases are then mixed and injected at the bottom of the burner (for more details, see /6/). The flame is dynamically stabilized by adjusting the mean flow velocity /7/.

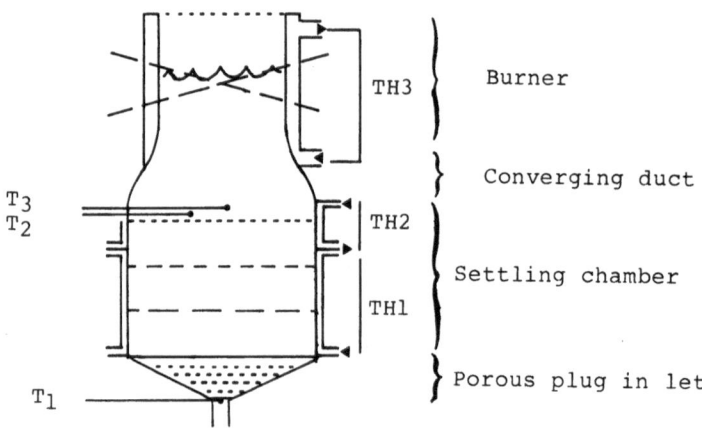

Fig.3 - Schematic view of the burner
T_1, T_2, T_3 : thermocouples
TH1, TH2, TH3 : Thermostated water flows

This experimental study is principally concerned in checking relation (3) which requires the measurement with non-perturbative methods of the normal flame velocity, the expansion ratio (obtained through the measurement of the burnt gas velocity as the flow may be considered to be incompressible at low Mach numbers) and the characteristic length of the structures. The first two quantities are measured by laser Doppler anemometry (the fresh mixture is seeded with oil droplets or alumina particles which have no influence on the dynamical properties of the flames presently studied).

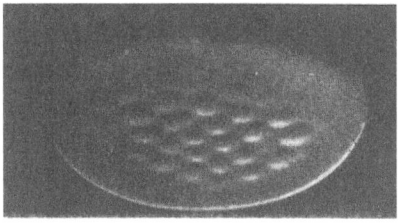 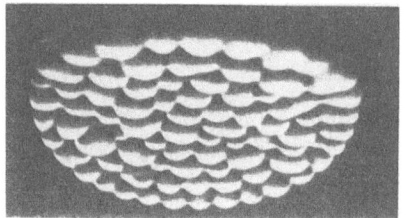

a   b

Fig.4 - Flame front structures viewed from below at different relative
distances from threshold.
a. V_f= 8.7 Λ_C= 1.1cm - $\dfrac{a}{\Lambda_C}$ = 5%
b. Λ_C= 1.1cm Λ_C= 1.1cm - $\dfrac{a}{\Lambda_C}$ = 30%

Concerning the cell size, tomographic methods /8/ are not suitable
here because of the small amplitude of the structures at the threshold
and because a single cut of the front is not sufficient to obtain the
characteristic length. However, classical high contrast photography
allows us to detect flame shape deformation with a height of about 5%
of their wavelength (fig.4). Such views are analyzed to give the mean
characteristic length for a given composition, and experience suggests
that a good estimation of the critical wavelength is obtained as Λ_C
remains relatively constant when we increase the ratio of cell amplitude to cell size from 5% to 50% (see figure 5).

4. RESULTS

These experiments were performed with two hydrocarbon fuels :
propane (C_3H_8) and methane (CH_4), which have different diffusive
properties in the mixture. To attain the low flame speed ($\simeq$ 10cm/s)
where the instability appears, a dilutant (N_2) is needed to decrease
the flame temperature T_b. The thermal diffusivity of the mixture is
then approximatly that of nitrogen.
For a given dilution $D = \dfrac{O_2}{O_2+N_2}$ (in volume), the flame speed follows
a parabolic curve when the equivalence ratio grows from values smaller
than 1 to higher values due to the joint effect of the variations of
the heat of combustion and of the species concentrations (fig.6 and 7).
Below 6cm/s, the flame front is quenched by heat losses to the walls.
The stability limit is determined when the flame front shows the first
perceptible signs of structure, but the cell size is measured with
sufficient precision only when the flame speed is 1 or 2% larger than

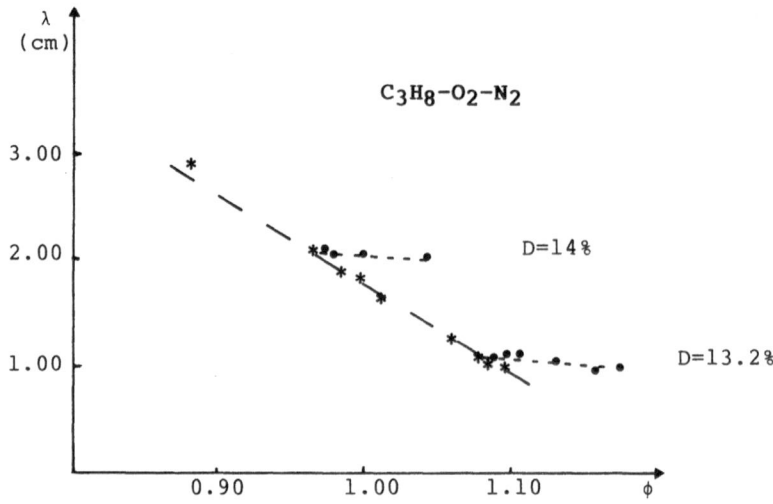

Fig.5 - Evolution of the mean wavelength with distance from threshold
for two dilution D of the mixture.
ϕ =equivalence ration ; * Critical wavelength ;
● mean wavelength

its critical value at the threshold.

For propane flames, the stability limit, see fig.6 is relatively
well explained by considering the competition between diffusive and
hydrodynamic effects : when $\phi<1$, the limiting component is C_3H_8 with
a Lewis number of about 1.7 ; Diffusion of heat is thus predominant
and stable fronts can be obtained with velocities up to 12 cm/s.

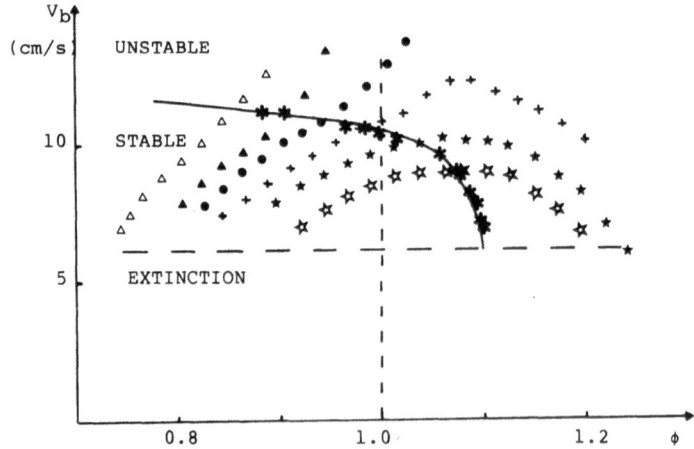

Fig.6 - Flame velocity of $C_3H_8/O_2/N_2$ mixtures.
Dilutions are : △ =15.4% ; ▲ = 14.8% ; ● = 14.4% ;
+ = 14% ; ★ = 13.6% ; ✿ = 13.3%
※ marginal stability limit ; ϕ = equivalence ratio.

As ϕ is increased, the effective Lewis number decreases towards a
value of 0.89 : here, the front is less stabilized by diffusive
effects leading to lower critical velocities at the threshold. Theoretically, it is anticipated that with this value of the Lewis number,
stable flames should be observed with flame speed smaller than 5cm/s
/9/. Unfortunatly, extinction phenomena hide this occurrence.

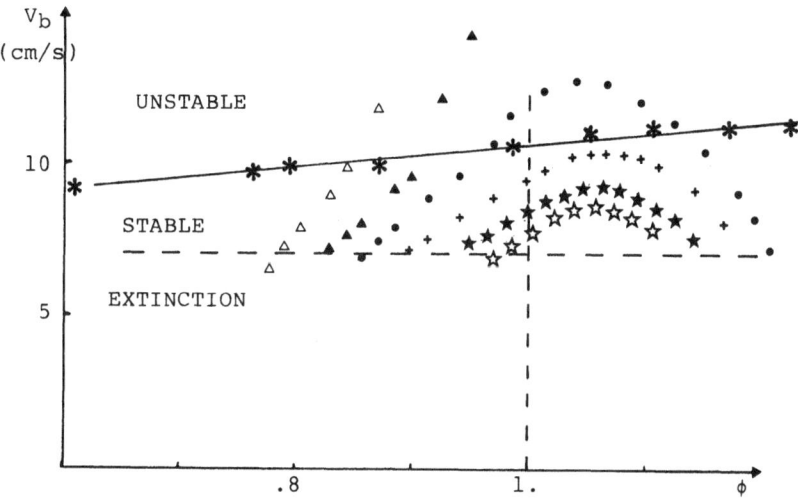

Fig.7 - Same as fig.6 for $CH_4/O_2/N_2$ mixtures
Dilution are : $\triangle$ = 16% ; $\blacktriangle$ = 15.4% ; $\bullet$ = 14.8% ;
+ = 14.4% ; $\bigstar$ = 14% ; $\maltese$ = 13.9%

For methane flames, see figure 7, the Lewis number should take limiting values of .85 and .89 for $\phi\ll1$ and $\phi\gg1$ respectively : this may
explain the small change of the critical velocity as a function of the
equivalence ratio. However, on the rich side ($\phi>1$) where O_2 is the
limiting component the critical velocity is not the same for CH_4 and
C_3H_8 flames. This points out the difficulty in evaluating an effective Lewis number considering only the diffusivities of the initial
components. In reality the chemical reactions involve may intermediate
species with different diffusion coefficients and which many drastically affect the effective Lewis number as the equivalence ratio is
changed or when different fuels are compared.

Figure 8 presents the critical wavelength as a function of the
critical velocity for both fuels. The theoretical prediction is given
by relation 3 with the measured value of the expansion ratio (γ remains relatively constant in the experimental range).

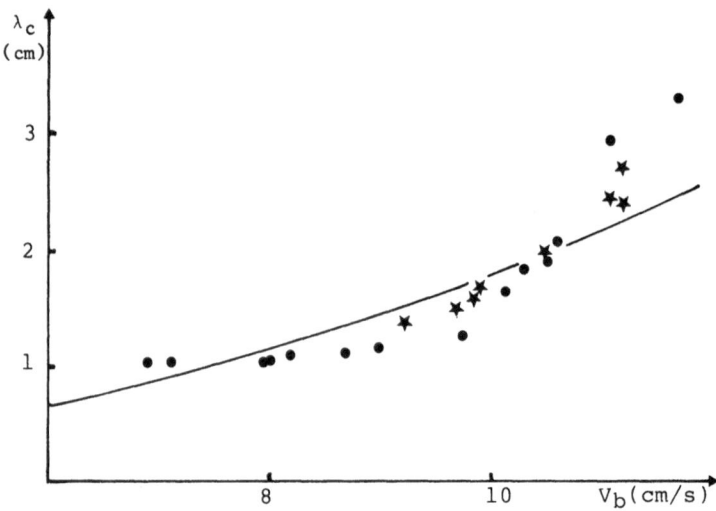

Fig.8 - Critical wavelength as a function of the critical flame
speed with $C_3H_8/O_2/N_2$ mixtures (●) and $CH_4/O_2/N_2$ mixtures(✷)
The full line is the theoretical prediction. Expression 3.

The order of magnitude of experiment and theory is quite comparable and the cell size increases effectively with the critical flame
velocity, but the parabolic dependence is not well verified. More
realistic analyses have been performed including heat loss effects /9/
or temperature dependent diffusivities /10/, but they are not likely
to change significantly the critical wavelength.

However chemical kinetic effects could in fact be responsible for
the quantitative change in behavior of the experimental curve on fig.8
which coincides with the intersection of the marginal stability curve
and the locus of the maximum flame speed for various isodilutions.
Two other phenomena may also affect the critical wavelength :

- The wavenumber increases as the flame speed decreases, and
stabilizing higher order effects would contribute towards a larger
cell size for low critical velocities.

- At large velocities, the cell size is no longer very small
compared to the burner size (∅= 11cm) and boundary effects may influence the critical cell size

5. CONCLUSION

These results show a good qualitative agreement with the model of
a threefold instability mechanism. The marginal stability is crossed
when the strength of the hydrodynamic instability is increased

following an increase in the normal flame speed. The critical velocity is lowered by decreasing the Lewis number due to the destabilizing influence of transverse mass fluxes. As expected, the critical wavelength is principally related to the critical flame speed and does not vary significantly from one fuel to another. This analysis has also led to a good estimation of the behavior of stable flame fronts in a weakly turbulent flow /11/.

However, the quantitative discrepancy evident in figure 8 may result from an over-simplification of the chemical kinetic scheme in the model (overall one-step reaction), from finite size effects in the burner or even from the neglect of non linear effects which appear as soon as the threshold is crossed /12/.

REFERENCES

1 - A. Smithells and H. Ingle, "The structure and chemistry of flames"; J. Chem. Soc. 61, 204 (1892).

2 - W.B. Bush, F.E. Fendell, "Asymptotic analysis of laminar flame propagation for general Lewis number", Combustion Sci. and Techn., 1, 421-428 (1970)

3 - L. Landau, "On the theory of slow combustion". Acta Physico-chimica URSS, 19, 77-85 (1944)

4 - L. Landau, E. Lifchitz, "Mécanique des Fluides", ed.Mir, Moscou (1971).

5 - P. Pelcé, P. Clavin, "Influence of hydrodynamics and diffusion upon the stability limits of laminar premixed flames", J. Fluid Mech., 124, 218-237 (1982)

6 - J. Quinard, G. Searby, L. Boyer, "The stability limits and critical size of structures in premixed flames". To appear in the proceedings of the 9th ICODERS, Poitiers (1983)

7 - L. Boyer, G. Searby, J. Quinard, "Some recent applications of Laser Tomography in combustion". Proceedings of the XV ICHMT Symposium on Heat and Mass Transfer Measurement Techniques. Dubrovnik (1983) to appear.

8 - L. Boyer, "Laser tomographic method for flame front movement studies", Combustion and Flame, 39, 321-323 (1980)

9 - P. Clavin, C. Nicoli, "Effect of the heat-losses on the limits of stability of premixed flames propagating downwards" to appear in Comb. and Flame.

10 - P. Clavin, P. Garcia, "The influence of the temperature dependence of diffusivities on the dynamics of flame fronts ; J. Mécanique Théo. et Appl. 2, 245-263 (1983)

11 - G. Searby, F. Sabathier, P. Clavin, L. Boyer, "Hydrodynamical coupling between the motion of a flame front and the upstream gas flow" Phys. Rev. Lett. 51, 1450-1453 (1983)

12 - P. Pelcé, "Influence du champ de pesanteur sur la propagation d'un front de flamme courbé dans un tube", Colloque de Mécanique appliquée, Ecole INRIA - CEA - EDF, Session combustion (18-21 oct. 83).

ENERGY IN THE BENARD-RAYLEIGH PROBLEM

M. Zamora, A. Rey de Luna

Departamento de Termología
Universidad de Sevilla (Spain)

1. INTRODUCTION

The classical experimental methods used to investigate the properties
of dissipative structures in fluids {1} do not include the measurement
of an essential quantity in physics, namely energy. The same situation
is observed in theoretical and numerical treatments of the problem {2}.
This paper presents a new experimental method to determine the formation
energy of dissipative structures, referred to the conductive state, in
the Benard-Rayleigh problem. This new method is based on the energy flux
balance and uses the heat flux measuring techniques developed by the
authors {3,4}.

It is generally accepted that every steady state in a thermodynamic
system has the energy content, U, {5,6}, that is an extrapolation of the
concept of internal energy in equilibrium. As with the latter the former
does not have a natural origin and, therefore, it can be measured only
relatively, i. e. the difference of energy between an initial steady
state (i) and a final steady state (f) can be measured. In the absence
of external work, we can express for the Benard- Rayleigh problem

$$U(f) - U(i) = \int_i^f \left(\oint_\Omega \vec{q} . d\vec{S} \right) dt$$

where $\vec{q}$ is the heat flux vector, Ω is the external surface of fluid and
t is the time. In our case, the initial state of fluid can be the conduc
tive state (the fluid is heated from above) and the final state can be
the convective state (fluid heated from below), or viceversa. Both
states support the same difference of temperature and have the same mean
temperature. The increase in energy in the process corresponds exclusi-
vely to the formation of convective cells, and so we shall call it the
structural energy or formation energy of the convective structures.

There are several reasons for attaching special importance to the
structural energy and among them the following:

a. It will permit a more effective thermodynamic approximation to a
   system far from equilibrium, as is the Benard-Rayleigh problem.

b. It represents a new physical parameter to detect the structural exchanges in convective cells, perhaps more sensitive than the Nusselt number and others, in superior transitions.

c. The continuous measurements of heat fluxes give a thermograph that provides information about the dynamics of cell formation.

d. In the oscillatory states, prior to chaotic turbulence, it is possible to know both the mean energy of structures and the characteristic energy of each oscillator, thus making it possible to study the energetic interactions between different oscillators.

e. In chaotic turbulence, the energy can be another quantitive parameter of state.

2. EXPERIMENTAL

The experimental device is shown schematically in Fig. 1, {7}. It is formed by two parts symmetrical to the horizontal axis FF´, around which the whole system can turn. Each part is formed by two isothermal pieces, A and A´, whose temperatures are regulated by thermostats, two heat flux meters, B, formed by 133 thermocouples of chromel-constantan, the sample, E_1 or E_2, and a lateral guard ring, C, to prevent horizontal heat losses, which is separated from the sample and the heat fluxmeters by an air chamber, D.

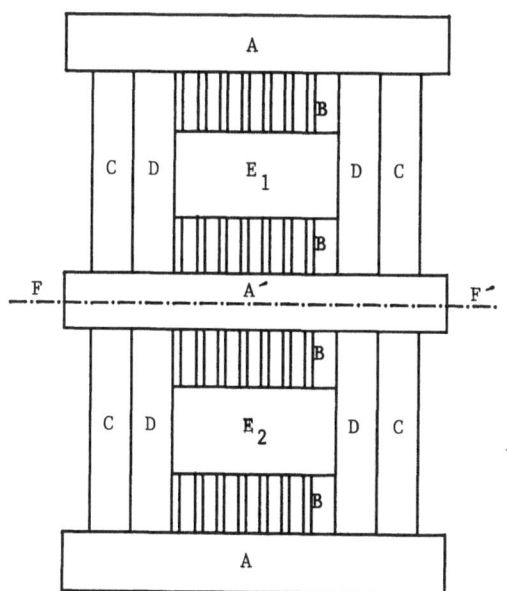

Fig. 1. Schematic representation of experimental apparatus.

The thermal connection is made by regulating the external pieces symmetrically with the same thermostat, at a temperature lower than that of the central piece. Then, if both samples contain fluid and the temperature difference, ΔT, applied is higher than the critical temperature, the upper fluid sample will be in convection and the lower one will transfer heat only by conduction. Turning the system 180° around the axis FF´ involves destroying the structures in E_1 and forming new structures in E_2. The study of the continuous response of heat fluxmeters during the process gives the desired information about the energy.

Each heat fluxmeter gives the measurement of the heat flux transferred by itself, thus the heat fluxes that enter and depart from the sample are known. The difference gives the lateral losses. If the thermal characteristics of the fluid and its box are known, the heat flux transferred vertically by the air chamber can be obtained.

Connecting the heat fluxmeters of each sample electrically in opposition, the difference in heat flux can be measured directly in function of time. Both measurements will be disturbed by the oscillation of temperature of each thermostat which acts on the reference junctions of the fluxmeters. To avoid this effect both setups are connected in opposition too, thus removing the influence of the external baths. The heat flux exchange in both samples

$$W = (q_2 - q_1) - (q_3 - q_4)$$

is measured directly and simultaneously. To desmonstrate the efficiency of the connection in opposition, Fig. 2 shows the error distribution of the heat fluxes q_1 and q_2 as well as W measuring in steady state.

The experimental system has been calibrated to determine the relation between the heat flux and electrical response of the heat fluxmeters and the lateral heat losses. Similarly, the system has been studied with solid samples to determine its physical response. An auxiliary experimental system formed by a differential interferometric laser device was included to observe the cells formed in samples.

The energetic behaviour of a silicon oil (Rhodorsil 47V350) in a little box of polymetacrilate of metile, P.M.M., (39.2 x 23.5 x 19.6 mm) has been studied. The box is hermetically sealed; it has an expansion system in a corner and four thermocouples in series, conveniently isolated and arranged on the top and bottom copper sheets to measure the tempera-

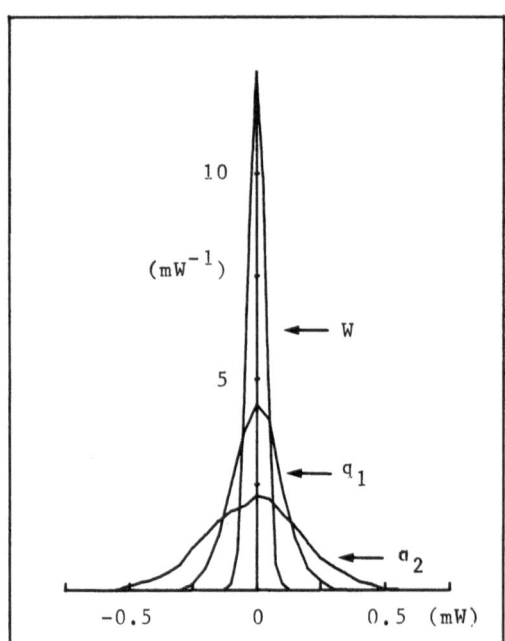

Fig. 2. Histogram of the heat fluxes and its difference.

ture difference applied.

Three series of measurements have been made. The first with fluid in both samples and the other two with fluid in one sample and solid P.M.M. in the other. The first series was made to obtain simultaneously a thermograph of the destruction and the formation of structures and to find their sum, thus finding ΔU with a lower relative error. The results have shown that the formation and the destruction of structures are not simultaneous, there being an important time difference between them. In the second series, only one fluid sample was studied and the following result was obtained: the vertical heat flux transfer through the lateral air chamber is larger when the sample is up than when it is down. We think that this result is the consequence of air convection, and to confirm it we plotted

$$(Nu)_{air} = \frac{q_{up}}{q_{down}}$$

versus ΔT (Fig. 3). The figure shows that, effectively, for differences of temperature greater than 5 K there is convection in the air. The final fall of experimental points in the graph can be justified by the induction of convective motion of the air in the heat fluxmeters, which changes its thermal properties. At the same time we observed the formation of three rolls in the fluid sample which remained stable for several hours. This effect does not appear in literature, which establishes two rolls as stable state {8}.

To avoid air convection, the guard ring was modified to remove part

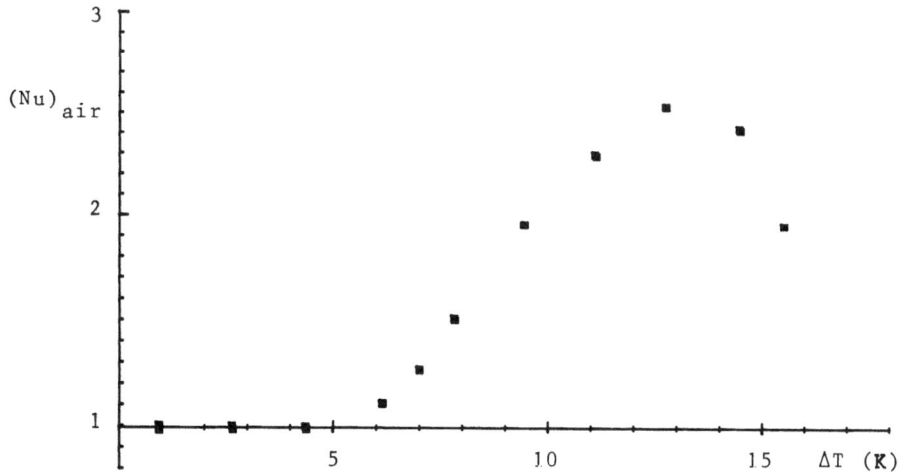

Fig. 3. Nusselt number of air chamber versus difference of temperature

of the air and to increase friction with new lateral walls. Thus, the
third series of measurements was made with no lateral air convection.
In the formation of the structures we observed three deformed rolls
which finally became two stable rolls. A typical thermograph is shown
in Fig. 4. The curves q_3 and q_4 are the heat fluxes in the solid sample,
and are not modified by changes in the position of the sample. The curves q_1 and q_2 are the heat fluxes in the fluid sample, the heat flux in
convection being higher than in the conduction state; the two transient
situations are not similar. The last curve represents W and clearly
shows the transitory states. As can be observed in the W graph, the
lines of the different steady states do not have the same values. This
effect is due principally to the change of the equivalent thermal con-

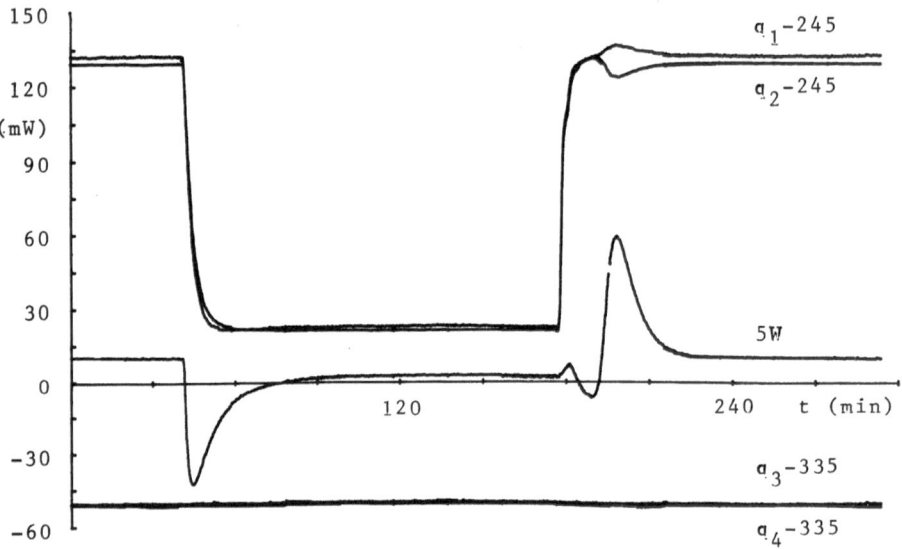

Fig. 4. Thermograph of a typical energetic experience.

ductivity of the fluid in both states, which affects both the vertical
heat conduction and the horizontal losses. Moreover, there are other
effects produced by the fluid and the measuring system. All these
effects must be corrected to equalize the steady values and to be able
to integrate the curve. The area measures the desired energy. A corrected
thermograph is shown in Fig. 5.

3. RESULTS AND DISCUSSION

Several quantitative results can be obtained from the last series
of measurements: The Nusselt number and the structural energy. In the

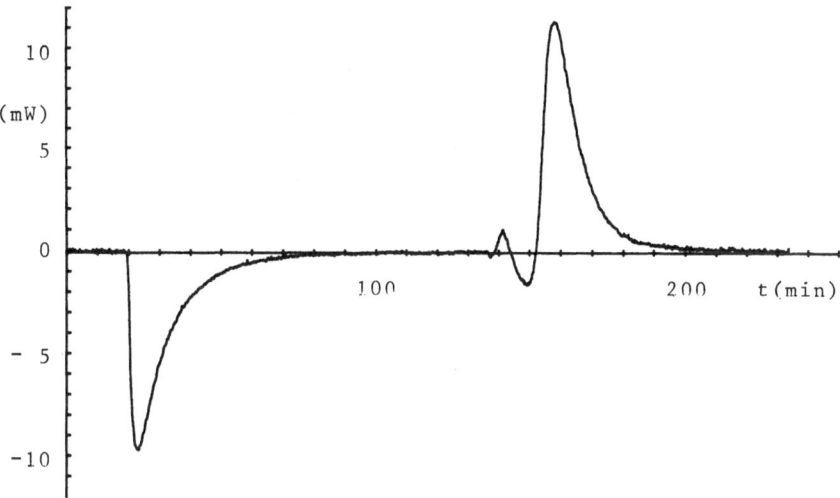

Fig. 5. Corrected thermograph.

same way, several qualitative results can be discussed from the corrected thermograph and the first experiences.

The Nusselt number versus the Rayleigh number is shown in Fig. 6.
The experimental points have been fitted with the mathematical expression

$$\frac{(Nu - 1) \, Ra/Ra_c}{(Ra/Ra_c - 1)} = \left\{ \frac{Ra/Ra_c - 1}{1.125} \right\}^{0.298}$$

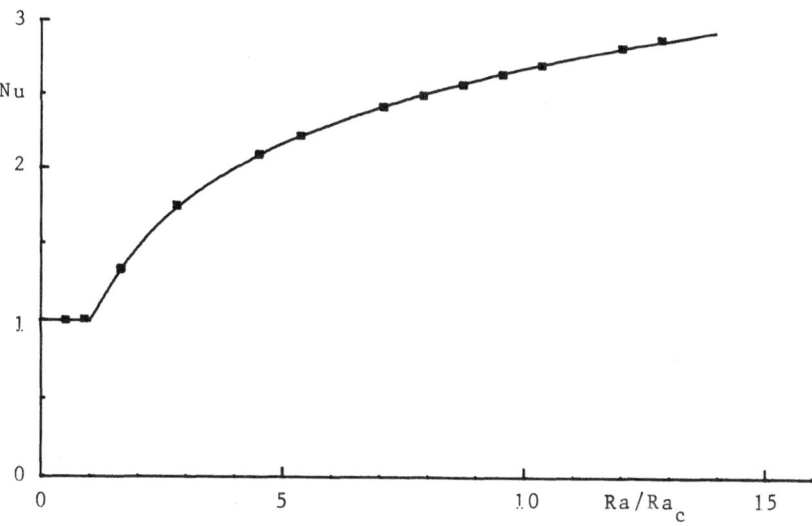

Fig. 6. Nusselt number versus Ra/Ra_c.

proposed by Busse {9}, with Ra_c = 2496 corresponding to ΔT_c = 0.95 K. The exponent 0.298 is within the interval predicted by numerical computations {10}.

The structural energy is shown in Table I and it is plotted versus ΔT in Fig. 7. As can be seen, the structural energy agrees with the principle of conservation and depends linearly on the difference of temperature applied. The straight line fitted to the experimental points (E (J) = 1.005 ΔT (K) – 1.204) cuts the abscissa axis at ΔT = 1.2 K, higher than ΔT_c; this would suggest that the line tends asymptotically to the critical value. It has not been possible to confirm this result because the energy values in this region are too small. The other side of the straight line, for high values of ΔT, is not well defined because we have not yet applied the temperature difference corresponding to the bimodal transition. The high critical Rayleigh number – Ra_c = 2496 >> 1708 – and the lateral air convection have hitherto rendered difficult to study the bimodal structures. It may be thought that the slope of the straight line of the energy versus the difference in temperature is a characteristic of the type of structure formed, but as yet it has not been possible to verify this point.

The form of the corrected typical thermograph, Fig. 5, shows several aspects that must be discussed. Firstly, the relation between the thermograph and the thermogenesis in the physical problem must be studied. If we impose the boundary conditions of the temperature on the thermostated pieces, A and A´ in Fig. 1, the temperature distribution in heat flux–

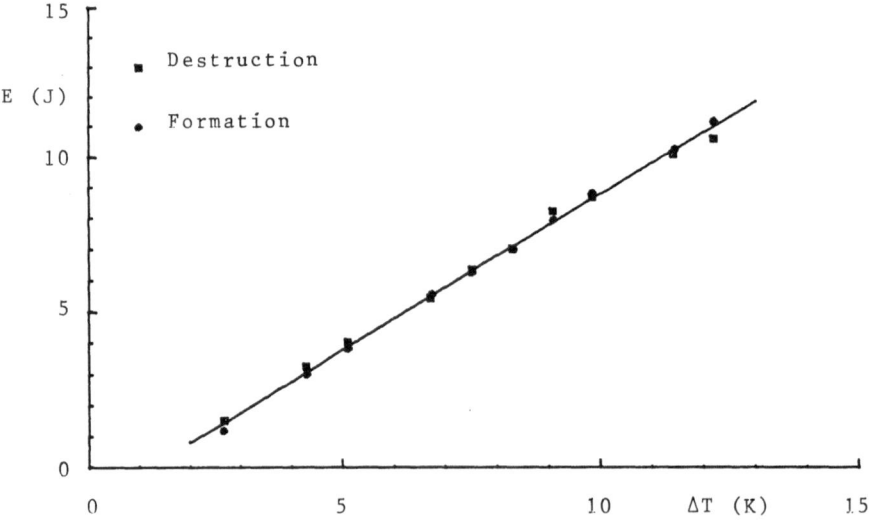

Fig. 7. Structural energy versus temperature difference.

Temperature difference (K)	Formation energy (J)	Temperature difference (K)	Destruction energy (J)
2.67	1.2	2.68	1.5
4.30	3.0	4.28	3.3
5.11	3.9	5.10	4.1
6.74	5.6	6.72	5.5
7.51	6.3	7.52	6.4
8.30	7.0	8.30	7.1
9.09	8.0	9.08	8.3
9.84	8.9	9.84	8.7
11.43	10.3	11.41	10.1
12.19	11.2	12.19	10.6

Table I. Experimental values of the structural energy.

meters changes when the experimental system turns, because the thermal
conductivity of the fluid is modified. For this reason, the thermograph
is disturbed by the fluid and the heat fluxmeters, but as the response
time of the fluxmeters is about 2 % of the mean characteristic time of
the fluid, it can be accepted that the thermograph directly represents
the thermogenesis. In the second place, the destruction of structures
is a continuous relaxation process, with a well defined characteristic
time. The formation of structures is a more complicated process. Initia
lly there is an effect with little energy corresponding to the formation
of the transient three deformed rolls. This confirms the qualitative
result obtained in the preceding series of measurements: the three roll
formation is made with little energy. The higher energetic effect begins
with the reorganization of the bulk of the fluid to two rolls and it
reaches the maximum when the boundary layer of the fluid is organized
definitively.

Another qualitative fact was obtained from the first two measurement
series. It has already been said that above 5 K of applied temperature
difference the convection of the air chamber was observed. In this
situation and for $\Delta T > 7$ K the existence of a dissipative structure
formed by three symmetrical stable rolls is seen. This perfectly formed
structure at times persisted indefinitely and other times was transformed
in two rolls after several hours. Comparing this with the results obtained without air convection, we must conclude that air convection in the
chamber seems to stabilize the three rolls.

4. REFERENCES

{ 1} E. L. Koshmieder, Adv. Chem. Phys. <u>26</u>, 177 (1974)

{ 2} E. Palm, Ann. Rev. Fluid Mech., 7, 39-61, (1975)

{ 3} J. Moreno, J.Jiménez, A. Córdoba, E. Rojas and M. Zamora, Rev. Sci. Instrum., <u>51</u>, 82, (1980)

{ 4} M. Zamora, A. Córdoba and J. Moreno, PhysicoChem. Hydrodynam.,<u>4</u>, 65, (1983)

{ 5} R. J. Tykodi, "Thermodynamics of Steady States" Macmillan, New York, 1967

{ 6} S. J. Benofy and P. M. Quay, J. Chem. Phys., <u>78</u>(6), 3177, (1983)

{ 7} M. Zamora and A. Rey de Luna, Rev. Sci. Instrum., to be published

{ 8} M. Dubois and P. Berge, J. Physique, <u>42</u>, 167, (1981)

{ 9} F. H. Busse, J. Math. Phys., <u>46</u>, 140, (1967)

{10} F. H. Busse, Rep. Prog. Phys., <u>41</u>, 1929, (1978)

MODELISATION OF THERMOCONVECTIVE INSTABILITIES IN SATURATED POROUS MEDIA- LATERAL BOUNDARY INFLUENCE

A.R. Deltour

Institut de Mécanique des Fluides de Toulouse, L.A. 005
2, rue Charles Camichel
31071 TOULOUSE CEDEX (FRANCE)

1. INTRODUCTION

We study thermoconvective instability evolution in saturated porous media within a vertical cylindrical cell bound   by two isothermal horizontal surfaces.

In fact, stability condition of the fluid and flow pattern and corresponding heat transfer are very dependent on geometrical dimensions and thermal boundary conditions.

Even though we know some recent investigations about aspect ratio, /1,2,3/ we have not found papers including thermal lateral condition and geometrical one.

The present study describes theoretical and experimental results about stability criterion and modelisation of finite amplitude convection.

2. FORMULATION OF THE PROBLEM

The physical system is composed of a cylindrical cell. The encloser is of height H and circular cross section D, filled with an isotropic, homogeneous saturated porous media. The lateral side wall is of external diameter D_1 with a thermal conductivity λ_p. A destabilizing vertical thermal gradient is applied to the horizontal isothermal boundaries.

According to current assumptions generally used to describe heat and mass transfer phenomena in porous media, thermal equilibrium between fluid and solid, Boussinesq approximation for the fluid, we may write the well-known following system.

In the fluid

Continuity equation :

$$\nabla v' = 0$$

Momentum equation :

$$\frac{\rho}{\varepsilon} \frac{\partial v'}{\partial t} = - \nabla P - \frac{\mu}{K} v' + \rho g$$

Energy equation :

$$(\rho C)^{\textstyle *} \frac{\partial T'}{\partial t} + (\rho C)_f \; V' \; . \; \nabla T' = \lambda^{\textstyle *} \; \nabla^2 \; T'$$

Fluid equation of state :

$$\rho = \rho_o (1 - \alpha(T' - T_o))$$

In the side wall :

Energy equation :

$$(\rho C)_p \; \frac{\partial T'_p}{\partial t} = \lambda_p \; \nabla^2 \; T'_p$$

We complete with the boundary conditions : impervious surfaces, thermal and heat flux continuity.

3. STABILITY ANALYSIS

It will be based on the assumptions of linear theory. The preconvective state will be characterised by the following temperature distribution, $T'(z') = T_2 - \dfrac{\Delta T}{H} \, z'$, and a velocity field $V_o = 0$ in the whole cell. Let V', θ', θ'_p and p' be velocity, temperature and pressure perturbations at the onset of thermoconvective instability.

According to the classical assumptions of linearisation we obtain for the perturbations.

In the porous medium :

$$\nabla V' = 0$$

$$\frac{\rho}{\varepsilon} \frac{\partial V'}{\partial t} = - \nabla p' - \frac{\mu}{K} \; \theta' + \rho \; g \; \alpha \; \theta'$$

$$(\rho C)^{\textstyle *} \frac{\partial \theta'}{\partial t} = \lambda^{\textstyle *} \; \nabla^2 \; \theta' + (\rho C)_f \; W' \; \frac{\Delta T}{H}$$

In the side wall :

$$(\rho C)_p \; \frac{\partial \theta'_p}{\partial t} = \lambda_p \; \nabla^2 \; \theta'_p$$

By elimination of velocity and pressure we look for an explicit solution of the thermal perturbations. The principle of exchange of stability may be assumed in the configuration. Then we specify the existence of stationary solutions.

By the introduction of adimensional variables r, z, θ, θ_p, the previous system reduces to :

$$\nabla^4 \theta = Ra^{\textstyle *} \; \nabla_1^2 \; \theta \qquad \text{with } \nabla_1^2 = \nabla^2 - \partial^2/\partial z^2$$

$$\nabla^2 \theta_p = 0$$

For the boundary conditions :

$$\theta = \theta_p, \quad \Lambda \frac{\partial \theta}{\partial r} = \frac{\partial \theta_p}{\partial r} \quad \text{as} \quad r = \frac{1}{2R_0} \; ; \; \frac{\partial \theta_p}{\partial r} = 0 \text{ or } \theta_p = 0 \text{ as } r = \frac{1}{2R_1}$$

with $R_0 = H/D$, $R_1 = H/D_1$, $\Lambda = \lambda^*/\lambda_p$.

So the perturbations θ and θ_p seem to be only dependent on the adimensional parameters Ra^*, R_0, R_1 and Λ ; Ra^* as the Rayleigh number, R_0 and R_1 as aspect ratio, Λ as the thermal porous media - side wall conductivity ratio.

According to the boundary conditions we look for periodic solutions with horizontal wave number c and $k\pi$ vertical one. The previous system allows to obtain the critical value of control parameter Ra^*. This leads to real solution :

$$Ra^* = (\frac{(k\ \pi)^2 + c^2}{c})^2$$

The marginal condition of stability will be satisfied for c and integer value of k, solution of the following expression with an adiabatic lateral external condition:

$$\begin{vmatrix} J_n(c/2\ R_0) & - I_n(k\pi/2R_0) & - K_n(k\pi/2R_0) \\ \Lambda c\ J_n'(c/2R_0) & - k\ I_n'(k\pi/2R_0) & - k\ K_n'(k\pi/2R_0) \\ 0 & I_n'(k\pi/2R_1) & K_n'(k\pi/2R_1) \end{vmatrix} = 0$$

Each solution involves a corresponding control parameter value, the smallest is selected as the critical value Ra_c^*, function of Λ , R_0 an R_1. We plot figure 1 $Ra_c^*(\Lambda)$ for several aspect ratio and thick side walls. More precisely with $R_0 = 3$, figure 2, we find the influence of the thickness of the wall side for two lateral external conditions and several conductivity ratio. The asymptotic value of the critical parameter for the thick side wall is obtained equal to the previous one calculated for the same conductivity ratio whatever the nature of the external conditions may be. On the contrary, whatever the conductivity ratio with thin side wall may be, the external condition is the director parameter, so we obtain values corresponding to an adiabatic or conductive condition.

4. FINITE AMPLITUDE CONVECTIVE FLOW

Let us assume that the arising disturbance is developed as the perturbation term in the linear analysis. Further, the successive modes are summerised above the corresponding control parameter values. Theoretical critical values are computed from marginal conductive state. With the power integral technic of Malkus, by integrating over the porous volume the weighted energy equation, we obtain the theoretical temperature distribution in the cell.

Comparison between theoretical values and experimental results (6) are displayed for high geometrical aspect ratio, with temperature field as :

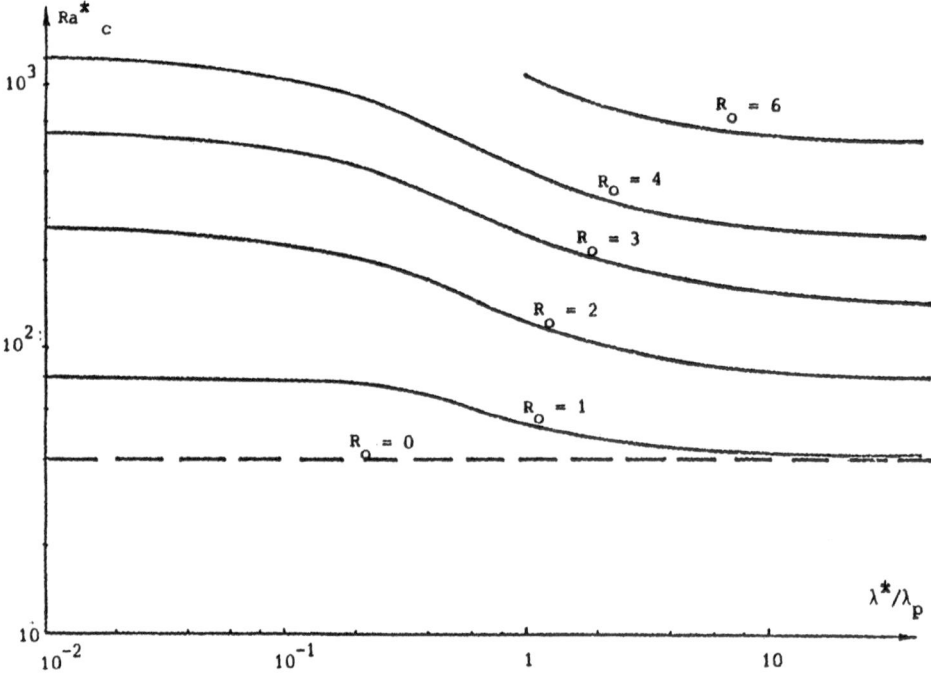

Figure 1 - Lateral side wall conductivity influence

2e/D adimensional side wall thickness

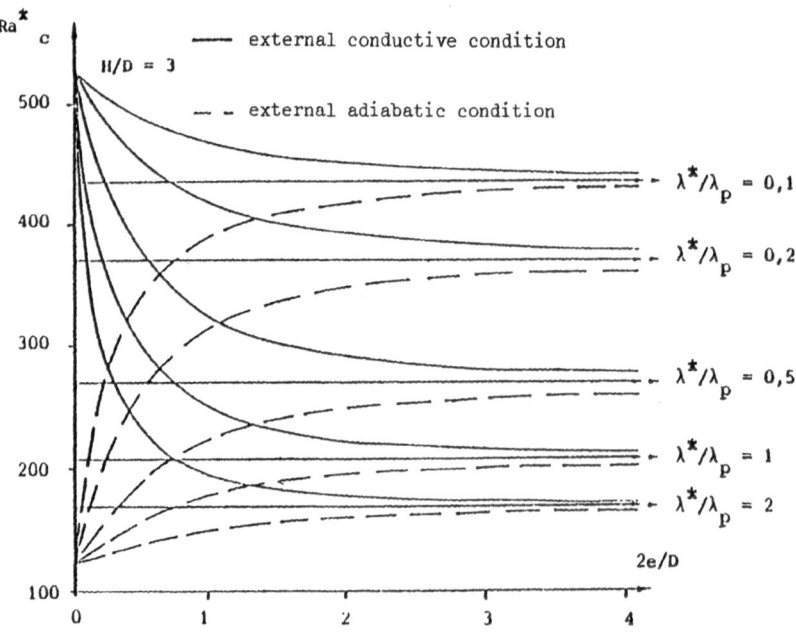

Figure 2 - Lateral boundary thickness influence

$$T(r,\phi,z) = 1 - z + \sum_k \frac{1}{k} (1 - \frac{Ra^*_{ck}}{Ra^*}) \sin 2k\pi z + \sum_k A_k J_1(cr) \cos\phi \sin k\pi z$$

with :

    $1 - z$ conduction part

    $\sum_k \frac{1}{k} (1 - \frac{Ra^*_{ck}}{Ra^*}) \sin 2k\pi z$ coupling velocity and temperature part

    $\sum_k A_k J_1(cr) \cos\phi \sin k\pi z$ succeding modes

Experimentally we observe axial temperature distribution in a porous medium composed of glass balls saturated with water (6). On these points theoretical field is the sum of conductive and coupling terms. Fourier analysis of experimental recordings and theoretical decomposition are plotted, figure 3, for $R_o = 3$.

5. CONCLUSION

Finite amplitude steady convection in large aspect ratio configuration is studied with drastic assumptions (summerised modes, stability analysis). We found intersting indications on the temperature and velocity distribution within several critical values for the control parameter. This paper will help to select a trial function in a further numerical approach.

NOMENCLATURE

C   specific heat
c   horizontal wave number
g   acceleration of gravity
k   integer
p   pressure perturbation
r   radial coordinate
t   time
z   vertical coordinate
D   porous media cell diameter
D_1 external experimental cell diameter
H   vertical distance between the horizontal boundaries
K   permeability
P   pressure
R_o $= H/D$ porous medial cell aspect ratio
R_1 $= H/D_1$ experimental cell aspect ratio
T   temperature
V   velocity

Greek Symbols

α thermal expansion coefficient

ϕ angular coordinate

ε porosity

λ thermal conductivity

θ temperature perturbation

ρ density

ν kinematic viscosity

$$Ra^* = g\alpha \, \frac{(\rho C)_f}{\nu} \, \frac{K}{\lambda^*} \, H \, \Delta T \quad \text{Rayleigh number}$$

In subscript

$*$ fictitions continuous medium equivalent to saturated porous medium

f fluid

p lateral wall side

In superscript

$'$ dimensional variable

REFERENCES

/1/ BECK J.L.
Convection in a box of porous material saturated with fluid
Phys. Fluids, 15, N° 3, pp. 1377-1383, (1972)

/2/ LOWELL R.P., SHYU C.T.
On the onset of convection in a water-saturated porous box : effect of conducting walls
Lett. Heat Mass Transfer, 5, pp. 371-378, (1978)

/3/ TEWARI P.K., TORRANCE K.E.
Onset of convection in a box of fluid saturated porous material with a permeable top
Phys. Fluids, 24, N° 5, pp. 981-983 (1981)

/4/ COMBARNOUS M., BORIES S.
Hydrothermal convection in saturated porous media
Advances in Hydroscience, Vol. 10, pp. 231-301, Academic Press, New-York (1975)

/5/ MALKUS W.V.R., VERONIS G.
Finite amplitude cellular convection
J. Fluid Mech., 4, pp. 225-260 (1958)

/6/ DELTOUR A.
Convection naturelle au sein d'un milieu poreux saturé confiné dans un domaine cylindrique vertical
Thèse, Toulouse, (1982)

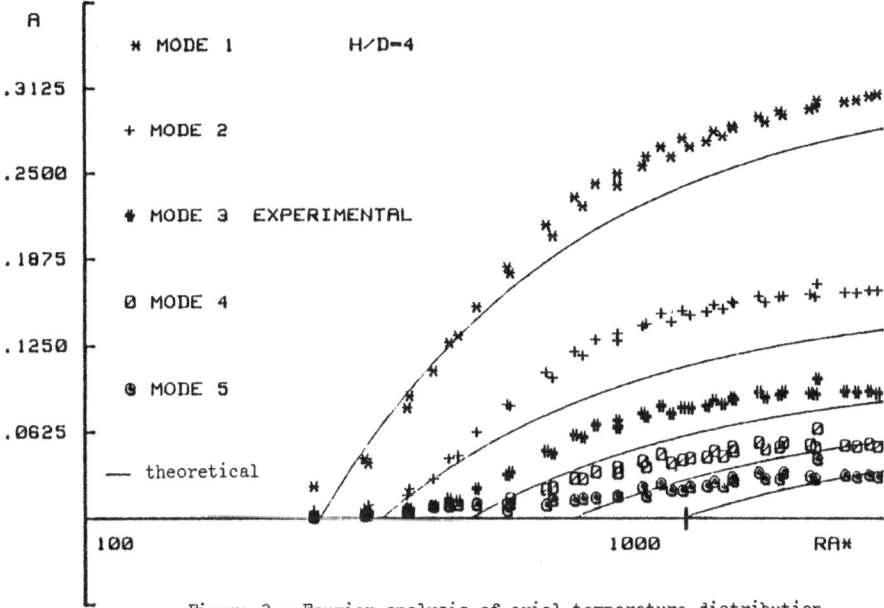

Figure 3 - Fourier analysis of axial temperature distribution

COMPETITION BETWEEN MARANGONI AND ARCHIMEDEAN FORCES, TO DETERMINE
THE SURFACE PROFILE OF A LIQUID HEATED, OPEN TO AIR

J.C. LOULERGUE

Institut d'Optique Théorique et Appliquée
Université de Paris-Sud B.P. 43
91406 ORSAY - FRANCE

1. Introduction

Convection in the absence of imposed velocities or pressure
gradients in fluid systems with interface may be classified as
natural, Marangoni or combined free convection according to wether
the motion is caused by buoyant forces, Marangoni stresses or both.

Problems of this type have recently witnessed renewed interest
also in view of their relevance in several fields of microgravity
sciences and space processing [1] and instabilities of an interface
between two phases, in the presence of gradients of chemical poten-
tials [2,3].

However, although many progresses have been made, especially in
the study of Marangoni convection [4], the state of the art is still
somewhat unsatisfactory. Thus for instance, it is not yet well
known how the problem's data determine the type of convection
prevailing, the coupling of the two flow fields the nature of flow
regimes and so on. [5]

The relevance of these questions is being able to answer them
will greatly help in the analysis of the deformation of the free
surface : there is an opposite behaviour of the surface relief for
the two driving forces [6] (buoyant and Marangoni forces).

The analysis proposed in the present paper, will be to investigate the surface profile of a liquid heated in which the extension
(L) of the interfacing fluids is larger than their height, and the
imposed temperature difference (δT) very small. We shall discuss here
the sign of the deformation, i.e. if the buoyant forces or Marangoni
stresses provoke a local depression or elevation at the free surface,
relative to the upwelling flows, in function of the depth (h) of the
fluid and wavenumber (k) of the imposed periodic perturbation.

2. STATEMENT OF THE PROBLEM

a. General formulation

We consider a layer of fluid in the gravity field (g), submitted to a spatially periodic modulation of temperature of wavenumber k, on its free surface, produced by absorption of radiation [7] in a thin skin at the top of the fluid (Figure 1). This temperature distribution induces in turn density and surface-tension $\rho(T)$, $\alpha(T)$:

$$\rho(T) = \rho_o + (\frac{\partial \rho}{\partial T})_{T = T_o} (T - T_o) \tag{1}$$

$$\alpha(T) = \alpha_o + (\frac{\partial \alpha}{\partial T})_{T = T_o} (T - T_o) \tag{2}$$

These are, of course, first order approximations. The quantities (ρ_o, α_o) are the initial values of (ρ, α) with respect to the temperature. $\frac{\partial \alpha}{\partial T}$ and $\frac{\partial \rho}{\partial T}$ are in general negative quantities. As a result of competition between the density and surface-tension variation [8] the free surface adopts a new profile (Figure 1).

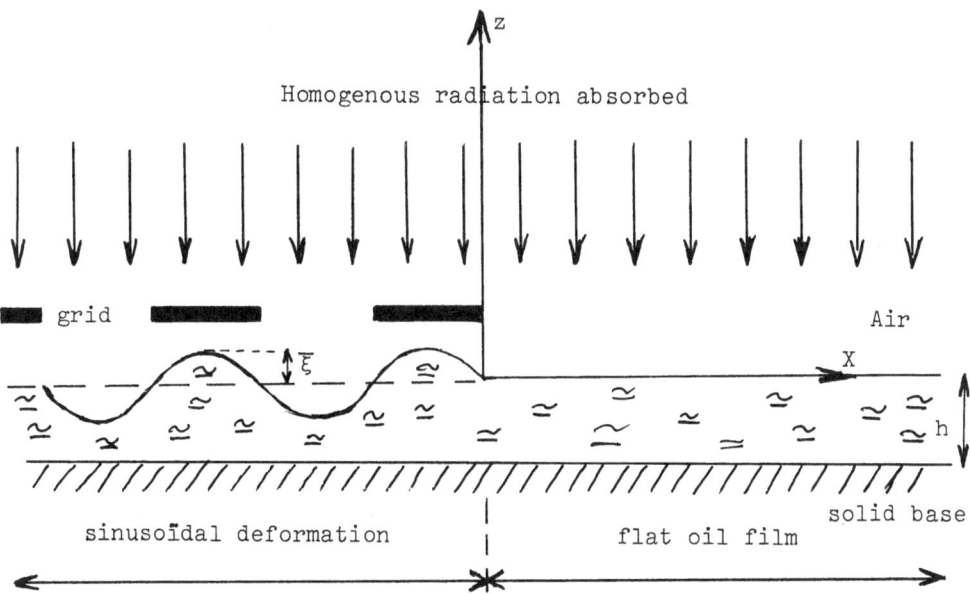

Figure n° 1 - Thermogravity and thermocapillarity induced by absorption of an homogenous radiation, spatially modulated by a grid of periodic transparent and opaque bars, of period p. Without grid the free surface remains level. With the grid, it appears a sinusoïdal deformation on the free surface, of period p, and amplitude ξ.

The problem is to know the response - the sign of the deformation - of the free surface at the predominancy of one or the other
flow related to the depth of the fluid. We shall make a certain
number of simplifying assumptions, in order to make the problem more
tractable, while retaining most of the physics.

b) Approximations

. We shall neglect the dynamics of the physical quantities which
produce the thermal modulation. We shall restrict ourselves under to
the static response D.C. excitation. Thus, flows will be described
by Navier-Stokes equations in the usual form for incompressible
fluids. However a theoretical analysis of the non-stationary stage
is lacking. Also from an experimental view point this initial
transient stage is practically unexplored.

. In the case of small perturbations, $T - T_o = \delta T \ll T_o$ one
make take the temperature modulation on the free surface as a
boundary condition given, this field being the result of a balance
between the non-uniform heating and the tendency to uniformity by
molecular heat-conduction, when the flow pattern is fully developped.

. As the heating of the fluid is very small we shall research
only the linear response of the free surface to an infinitesimal
excitation. In the following we shall retain only terms of first
order in the hydrodynamic variables and interface deformations.

3. THEORETICAL CALCULATION OF THE SURFACE SHAPE

We assume a temperature perturbation T at the interface, periodic
in the X direction, independant of y and also of z, the thickness h
being much smaller than the lateral extension L of the fluid
(Figure 1).

$$T(x) = T_o + \delta T \sin kx \qquad (3)$$

with $\delta T \ll T_o$ and $k = \frac{2\pi}{p}$ is the wavenumber of the thermal perturbation
of step p. As usual for linear problems, more general excitation can
be decomposed in such modes, and the total response is a superposition
of individual responses for each mode. The calculation of the free
surface deformation (sign and amplitude) involves mainly two steps :
the first is to solve the velocity field taking into account boundary
conditions, the second is to derive the pressure field and from it
the shape of the free surface.

This particular thermal distribution induces in turn density and surface-tension distribution $\rho(x)$ and $\alpha(x)$:

$$\rho(x) = \rho_o + \delta\rho \; \sin kx \qquad\qquad (4)$$

$$\alpha(x) = \alpha_o + \delta\alpha \; \sin kx \qquad\qquad (5)$$

with in general

$$\left|\delta\rho\right| = \left|\frac{\partial\rho}{\partial T}\right| \delta T \ll \rho_o \qquad \text{and} \qquad \left|\delta\alpha\right| = \left|\frac{\partial\alpha}{\partial T}\right| \delta T \ll \alpha_o$$

and a new profile of the free surface that we take of the form

$$\xi(x) = h + \overline{\xi} \; \sin kx$$

and as the fluid is submitted to an infinitesimal excitation we have $\overline{\xi} \ll h$. Moreover, the surface relief will be considered, with a good approximation, as sinusoïdal.

a) <u>Boundary conditions</u>

The problem being two-dimensional in the (x, z) plane, there is no velocity component in the y direction and we shall denote u and w the horizontal and vertical velocity components in the fluid.

. At the bottom of the vessel $(z = 0)$ there is the usual no-slip condition :

$$u(z = 0) = w(z = 0) = 0 \qquad\qquad (6)$$

Other boundary conditions describe the free interface :

$$w \; (z = h) = 0 \qquad\qquad (7)$$

which describes the impenetrability of the steady interface.

. and the continuity equation

$$\left. \frac{\partial}{\partial z} \; w \; \right|_{z \, = \, h} = 0 \qquad\qquad (8)$$

. continuity of the normal stresses generalises the Laplace law.

$$p_a = \left[p - 2\eta\left(\frac{\partial w}{\partial z}\right) \right]_{z \, = \, h} + \alpha \; \frac{\partial^2}{\partial x^2} \; \xi \qquad\qquad (9)$$

where $p(x,z)$ and p_a are the pressure field in the fluid and in the air, η the viscosity.

. Continuity of the tangential stresses involves the supplementa--ry tractions due to the variation of the surface-tension (Marangoni forces).

$$\eta \frac{\partial}{\partial z} u \bigg|_{z\,=\,h} = \frac{\partial \alpha}{\partial x} \tag{10}$$

b) Determination of the velocity fields

Linearised time-dependent NAVIER-STOKES equations are

$$- \partial_x p + \eta (\partial^2_{xx} + \partial^2_{zz}) u = 0 \tag{11}$$

$$- \partial_z p + \eta (\partial^2_{xx} + \partial^2_{zz}) w = \rho(x) g \tag{12}$$

to which we add the continuity equation for incompressible fluids

$$\partial_x u + \partial_z w = 0 \tag{13}$$

As regards the horizontal dependance of the velocity components one sees from the boundary condition (10) that u is in phase with $\frac{\partial T}{\partial x}$ and from the continuity equation (13) that w is in phase with $\partial_x u$, so that as $T(x)$ is of the form (3), we shall find

$$w(x,\,z) = \bar{w}(z)\, sinkx \tag{14}$$

$\bar{w}$ being still unknown functions of the variable z, to be determined from equation (12) and boundary conditions.

Elimination of the pressure p and the velocity component u leads to :

$$\eta(\partial^2_{xx} + \partial^2_{zz})^2 w = g\partial^2_{xx} \rho(x) \tag{15}$$

with equation (4) and (14), the equation (15) leads to

$$(\partial^2_{zz} - k^2)^2 \bar{w}(z) = - g \frac{k^2}{\eta} \delta\rho \tag{16}$$

so that $\bar{w}(z)$ is a combination of exp $(\pm kz)$ and z exp $(\pm kz)$. It is easily seen that $\bar{w}(z)$ is most conveniently written in the form :

$$\bar{w}(z) = (A + B \frac{z}{h}) \cos h(kz) + (C + D \frac{z}{h}) \sin h(kz)$$

introducting three unknown constants B, C, D. Denoting δ_o = kh, c = cos h(δ_o), s = sin h(δ_o), boundary conditions (6) on w give :

$$A \simeq \frac{g\delta\rho}{\eta k^2} \qquad (17)$$

$$Bc + (C + D) s = A(1 - C) \qquad (18)$$

and the continuity equation (8)

$$\delta_o C + B = 0 \qquad (19)$$

At the upper interface the boundary condition (10) reads :

$$\eta \left. \partial^2_{zz} \overline{w} \right|_{z = h} = k^2 \delta\alpha \qquad (20)$$

Replacing $\overline{w}$ in (20) one easily deduces

$$(A + B) c\delta_o + (C + D) s\delta_o + 2(Bs + Dc) = \frac{\delta_o \delta\alpha}{2\eta} = E \qquad (21)$$

After simplification using equation (18), the equation (21) reads

$$Bs + Dc + \frac{1}{2} A\delta_o = E \qquad (22)$$

we have a system of three equations (18,19,22) with three unknowns
B, C, D, non homogeneous due to the source terms :

$$A = \frac{g\delta\rho}{\eta k^2} \qquad \text{and} \qquad E = \frac{\delta_o \delta_\alpha}{2\eta}$$

in equations (18) and (22).

Solving this system ends the first step. Only C will be needed
below.

$$C = \frac{A}{(sc - \delta_o)} \left[(1-c) c + \frac{s\delta_o}{2} \right] - \frac{s E}{(sc - \delta_o)}$$

where can have two interesting limits :

E → 0 buoyancy controlled convection

A → 0 surface-tension controlled convection for k ≠ 0

c) Determination of the pressure fields

We have now to determine the pressure field from the Navier-
Stokes equation (12)

$$\partial_z p = - \rho(x) g + \eta(\partial^2_{zz} - k^2) w \qquad (24)$$

by integration we have

$$p(x, z) = -\rho(x)\, gz + \eta(-k^2 \int w\, dz + \partial_z w) + c^{ste} \tag{25}$$

The unknown constant is given by the boundary condition at the free surface :

$$c^{ste} = p_a + \bar{\rho} gh \tag{26}$$

with p_a atmosphéric pressure .

In the fluid the pressure is

$$p(x,z) = -\rho(x)\, gz + \eta\left[-k^2 \int w\, dz + \partial_z w\right] + p_a + \bar{\rho} gh \tag{27}$$

d) The free surface deflection

The amplitude ξ of the interface is determined using equations (9) and (27).

$$\left[\rho g + \alpha k^2\right]\xi = -\eta k^2 \left\{Ah + \left[\int \bar{w}\, dz + \frac{1}{k^2} \partial_z \bar{w}\right]\right\}_{z = h} \tag{28}$$

Without thermal perturbation, the free surface is level, and its height is z = h. With an infinitesimal perturbation, the free surface is slightly deformed. Its new height is z = h + ξ sinkx with $\xi \ll$ h. Strictly speaking, we must integrate at z = h + ξ sinkx However, as ξ is much smaller than h, we have calculated as the position of unperturbed free surface.

The calcul of the integral, in equation (28) for z $\approx$ h give, using equation (29)

$$Bs + cD = -\frac{1}{2} A\delta_o + E \tag{29}$$

By using equations (23, 28, 29) we get the complete expression of

$$\xi = \frac{k^2 \delta_o^2\, \delta\alpha}{\bar{\rho} g(1 + \delta_o^2 \Lambda_c^2)\, k(sc - \delta_o)} - \frac{g\delta\rho\phi(\delta_o)}{\bar{\rho} g(1 + \delta_o^2 \Lambda_c^2)\, k(sc - \delta_o)} \tag{30}$$

where $\phi(\delta_o)$ is a function of the non dimensional number δ_o equals at

$$\phi(\delta_o) = 2c\,(c - 1) - 2\, s\delta_o + \delta_o^2 \tag{31}$$

we define the capillary length $\lambda_c = \left(\frac{\alpha}{\rho g}\right)^{1/2}$ (32) and its dimension-less expression in units of h by $\Lambda_c = \frac{\lambda_c}{h}$

4. DISCUSSION

In this section we examine the main features of this study : the question of the sign of the reponse of the interface. The discussion on the value of the amplitude, relative to the reduced wave-vector δ_0, is beyond the scope of this paper. Such a study will be done in another article. We note that the equation (30) is the sum of two factors : * the first factor describes the action of the surface-tension gradients or Marangoni-driven flows $\frac{\partial \alpha}{\partial x}$.

    * the second factor describes the action of the density gradients or buoyancy-driven flow $g \, \delta\rho$.

The sign of the amplitude (30) depends of the expression

$$g \left| \delta\rho \right| \phi(\delta_0) - k^2 \, \delta_0^{\,2} \left| \delta\alpha \right| = 0 \qquad (33)$$

Figure 2 - Surface profiles defined from the sign of the equation (33) for a fluid given and δ_0 variable.

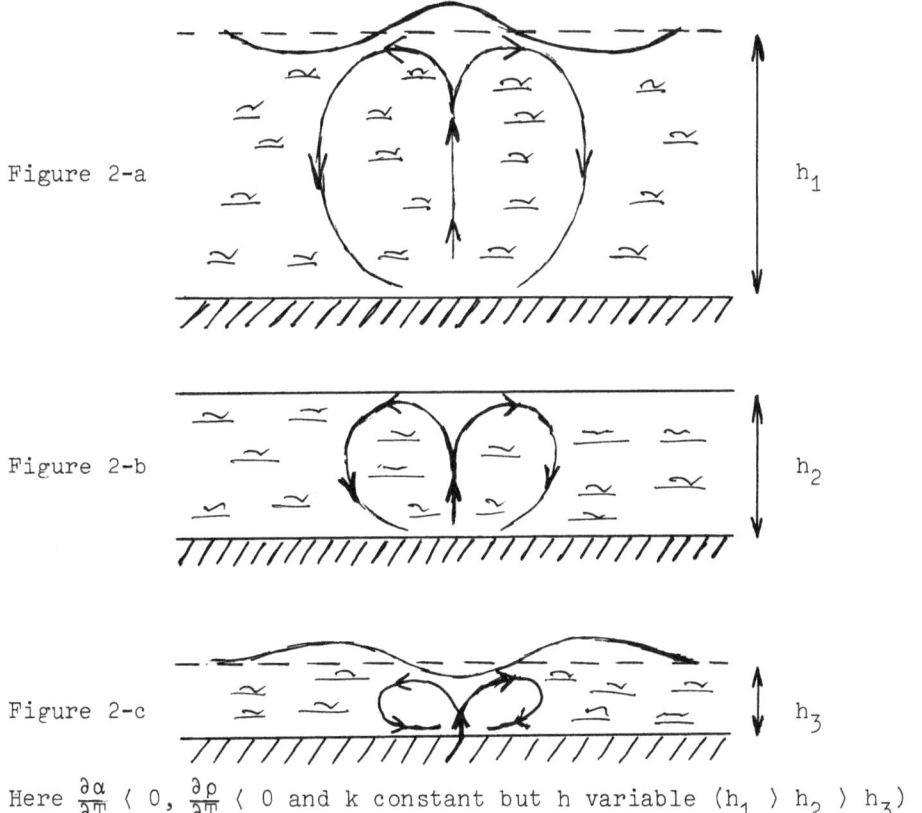

Figure 2-a h_1

Figure 2-b h_2

Figure 2-c h_3

Here $\frac{\partial \alpha}{\partial T} \langle 0$, $\frac{\partial \rho}{\partial T} \langle 0$ and k constant but h variable $(h_1 \rangle h_2 \rangle h_3)$

There exist three possibilities :

a) $$g \left| \delta\rho \right| \left| \phi(\delta_o) - k^2 \delta_o^2 \right| \left| \delta\alpha \right| \rangle 0$$

The amplitude $\bar{\xi}$ is positive = the free surface has an elevation at
the hotter areas (Figure 2a with $\frac{\partial\alpha}{\partial T}$ and $\frac{\partial\rho}{\partial T}$ $\langle$ 0).

b) $$g \left| \delta\rho \right| \left| \phi(\delta_o) - k^2 \delta_o^2 \right| \left| \delta\alpha \right| \langle 0$$

The amplitude is negative : the free surface has an opposite structure. It is depressed at the hotter areas near the points of upwelling
flow (Figure 2c with the same fluid).

c) The two previous cases are separated by the situation represented by equation (33). Then the free surface of the liquid sample in
convection remains level ; there is no deformation (Figure 2 b
with the same fluid).

Replacing equation (31) in the equation (33) one easily deduces :

$$g \left| \delta\rho \right| \left\{ 2 \left[c(c-1) - s\delta_o \right] + \delta_o^2 \right\} = k^2 \delta_o^2 \left| \delta\alpha \right| \qquad (34)$$

or

$$2 \left[c(c-1) - s\delta_o \right] = \lambda_T^2 (k^2-1) \delta_o^2 \qquad (35)$$

where $\lambda_T = \left(\frac{\delta\alpha}{g\delta\rho} \right)^{1/2}$ is a characteristic length. It depends of thermal
variation of ρ and α of the fluid. By comparaison at the hydrostatic
capillary length (32) or hydrostatic Bond length [9], we shall call
it "thermal_capillary_length" - or dynamic Bond length [4] - This new
parameter should be considered with some care later.

It is easily seen that equation (35) is satisfied for one particular value of δ_o. For example, the table I give several values or
δ_o which satisfy the equation (35) for a silicon oil whose characteristics are given in table II. The figure 3 shows a graph, giving the
sign of $\bar{\xi}$ as a function of the wavelength p for several heights of
a silicon oil.

Let us take this liquid sample with h = 3 mm heated by irradiation
with a two-beams interference fringes of variable spatial wavelength [10]. Then the motion of the fluid is caused by buoyant forces
(bulk forces) and Marangoni stresses (surface forces). If the flow
regime is principally controlled by buoyant forces, the free surface

- TABLE I -

h(mm)	p(mm)	$\delta_o = kh$
0,1	0,08	810^{-3}
0,2	0,18	$3,610^{-2}$
0,3	0,30	910^{-2}
0,4	0,43	0,172
0,5	0,57	0,285
0,6	0,78	0,468
0,7	0,88	0,616
0,8	1,05	0,84
0,9	1,23	1,107
1	1,42	1,42
1,5	2,5	3,75
2	4	8
2,5	5,7	14,25
3	7,8	23,4

- TABLE II -

Silicon oil (20073)

a) Data at room temperature

viscosity	$3\ 10^{-2}$ St
density	1 g.cm^{-3}
surface-tension	10 dyne cm^{-1}

b) Thermal data at 20°C

$$\frac{\partial \alpha}{\partial T} = 810^{-2} \text{ dyne cm}^{-1} \text{ }^\circ\text{C}^{-1}$$

$$\frac{\partial \rho}{\partial T} = 10^{-3} \text{ g. cm}^{-3} \text{ }^\circ\text{C}^{-1}$$

c) Characteristic lengths

$$\lambda_c = 1,43 \text{ mm}$$

$$\lambda_T = 2,85 \text{ mm}$$

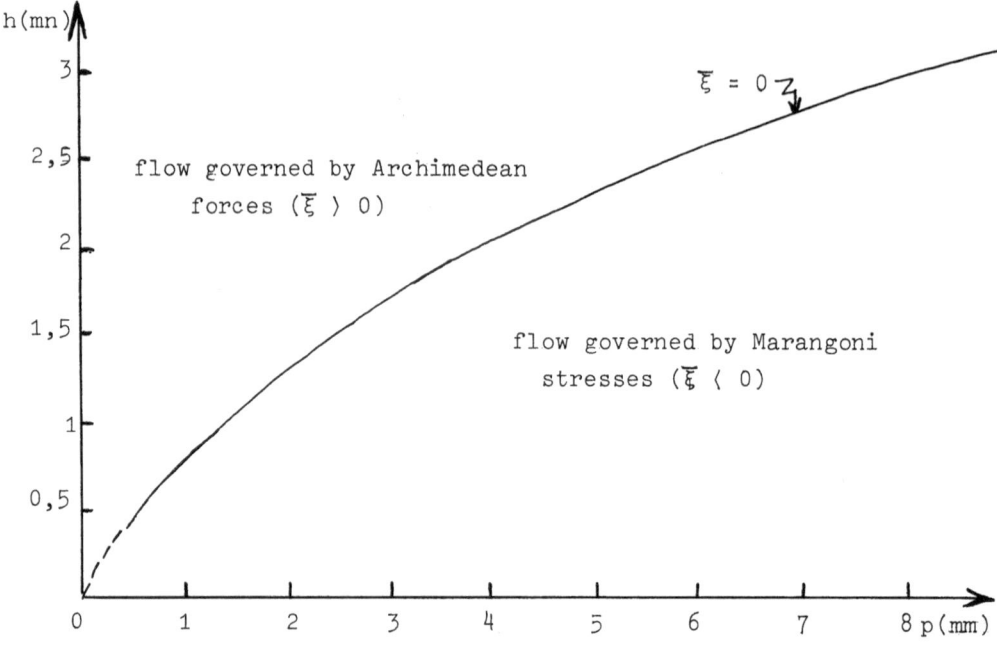

Figure n° 3 - No free surface deformation as a function of
the wavelength p for several depths h of the vessel : above
the curve, the deformation is governed by Archimedean
forces ; below the curve, the deformation is governed by
Marangoni forces.

is elevated at the hotter areas, near the points of up-welling flow
(figure 4-a). This happens in an interval of wavelengths, for the
thermal perturbation, situated between 0 and 7,8 mm. Now, if the flow
regime is principally controlled by Marangoni forces, then the free
surface is depressed near the points of up-welling flow (figure 4-c)
while keeping the same experimental set up. This happens, for a
thermal perturbation of wavelengths higher than 7,8 mm. The intermediate case corresponds at a coupling of the two-flow fields of same
magnitude. Then the horizontal layer of liquid heated non uniformly,
with a spatially periodic thermal perturbation of wavelength equals
at 7,8 mm has an upper free surface level (Figure 4b). The competition between Marangoni and buoyant forces over the surface profile
is opposite and of same intensity. There is no predominancy of one
over the other. Surface profiles showed on figure n° 4 are discussed
from equation (30), with a fluid of λ_c = 1,43 mm and λ_T = 2,85 mm.
Its height is 3 mm. The wavelength p of the periodic thermal perturbation increase from 4-a to 4-c.

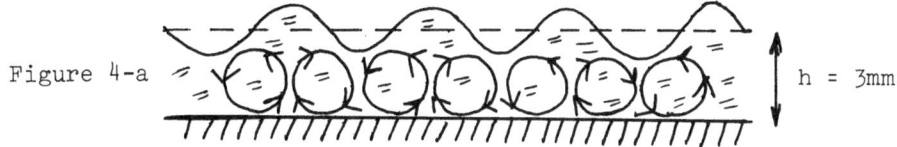

Figure 4-a   h = 3mm

p < 7,8 mm

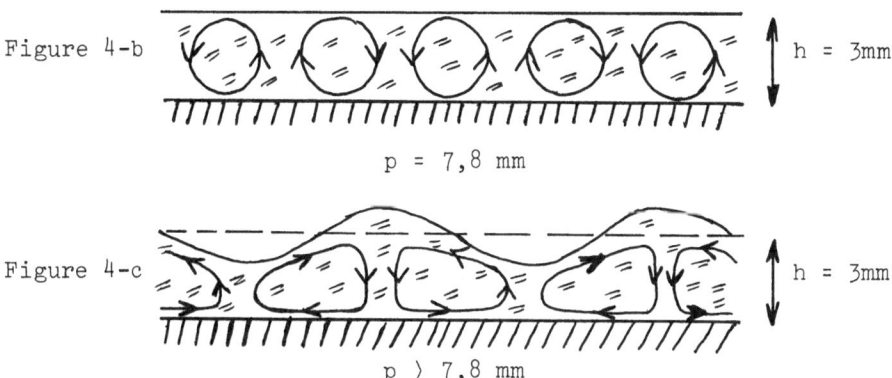

Figure 4-b   h = 3mm

p = 7,8 mm

Figure 4-c   h = 3mm

p > 7,8 mm

5. CONCLUSION AND REMARKS

In this paper we have developped the calculation of the linear
response of a free surface open to air, in a one-fluid system,
submitted to a surface-tension modulation, a configuration which has
already shown interesting applications[11, 13] for which a detailed
hydrodynamical study similar to the present was lacking. The
discussion on the sign of the free surface deformation has shown
clearly well that the concavity of the surface is determined by the
competition between surface-tension and buoyancy-driven flows, the
behaviour of the free surface being opposite for the two driving
mechanism. We can safely say that this study gives now an unambiguous
response, and there exists a critical depth to delineate the predominancy of one force over the other. The sign of the relief surface
depends on the relative importance of Marangoni and buoyancy forces,
which is governed by the depth of the pool. With λ_T called "thermal
capillary length" the effect of Marangoni stresses (surface forces)
is compared with the effect of gravity (bulk forces). If the depth
h is much smaller (h $\ll \lambda_T$) or much greater (h $\gg \lambda_T$) than λ_T of the
fluid, the relief surface is principally governed by Marangoni

stresses or Archimedean forces. Moreover, if the depth of the pool is equal at the "thermal capillary length" the relief surface is level. A model of a level surface with surface-tension stresses on it, is equivalent to assuming an infinite surface-tension with the consequent vanishing of the "crispation number"[6]. These predictions are in good agreement (for the case only where p = 2 h), with the ideas of critical thickness described by J.R.A. Pearson[5], and with the experimental results reported by P. Cerisier and J. Pantaloni[14] in Benard-Marangoni convection. It is clear that the study of the deformation delineate the surface effects over bulk effects. Such a study could in turn have interesting applications for the determination of physical and hydrodynamical interfacial quantities at a fluid-fluid interface.

ACKNOWLEDGEMENTS

The author thanks P. Manneville and Prof E. Guyon for the many enlightening discussions he has had with them.

REFERENCES

1 - Challenges and Prospectives of Microgravity research in space. ESA BR-05 October 1981

2 - J. Friedel - J. Physique Lett. 41 (1980) - L251 - L254

3 - E. Nakache, M. Dupeyrat, M. Vignes-Adler - J. Colloïd Interface Sci. 94, 187-200, (1983).

4 - L. G. Napolitano - Acta Astronautica - Vol. 9 n°4 pp. 199-215 (1982).

5 - J.R. A. Pearson - J. Fluid Mechanics, 4, 489-500, 1958.

6 - C. Normand, Y. Pomeau, M.G. Velarde 1977 - Rev. Mod. Phys. Vol. 49, n° 3 581-624.

7 - J.C. Loulergue, P. Manneville, Y. Pomeau - J. Phys. D : Appl. Phys. 14 (1981).

8 - G. Da Costa - J. Physique 43 (1982), 1503-1508.

9 - L. Landau, E. Lifchitz, Editions MIR, page 293.

10- J.C. Loulergue, Thin solid Films, 82 (1981) 61.

11- F. Mast, U. La Roche - Proc. Int. Electro-Optical Design. Brighton (1971).

12- F. Laeri, B. Schneeberger, T. Tschudi - Optics Comm. 34 (1980) 23.

13 - M. Cormier, M. Blanchard, M. Rioux, R. Beaulieu - Appl. Optics 17 (1978) 3622.

14- P. Cerisier, J. Pantaloni - Ann. New-York Acad. of Science 404, 1983

SURFACE RELIEF ACCOMPANYING NATURAL CONVECTION IN LIQUID LAYERS HEATED FROM BELOW.

P. Cerisier, J. Pantaloni

Laboratoire de Thermophysique
Université de Provence
Rue H. Poincaré
13397 MARSEILLE CEDEX 13

1.   INTRODUCTION

It is well known that a thin horizontal layer of liquid, with a free upper surface, uniformly heated from below, begins to convect when the temperature difference between the two limiting surfaces reaches a critical value ΔT_c. This convection is well organized : the liquid layer has a honeycomb structure made up of hexagonal cells with only a few structural defects. In each cell the liquid rises along the the axis and goes down along the dihedral angles. It is note worthy that the liquids of two adjacent cells do not mix (this has been proved by dropping coloring matter into a cell).

This unstability has been studied since the beginning of the century by many authors, but a number of problems still remain. In particular the sign of the surface deformation is still a subject of controversy. In 1901 Benard found that the surface of a convective cell is concave above the warm ascending streams. Other experiments were performed afterwards. But in some cases the sign of curvature was not determined (2) or the experimental conditions were not well defined. Various authors do not agree on the nature of the deformation. For instance Volkovisky (3) found the same observation as Benard.On the other hand Spangenberg (4) et Davidhazy (5) found the opposite : a convex surface. The theoretical studies have been made by different authors. To our knowledge the first was Jeffreys (6) who concluded that the surface must be convex and that Benard's experimental work should be repeated. Now we know that this theory is erroneous because Jeffrey only considered the buoyancy forces and neglected the surface tension terms. Later on Pearson (7), Scriven and Sternling (8) taking into account both driving forces concluded that the free surface must be concave at the cell centre in concordance with Benard's results. Then comparing their results to that of Jeffreys they proposed the following conclusion : "in steady cellular convection driven by surface tension, there is up flow beneath depressions and downflow beneath elevations of the free surface".

Some very simple models have been proposed to calculate the height h_m between the centre and the side of a cell (9-10). But they give only a rough estimation of h_m. It is impossible to predict the influence of the depth layer on h_m. Moreover the buoyancy forces are not taken into account.

Finally an interesting work of Kayser and Berg (11) must be reported. They studied a liquid layer heated from below not on the whole surface but using a straight wire. They found above the rising warm current a concave surface in the shallower pools, and a convex surface in the deeper pools. Their mathematical model is in qualitative agreement with experiment.

To our knowledge no systematic experimental study of the profile has been made on the influence of the various fluid properties and system parameters, such as depth e, temperature difference ΔT, cinematic viscosity ν, coefficient of thermal dilatation α, surface tension σ, thermal coefficient of surface tension σ', etc... In this essentially experimental paper we present the first results of such a study.

2. BASIC PRINCIPLES

Let us call, K the thermal diffusivity, g the gravity acceleration and ρ the density. It has been shown (12-15) that the dimensionless numbers which represent the density and the surface tension effects are the Rayleigh number $R = \alpha g e^3 \Delta T / \nu K$ and the Marangoni number $M = \sigma' e \Delta T / \rho \nu K$. When there is no surface tension gradient ($\sigma' = 0$) the convection starts for the critical value $R_{oc} \simeq 680$, and in the same way in weightlessness the critical Marangoni number is $M_{oc} \simeq 81$. In the general case where the two driving forces act together, the four numbers are related by :

$$(M/M_{oc}) + (R/R_{oc}) \simeq 1 + \varepsilon$$

At the onset of convection $\varepsilon = 0$, R and M take the values R_C and M_C which are the critical values for the physical conditions of the experiment. For a given experiment, ε characterizes the distance to threshold of the instability.

3. EXPERIMENTS AND RESULTS

The surface profile has been characterized by several methods :

　　1. measurement of deviation of a laser beam reflected by the surface (Poggendorf method)

　　2. interferences between the liquid free surface and a glass

　　3. Michelson method where the interferences are created bet-ween the liquid free surface to study and that of a motionless liquid.

The last two methods provide the amplitude of relief but not its sign. The latter has been obtained by the Foucault's method which is often used during the making of optical instruments.

The liquid studied is a silicon oil Rhodorsil 47V100. We focused our attention on the influence of two parameters : the depth e and the distance to threshold ε. The three methods provide concordants results and the following conclusions can be drawn :

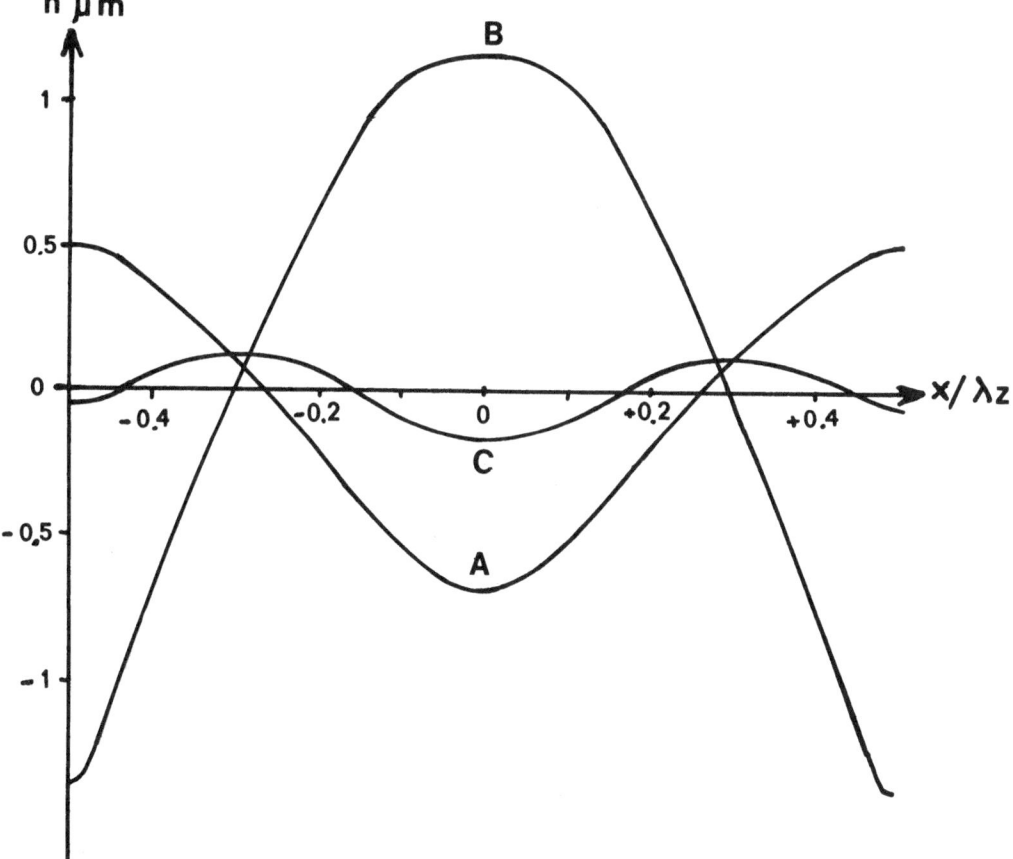

Figure 1. Profile of superficial relief. (A): concave relief, e=1.75mm, ε=0.98 - (B): convex relief, e=4.04mm, ε =3.26 - (C): hybrid relief, e=2.4mm, ε=0.60.

1) influence of the depth. For great depths(e>3mm) the surface
is convex at the cell centres. For small depths (e<2mm) it is concave.
For both cases the amplitude of h_m can amount to a few µm. For intermediate depths there is almost no deformation (about 0.2µ) and the profile is hybrid : there is a shallow depression at the centre of a shallow
bump which covers all the rest of the surface.

2) influence of the distance to threshold. Generally speaking
the amplitude $|h_m|$ first increases sharply with ε. For greater values
of ε a saturation phenomenon progressively appears, $|h_m|$ increases
very slowly and propably tends to a limiting value.

4. DISCUSSION

It has been shown (11,15) that the behaviour of the liquid can be
described by the graph in figure 2. The curve AB which is very close
to a straight line, parts the quadrant in two regions. The first one
(S) in OAB corresponds to the stability zone of the fluid, the second
one (U) represents the unstability zone. The line AB corresponds to
the onset of convection. For a given experiment (ρ,α,σ' and g well
definite) the representative point P moves along a straight line Oz
when ΔT varies. The slope of Oz is $(\sigma' Roc/\rho \alpha g Moc)e^{-2}$: it depends
strongly on e. So it is clearly seen that when e is small Oz is close
to the (M/Moc) axis. The surface tension forces are predominant.

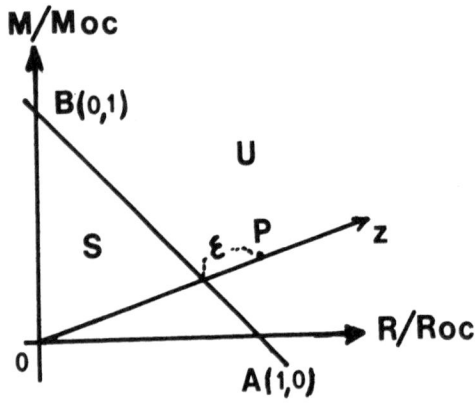

Figure 2. Hydrodynamic state of
the liquid as a function of Rayleigh and Marangoni numbers.
(S): Stable liquid - (U): Unstable liquid - (AB): line corresponding to the onset unstability.

Now the relief is concave (because on the cold points of the
surface of the liquid a greater
cohesion than above the warm
points of central surface of the
cells) is found. When e is large,
Oz is close to the (R/Roc) axis, the
buoyancy forces are predominant,
relief is convex (the liquid rises in the central

part of the cell and springs up at the surface).

This conclusion is in agreement with that of Loulergue (16) who very recently studied the two-dimensional forced Benard-Marangoni convection and with that of Kayser and Berg (11). For intermediate depths both driving forces have about the same importance and the relief profile is hybrid : a hollow in a bump. This is in disagreement with Loulergue who found a flat surface, an also with Kayser and Berg who announced the contrary result "a shallow crest rising from a shallow through". We estimated the aproximative value of critical depth. The linear theory gives :

$$e_c \wedge \sqrt{\frac{0.35\sigma'}{\alpha\rho g}}$$

Loulergue (16) in an independent way, obtained the same expression but without 0.35, and called e_c, the "thermal capillary length".

Using the data found in literature the calculated value $e_c \wedge 2 \mu m$ for the oil Rhodorsil 47V100 agrees with experimental value $(2 < e_c < 3mm)$. This expressions shows that the relief sign is essentially a function of σ', α and ρ. The other parameters such as ν, σ or κ have no influence on e_c. On the other hand it seems that σ has a major importance on relief amplitude : if σ is large the liquid is not much deformed, if σ is small the deformation can be more important.

REFERENCES

 (1) Benard, H., Rev. Gen. Sci. Pur. Appl. 11 (1900) 1261, Am. Chim.
     Phys. 23 (1901) 62, Thèse, Université de Paris (1901).
 (2) Block, M.J., Nature 178 (1956) 529.
 (3) Volkovisky, V., Publ. Sci. Tech. Ministère de l'Air, 151 (1939).
 (4) Spangenberg, W.B. and Rowland, W.R., Phys. Fluids 4, 6 (1961) 743.
 (5) Davidhazy, A., Photographic Sci. Engng. 13 (1969) 156.
 (6) Jeffreys, H. Quart. J. Mech. 4 (1951) 283.
 (7) Pearson, J.R.A., J. Fluid Mech. 4 (1958) 489.
 (8) Scriven, L.E. and Sternling, C.V., J. Fluid Mech. 19 (1964) 321
 (9) Hershey, A.V., Phys. Rev. 56 (1939) 204.
(10) Anand, J.N. and Karam, H.J., J. Colloid Interface Sci. 31 (1969)
     196.
(11) Kayser, W.V. and Berg, J.C., J. Fluid Mech. 57, 4 (1973) 739
(12) Nield, D.A., J. Fluid Mech. 19 (1964) 1941
(13) Scriven, L.E. and Sterling, C.V., J. Fluid. Mech. 19 (1964) 321
(14) Takashima, M., J. Phys. Soc. Japan 29 (1970) 531
(15) Pantaloni, J., Bailleux, R., Salan, J., Velarde, M.G., J. Non-
     Equili. Thermod. 4 (1979) 201
(16) Loulergue, J.C., Thèse, Université de Paris (1983).
     See also a communication in this symposium.

ON THE SUBJECT OF GORTLER VORTEX

H. Peerhossaini

Laboratoire d'Hydrodynamique et Mécanique Physique
E.S.P.C.I.
10, rue Vauquelin
75235 Paris, Cedex 5, France

INTRODUCTION

Stability of boundary layers on concave walls in the face of counter rotating longitudinal vortices has attracted much attention due to the increasing need for a better understanding of laminar-turbulent transition, and especially its tie with modern technology. A large part of modern gas turbine technology is dedicated to cooling of turbine blades with fluidmechanical methods such as film-cooling or transpiration. The boundary layer on the concave side of the turbine blade is unstable in the shape of longitudinal counter rotating Görtler vortices. Transition from laminar to turbulent due to Görtler vortices, and persistence of vortices even in turbulent regime, cause a drastic change in heat transfer rate and skin friction characteristics of the boundary layer.

Boundary layer instability, through its relation to boundary layer control methods, also plays a major role in the realization of the Short Take Off and Landing (STOL) aircraft. Sophisticated methods of boundary layer control are in demand to secure low takeoff and landing speed and its related high lift coefficient for STOL.

Görtler vortices are streamwise vortices having right and lefthand rotation (Fig. 1). They may appear in the boundary layer along a concave surface.

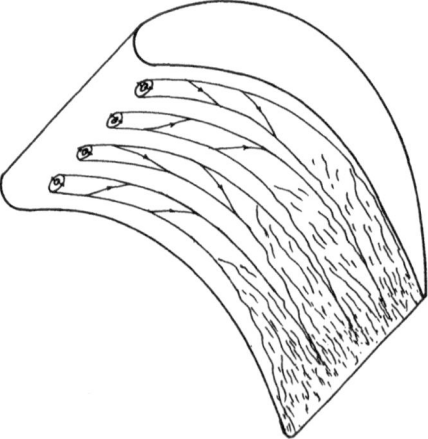

Fig. 1 : A schematic view of Görtler vortices on the concave side of a gas turbine blade.

Like as in the Taylor-Couette problem, the destabilizing mechanism in
the case of a concave boundary layer is due to the opposite effects of the
pressure gradient across the boundary layer, and the centrifugal force. Viscosity acts as a damper : attenuates the effect of destabilizing force.

In this sense, one can define the Görtler number (the equivalent of Reynolds
number for flat plate boundary layer) in terms of a stabilizing and a destabilizing "characteristic time".The centrifugal force can be formulated as :
$f = \frac{\rho U_\infty^2}{R}$, of which the penetration distance (with respect to the boundary
layer) can be assumed to be the momentum thickness. So one can define a
"destabilizing characteristic time " as : $\tau_d = \left[\frac{\theta}{f/\rho}\right]^{.5} = \frac{(\theta R)^{0.5}}{U_\infty}$
The " stabilizing characteristic time " of diffusive type can be formed as :
$\tau_s = \frac{\theta^2}{\nu}$. The ratio of the two characteristic times renders the Görtler number :

$$G = \frac{\tau_s}{\tau_d} = \frac{\frac{\theta^2}{\nu}}{\frac{(\theta R)^{.5}}{U_\infty}} = \frac{\theta U_\infty}{\nu}\left[\frac{\theta}{R}\right]^{.5}$$

Thus when the ratio of stabilizing characteristic time to that of destabilizing exceeds a certain value, the flow becomes instable.

For the first time the streamwise vortices were pointed out by Görtler (1)
in a theoretical investigation of centrifugal instability in boundary layer
flows over concavely curved walls. Görtler performed no experiment to support
his theory. Clauser and Clauser (2) found that boundary layers over concave
surfaces turn to turbulent in Reynolds numbers smaller than their corresponding
flat surfaces. Liepmann (3) in an experimental study of the effect of curvature
on transition of boundary layer, noticed that the Görtler number can correlate
the instability conditions.Subsequently existence of Görtler vortices and their
trace were demonstrated by Aihara (4) , Tani (5) and Wortmann (6) , using
different visualization methods : china-clay technique, colour liquids,
smoke threads, tellurium method and hydrogen bubble technique. Also hot wire
was used for direct measurements in vortices.

The very special case of Görtler type instability appears also as an interesting
tool to study the mechanism and successive states of laminar-turbulent transition
and interaction between large structures with other mechanisms such as Tollmien
Schlichting waves.

Eventhough a large body of investigation has been carried out in the
subject of Görtler vortex, the basis provided by linear theory (there is a
very limited nonlinear theory) is rather incomplete and to some extent not
sufficiently reliable. The experimental investigations are not only small in
number, but they are carried out in a different range than the theory. There
exist some basic open problems that play a fundamental role in the understanding of the subject. In the following we will examine some of them :

The complex nature of the phenomenon of the Görtler vortex, coupled with
experimental difficulties for verification of theory, has caused a plethora
of theoretical marginal stability curves : almost all in disagreement, if not
in contradiction with each other (see (7)). A brief review of the literature
reveals at least 11 different marginal stability curves.

A simple classification of the existing neutral stability curves (Fig 2)
signifies three distinguished categories as noted below :

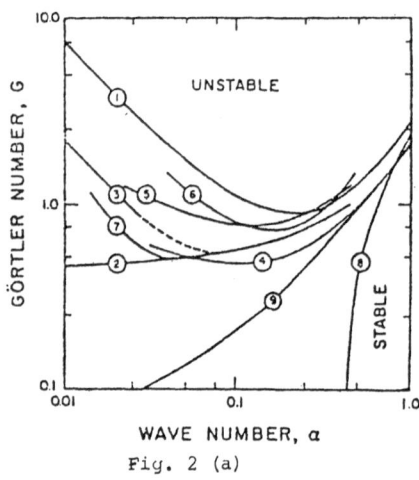

Fig. 2 (a)

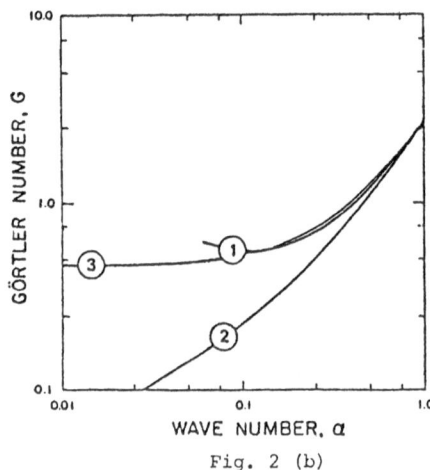

Fig. 2 (b)

Fig. 2 (a) : Various neutral curves resulted from different models of Görtler instability (after Floryan) 1) Görtler[1] model with original computation
2) Hämmerlin computation for Görtler model;;3) Hämmerlin computation for Görtler
model with correction for small wavenumbers; 4) Smith[10] model, original computation; 5) Hämmerlin model, based on streamlines and potential lines of the
flow over a wavy wall in the valley region; 6) Schultz-Grunow and Behbahani
model; 7) Kahawita and Meroney model, i.e; computation of Smith model with normal velocity component excluded;;8) Kahawita and Meroney ; Smith model with normal velocity component terms included; and 9) Floryan and Saric [9] computation
for Smith model.

Fig. 2 (b) :

Comparison of different computations of the neutral stability curves (after
Floryan and Saric[9]) ; 1) Smith's[10] original computations;
2) Floryan and Saric computation for Smith's model; 3) Floryan and Saric's
model.

- A majority of the neutral curves pass through a minimum for a finite wavenumber corresponding to a nonzero Görtler number in the stability diagram. That of H. Görtler (like those of the majority of other types of cellular instabilities) falls in this category.

- Some curves show a minimum (nonzero) Görtler number corresponding to a zero wavenumber. The curve due to Aihara (8) and recent work of Floryan et al (9) are those of this type.

- The last category comprises the curves which pass through or close to the origin of the stability plane. Floryan(9) resolved the set of equations due to Smith (10) in a new set of wall oriented coordinates and found a behaviour of this kind for the neutral stability curve.

Especially to include of the normal velocity component of the main flow, brings a deviation of the neutral stability curves from the first type to the second and third.

In fact, as long as normal component of the main velocity field is neglected in the system of Navier-Stokes and perturbation equations, should the critical Görtler number be reached; tne chance for fluid particles to move away or towards the wall (inside the boundary layer and in between the rolls) will be equal.On the contrary, to include of the normal component of the velocity in the system, is virtually identical to imposing an off-wall directed velocity field on the flow, thus breaking the symmetry in the flow field.

Inclusion of the normal component of the base velocity field encourages perturbationsof the disturbances into the region where viscous dissipation is small. Hence the pertinent "vortex-size" tends to large magnitudes.

It is reasonable to expect that suction at the wall induces an opposite effect, i.e. increasing the critical Görtler number and decreasing the corresponding critical "vortex-size". Theoretically it has been shown (11) that homogeneous suction from a concave permeable wall will increase tne critical Görtler number and also shifts the critical "vortex-size" to relatively smaller magnitudes; as an effect of suction, the vortices appear closer to the wall, where velocity gradient and consequently viscous dissipation near the center of the vortices becomes stronger. This shift of the center of the vortices to a region of higher viscous dissipation,forces them to choose an "optimum" vortex-size of smaller magnitude compared to the non suction case. Blowing has an opposite effect.

According to linear theory, in the unstable concave boundary layer, the possible longitudinal vortices can assume any wavelength within a certain range, which is covered by the neutral stability curve. On the other hand, this theory predicts nothing about the mechanism by which wavelength is selected. There are experimental studies of variation of vortex-size (physical wavelength) with radius of curvature and freestream velocity.

As an example, results taken from the work of Aihara (4) are reduced to the following table :

Radius of curvature (m)	Freestream velocity m/sec	Wavelength cm	$\dfrac{U_\infty R}{\nu}(\alpha R)^{-1.5}$
3	7	1.8	31
5	3	3.5	27
5	7	3.5	65
5	13	3.5	120
10	11	2.1	34
10	16	2.1	49

The table shows that for a fixed radius of curvature, prevailing "vortex-size" is insensitive to the freestream velocity.

Being suspecious that the wavelength (vortex-size) observed (above) is the one inherent in the experimental apparatus that has been amplified according to theory, Tani and Aihara (14) carried out the following positive experiment :

A row of wings (chord 0.8 cm, angle of attack 3 deg.,gap between wing tips 2 cm) were placed in the upstream flow at a height of 2 cm from the concave wall. The wavelength of 4 cm was observed.

To the above mentioned results we may add those of Bipps (12) as shown in the following table :

Radius of curvature (m)	Freestream velocity m/sec	Wavelength cm	$\dfrac{U_\infty R}{\nu}(\alpha R)^{-1.5}$
0.5	.075	1.5	12.3
0.5	.30	0.65	14.11
1.0	.075	2.0	13.5

Bipps produced the vortices by inducing an isotropic disturbance field, (using cloth screen with fine mesh size) hence no preferred vortex-size was imposed on the flow. Contrary to the former data, the latter shows that:the vortex-size is dependent on both radius of curvature and freestream velocity.

However in order to make sense out of the different results, we plot them on the G vs. $\alpha\Theta$ coordinates. Doing so, one can write the Görtler number in the following form just deduced from its definition :

$$G = \frac{U_\infty \, \Theta}{\Theta} \left[\frac{\Theta}{\nu} \frac{}{R} \right]^{0.5} \equiv \left[\frac{U_\infty \, R}{\nu} (\alpha R)^{-1.5} \right] (\alpha\Theta)^{1.5}$$

In the G vs. $\alpha\Theta$ coordinates (in log-log form), the above expression appears as a line with a slope of 3/2 for constant value of $\frac{U_\infty \, R}{\nu}(\alpha R)^{-1.5}$. Thus for a given radius of curvature, a change in V_∞ simply translates the distribution of experimental data from one $G \propto (\alpha\Theta)^{1.5}$ line to another parallel one with a different value of $\frac{U_\infty \, R}{\nu}(\alpha R)^{-1.5}$. For the experimental results of Aihara, the variation of G with $\alpha\Theta$ (for a constant wavelength and for a common radius of curvature but different freestream velocity) is caused only by the variation of Θ with X downstreem of the flow.

We have carried the results of Bipps experiment (second table)over the diagram of Aihara's results (Fig 3) They are concentrated around the line with $\frac{U_\infty \, R}{\nu}(\alpha R)^{-1.5}$ = 13,5 . Both Bipps and Aihara's experimental results lay far in the amplified region and are consistent with the expected $(\alpha\Theta)^{1.5}$ variation, but with different initial wavelengths.

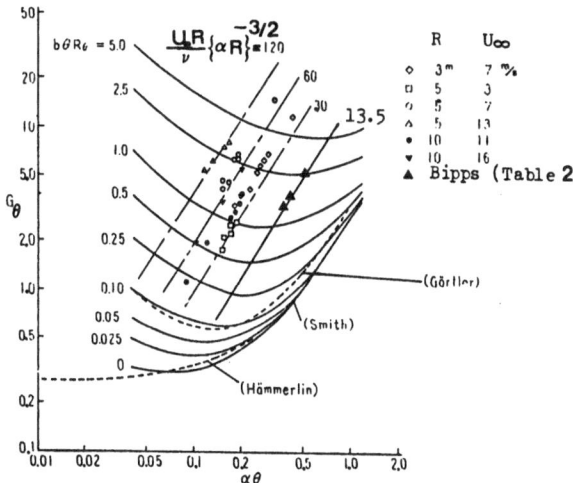

Fig.3 : Comparison of experimental results of Aihara (4) and Bipps (12) with theory. (After Aihara)

Most of the experimental results also agree in the observation of a constant vortex-size all along the flow, for a fixed radius of curvature.

This finding is rather unexpected in face of the fact that in analysis of the Görtler type instability and also in reduction of the experimental data (including stability diagram), momentum and/or boundary layer thickness is considered as the scaling length.

To have a meaningful scaling one expects that wavelength of the vortices also grows (as they run downstream) in a manner similar to the boundary layer thickness. Temporal or spatial growth of perturbations in conjunction with viscous diffusion also hints that vortex-size and consequently their wavelength should grow downstream of the onset point.

Constant "vortex-size" observation can be considered as an effect of the experimental apparatus walls on the flow. However, hydrogen bubble flow visualizations of Bipps (12) reveal the tendency of the vortices to converge or diverge, depending on the place of the bubble generator with respect to the vortex axis.

However, since the visualization photos show only the central vortices, verification of vortex growth from Bipps photos is not possible. Also, to verify the wavelength growth of the vortices, it is necessary to let them grow for a longer distance before nonlinear instability effects can interrupt their regular appearance.

In this respect the recent experimental results of Swearingen et al (13) are instructive. Their constant mean velocity contours reveal a two dimensional growth of the vortices. However, apparently due to experimental difficulties and breakdown of the regular structure of the vortices, the measurements have not been proceeded up to the point where the vortices grow enough to laterally interact with each other . It is from this point that wavelength variation (growth) should be expected.

NOMENCLATURE

Görtler number ; $G = \dfrac{U_\infty \, \Theta}{\nu} \left(\dfrac{\Theta}{R} \right)^{0.5}$

where :

U_∞	freestream velocity
Θ	boundary layer momentum thickness
ν	kinematic viscosity
R	radius of curvature of the concave wall
α	$\dfrac{2\pi}{\lambda}$ wavenumber
λ	wavelength

REFERENCES

1) Görtler, H.,"On the Three Dimensional Instability of Laminar
    Boundary Layers on Concave Walls", NACA Tech. Memo. 1375, (1954)

2) Clauser, M. and Clauser F., "The Effect of Curvature on Transition
    From Laminar to Turbulent Boundary Layer ", NACA TN-613, (1937)

3) Liepmann, H.W., "Investigation of Boundary Layer Transition on
    Concave Walls" , NACA War Time Report W-87, (1945)

4) Aihara, Y., "Transition in an Incompressible Bondary Layer Along
    a Concave Wall" , Bulletin of Aero.Res.Inst. Tokyo University, $\underline{3}$,
    195-240, (1962)

5) Tani, I.,"Production of Longitudinal Vortices in the Boundary Layer
    Along a Curved Wall" , J. Geophys.Res. $\underline{67}$, 3075-3080, (1962)

6) Wortmann, F.X., "Experimental Investigation of Vortex Occurance at
    Transition in Unstable Laminar Boundary Layer ", AFOSR Rep. 64-1280
    AF 61(052) 220, (1964)

7) Peerhossaini, M.H.,"Open Problems in the Subject of Görtler Vortex"
    Report N° 83/26; Laboratoire d'Hydrodynamique et Mécanique Physique,
    E.S.P.C.I, June 1983.

8) Aihara, Y. "Nonlinear Analysis of Görtler Vortices", Physics of Fluids,
    $\underline{19}$, 1655-1660.(1976)

9) Floryan, J.M. and Saric, W.S, " Stability of Görtler Vortices in
    Boundary Layers", AIAA Journal $\underline{20}$, 316-324,(1982)

10) Smith, A.M.O,"On the Growth of Taylor-Görtler Vortices Along Highly
    Concave Walls " , Quart.Appl. Math. $\underline{13}$, 233-262 , (1955)

11) Kobayashi, R;"Note on the Stability of a Boundary Layer on a Concave
    Wall with Suction", J. Fluid Mechanics, $\underline{52}$, 269-272, (1972)

./.

12) Bipps, H.,"Experimental Study of the Laminar-Turbulent Transition of
a Concave Wall in a Parallel Flow ", NASA TM-75243, (1978)

13) Swearingen J.D. and Blackwelder R.F.," Structure of Complex
Turbulent Shear Flow" , Ed. by R. Dumas and L. Fulachier,
IUTAM Symposium Marseille 1982 - Springer-Berlin p.10-19, (1983)

14) Tani, I., Aihara, Y, :"Görtler Vortices and Boundary Layer
Transition" , ZAMP, 20 , 609-618, (1969).

AIT AIDER O., Laboratoire d'Energétique et de Mécanique Théorique
et Appliquée,
2 rue de la Citadelle - 51011 NANCY (France)

ATTEN P., Laboratoire d'Electrostatique et de matériaux
diélectriques,
166 X - 38042 GRENOBLE CEDEX (France)

BERGE P., Service de Physique du Solide et Résonance Magnétique,
CEN-Saclay - Orme des Merisiers
91191 GIF-SUR-YVETTE (France)

BORCKMANS P., Service de Chimie Physique II
C.P. 231 - Campus Plaine, Boulevard du Triomphe
1050 BRUXELLES (Belgium)

BOUCIF M., Laboratoire d'Hydrodynamique et de Mécanique Physique,
E.S.P.C.I.
10 rue Vauquelin - 75231 PARIS CEDEX 05 (France)

CALTAGIRONE J.P., Laboratoire d'Energétique et Phénomènes de Transfert,
E.N.S.A.M.
Esplanade des Arts et Métiers - 33405 TALENCE CEDEX (France)

CAROLI B., Groupe de Physique des Solides de l'Ecole Normale
Supérieure,
Tour 23 - 2 Place Jussieu - 75231 PARIS CEDEX 05 (France)

CAROLI C., Groupe de Physique des Solides de l'Ecole Normale
Supérieure,
Tour 23 - 2 Place Jussieu - 75231 PARIS CEDEX 05 (France)

CERISIER P., Laboratoire de Dynamique et Thermo-Physique des Fluides
Centre St-Jérôme, rue H. Poincaré
13397 MARSEILLE CEDEX 13 (France)

CHAUVE M.P., Institut de Mécanique Statistique de la Turbulence,
12 avenue du Général Leclerc - 13003 MARSEILLE (France)

de CHEVEIGNE S., Groupe de Physique des Solides de l'Ecole Normale
        Supérieure,
        Tour 23, 2 Place Jussieu - 75251 Paris CEDEX 05 (France)

COGNET G.,   Laboratoire d'Energétique et de Mécanique Théorique
        et Appliquée,
        2 rue de la Citadelle - 51011 NANCY (France)

CROQUETTE V., Service de Physique du Solide et Résonance Magnétique,
        CEN-SACLAY - Orme des Merisiers
        91191 GIF-SUR-YVETTE (France)

DAMIL N.,   Laboratoire de Mécanique Théorique
        4 Place Jussieu - 75230 PARIS CEDEX 05 (France)

DELTOUR A.,   Institut de Mécanique des Fluides de Toulouse
        2 rue Ch. Carmichel - 31071 TOULOUSE CEDEX (France)

DUBOIS M.,   Service de Physique du Solide et Résonance Magnétique,
        CEN-SACLAY - Orme des Merisiers
        91191 GIF-SUR-YVETTE (France)

DUBOIS VIOLETTE E., Laboratoire de Physique du Solide,
        Bât. 510 - 91405 ORSAY CEDEX (France)

DEWEL G.,   Service de Chimie Physique II
        C.P. 231 - Campus Plaine, Boulevard du Triomphe
        1050 BRUXELLES (Belgium)

FAUVE S.,   Groupe de Physique des Solides de l'Ecole Normale
        Supérieure,
        24 rue Lhomond - 75231 PARIS CEDEX 05 (France)

 GIMENEZ M.,   Laboratoire d'Interaction Moléculaire et Réactivité
        Chimique et Photochimique,
        Université Paul Sabatier - 31063 TOULOUSE (France)
 GOLLUB J.P., Physics Department, Haverford College,
        HAVERFORD, PA 19041 (U.S.A.)
 GUAZZELLI E, Laboratoire de Physique des Systèmes Désordonnés,
        Centre St-Jérôme, rue H. Poincaré
        13397 MARSEILLE CEDEX (France)

GUTHMANN Claude, Groupe de Physique des Solides de l'Ecole Normale
          Supérieure,
          24 rue Lhomond - 75231 PARIS CEDEX 05 (France)

GUTHMANN Claudine, Groupe de Physique des Solides de l'Ecole Normale
          Supérieure,
          Tour 23, 2 Place Jussieu - 75251 PARIS CEDEX 05 (France)

GUYON E.,   Laboratoire d'Hydrodynamique et de Mécanique Physique,
          E.S.P.C.I.
          10 rue Vauquelin - 75231 PARIS CEDEX 05 (France)

HALDENWANG P., Laboratoire Héliophysique,
          Centre St-Jérôme, rue H. Poincaré
          13397 MARSEILLE CEDEX (France)

HOHENBERG P.C., Bell Laboratories,
          MURRAY HILL, N.J. 07974 (U.S.A.)

JOETS A.,   Laboratoire de Physique du Solide, Univ. Paris Sud
          Bât. 510 - 91405 ORSAY CEDEX (France)

KRAMER L.,   Physikalisches Institut der Universitat Bayreuth
          D - 8560 BAYREUTH (West Germany)

LALLEMAND P., Laboratoire de Spectroscopie Hertzienne de l'Ecole
          Normale Supérieur,
          24 rue Lhomond - 75231 PARIS CEDEX 05 (France)

LEGAL J.   DRET,
          26 Boulevard Victor - 75015 PARIS (France)

LOULERGUE J.C., Institut d'Optique,
          Bât. 503, Université Paris-Sud, B.P. N°43 -
          91405 ORSAY (France)

MANNEVILLE P., Service de Physique du Solide et Résonance Magnétique,
          CEN-SACLAY - Orme des Merisiers
          91191 GIF-SUR-YVETTE (France)

MASSAGUER J., E.T.S.I. Caminos - Universidad Politecnica Barcelona,
          Jorge Girona Salgado 31 - BARCELONA 34 (Spain)

MICHEAU J.C.,   Laboratoire d'Interaction Moléculaire et Réactivité
                Chimique et Photochimique,
                Université Paul Sabatier - 31063 TOULOUSE (France)

MULLIN T.,   Mathematical Institute,
                24/29 St Giles - OXFORD (Great Britain)

NORMAND C.,   Service de Physique Théorique
                CEN-SACLAY - Orme des Merisiers
                91191 GIF-SUR-YVETTE (France)

MARTINET B.,   Laboratoire d'Héliophysique,
                Centre St-Jérôme, rue H. Poincaré
                13397 MARSEILLE CEDEX (France)

OCELLI R.,   Laboratoire de Dynamique et Thermo-Physique des Fluides,
                Centre St-Jérôme, rue H. Poincaré
                13397 MARSEILLE CEDEX 13 (France)

PANTALONI I.,   Laboratoire de Dynamique et Thermo-Physique des Fluides,
                Centre St-Jérôme, rue H. Poincaré
                13397 MARSEILLE CEDEX 13 (France)
PELCE P.,   Laboratoire de Dynamique et Thermo-Physique des Fluides,
                Centre St-Jérôme, rue H. Poincaré
                13397 MARSEILLE CEDEX 13 (France)

PEREZ GARCIA C., Departemento de Termologia, Universidad Autonoma de
                Barcelona
                BELLATERRA - BARCELONA (Spain)

PERRIN B.,   Groupe de Physique des Solides de l'Ecole Normale
                Supérieure,
                24 rue Lhomond - 75231 PARIS CEDEX 05 (France)

PEERHOSSAINI H., Laboratoire d'Hydrodynamique et de Mécanique Physique,
                E.S.P.C.I.
                10 rue Vauquelin - 75231 PARIS CEDEX 05 (France)

PIQUEMAL J.M. Service de Physique du Solide et Résonance Magnétique,
                CEN-SACLAY - Orme des Merisiers
                91191 GIF-SUR-YVETTE (France)

POMEAU Y.,   Service de Physique Théorique
                CEN-SACLAY _Orme des Merisiers
                91191 GIF-SUR-YVETTE (France)

POTIER FERRY M., Laboratoire de Mécanique Théorique,
                4 Place Jussieu - 75230 PARIS CEDEX 05 (France)

QUINARD J.,   Laboratoire de Dynamique et Thermo-Physique des Fluides
                Centre St-Jérôme, rue H. Poincaré
                13397 MARSEILLE CEDEX 13 (France)

RIBOTTA B.,   Laboratoire de Physique du Solide, Univ. Paris Sud
                Bât. 510 - 91405 ORSAY CEDEX (France)

ROULET B.,   Groupe de Physique des Solides de l'Ecole Normale
                Supérieure,
                Tour 23 - 2 Place Jussieu - 75251 PARIS CEDEX 05 (France)

SEARBY G.,   Laboratoire de Dynamique et Thermo-Physique des Fluides
                Centre St-Jérôme, rue H. Poincaré
                13397 MARSEILLE CEDEX 13 (France)

TABELING P.,   Laboratoire de Génie Electrique de Paris,
                Plateau du Moulon - 91190 GIF-SUR-YVETTE (France)

TRAKAS C.,   Laboratoire de Génie Electrique de Paris,
                Plateau du Moulon - 91190 GIF-SUR-YVETTE (France)

VIET O.,   Laboratoire d'Hydrodynamique et de Mécanique Physique,
                E.S.P.C.I.
                10 rue Vauquelin - 75231 PARIS CEDEX 05 (France)

WESFREID J.E,   Laboratoire d'Hydrodynamique et de Mécanique Physique,
                E.S.P.C.I.
                10 rue Vauquelin - 75231 PARIS CEDEX 05 (France)

ZALESKI S.,   Groupe de Physique des Solides de l'Ecole Normale
                Supérieure,
                24 rue Lhomond - 75231 PARIS CEDEX 05 (France)

ZAMORA CARRANZA M., Departamento de Termologia, Universidad de Sevilla,
                B.P. 1065 - SECTOR SUR SEVILLA (Spain)

Y. Kuramoto

Chemical Oscillations, Waves, and Turbulences

1984. 41 figures. VIII, 156 pages. (Springer Series in Synergetics, Volume 19). ISBN 3-540-13322-4

H. Haken

Synergetics

An Introduction
Nonequilibrium Phase Transitions and Self-Organisation in Physics, Chemistry, and Biology
3rd revised and enlarged edition. 1983. 161 figures. XIV, 371 pages. (Springer Series in Synergetics, Volume 1) ISBN 3-540-12356-3

H. Haken

Advanced Synergetics

Instability Hierarchies of Self-Organizing Systems and Devices
1983. 105 figures. XV, 356 pages. (Springer Series in Synergetics, Volume 20). ISBN 3-540-12162-5

Chaos and Statistical Methods

Proceedings of the Sixth Kyoto Summer Institute, Kyoto, Japan, September 12–15, 1983
Editor: **Y. Kuramoto**
1984. 123 figures. XI, 273 pages. (Springer Series in Synergetics, Volume 24). ISBN 3-540-13156-6

Hydrodynamic Instabilities and the Transition to Turbulence

Editors: **H. L. Swinney, J. P. Gollub**
With contributions by numerous experts
2nd edition. 1985. Approx. 81 figures. Approx. 300 pages. (Topics in Applied Physics, Volume 45)
ISBN 3-540-13319-4
In preparation

Statics and Dynamics of Nonlinear Systems

Proceedings of a Workshop at the Ettore Majorana Centre, Erice, Italy, July, 1-11, 1983
Editors: **G. Benedek, H. Bilz, R. Zeyher**
1983. 117 figures. VIII, 311 pages. (Springer Series in Solid-State Sciences, Volume 47). ISBN 3-540-12841-7

Springer-Verlag
Berlin
Heidelberg
New York
Tokyo

Lecture Notes in Physics

Selected Issues from
Lecture Notes in Mathematics